C0-AVN-308

ELECTROMIGRATION AND ELECTRONIC DEVICE DEGRADATION

ELECTROMIGRATION AND ELECTRONIC DEVICE DEGRADATION

Edited by

ARIS CHRISTOU

CALCE Electronic Packaging Research Center,
University of Maryland

A Wiley-Interscience Publication

JOHN WILEY & SONS, INC.

New York • Chichester • Brisbane • Toronto • Singapore

This text is printed on acid-free paper.

Library of Congress Cataloging in Publication Data:

Electromigration and electronic device degradation / edited by
Aris Christou.
p. cm.
Includes bibliographical references and index.
ISBN 0-471-58489-4 (cloth : alk. paper)
1. Integrated circuits—Deterioration. 2. Semiconductors-
—Failures. 3. Electrodiffusion. I. Christou, A.
TK7874.E477 1993
621.3815—dc20 93-16841
CIP

Printed in the United States of America

10 9 8 7 6 5 4 3 2 1

CONTRIBUTORS

Aris Christou, CALCE Electronic Packaging Research Center, University of Maryland, College Park, Maryland 20742

A. Georgakilas, Foundation for Research and Technology–Hellas, Research Center of Crete, Institute of Electronic Structure and Lasers, 711 10 Heraklion, Crete, Greece

N. Kornilios, CALCE Center for Electronics Packaging, University of Maryland, College Park, Maryland 20742

Simeon J. Krumbein, AMP Incorporated, P.O. Box 3608, Harrisburg, Pennsylvania 17105

Pradeep Lall, CALCE Electronic Packaging Research Center, University of Maryland, College Park, Maryland 20742

P. Panayotatos, Department of Electrical and Computer Engineering, Rutgers, The State University of New Jersey, Piscataway, New Jersey 08855-0909

Michael Pecht, CALCE Electronic Packaging Research Center, University of Maryland, College Park, Maryland 20742

M. C. Peckerar, Naval Research Laboratory, Surface and Interface Science Branch, Electronics Science and Technology Division, Washington, DC 20375

Pin Fang Tang, IBM Burlington, Essex Junction, Vermont 05452

Jian Hui Zhao, Department of Electrical and Computer Engineering, Rutgers, The State University of New Jersey, Piscataway, New Jersey 08855-0909

PREFACE

Since 1970 the major failure phenomenon in discrete solid state devices and integrated circuits has been electromigration. The present book investigates the role of electromigration and related metal migration effects in the degradation of metal thin film interconnects. Therefore the classical definition of electromigration refers to the structural damage caused by ion transport in metal thin films as a result of high current densities. In integrated circuit metallizations, the high current-density induced mass transport manifests itself as voids, open circuits, and hillocks resulting in shorts between adjacent interconnect stripes.

The phenomenon of electromigration has become a critical problem both in VHSIC and MMIC circuits. As a result of the continuing device dimension scaling to submicrometers, as in ultra-large-scale integrated circuits, current densities of 1×10^6 A/cm^2 or higher exist in metallizations, although the total current may only be a few tenths of milliamperes. The threshold current density is exceeded even in digital ICs.

The physical models have not changed significantly since 1980 and are still based on the magnitude of the electric field, grain boundary diffusivity, and grain boundary structural factors that determine the atomic flux distribution and the distribution of flux divergence. The mass accumulation and depletion caused by the nonuniform flux divergence directly influences the metallization resistance and the current density distribution. The change in current density changes the surface temperature, which in turn changes the flux distribution. The process continues until mass depletion or accumulation becomes severe enough for circuit failure.

Since 1985, electromigration has been shown to be especially important in microwave monolithic integrated circuits (MMICs) due to the large power dissipation and the high current density in GaAs power MESFETs. In addition, the mass transport in GaAs not only takes place in the metallization, but also at the interfaces of the metal and semiconductor. Electromigration induces voids and openings in the gate fingers, which result in loss of control of device drain current.

This book has been written to review in a single volume the status of electromigration physics in microelectronics. The phenomenon of electromigration is discussed both for the silicon and GaAs technologies. Therefore the significant differences between GaAs ICs and silicon VLSI circuits as affected by the electromigration failure mechanisms are presented. Electromigration failure modes in electronics covering both theory and experiments are presented in this volume with the purpose of educating students and researchers in reliability. Thin film conductors carrying high current densities are often observed to fail by mass depletion effects. This process of metal migration has been a dominant failure mechanism in microelectronics. In addition to reviewing the experimental results, the theory and simulation models for electromigration are also presented. Finally, various rate controlling details are summarized, including an investigation of temperature dependence. The book also summarizes the metallization systems available for both GaAs and silicon integrated circuits and devices.

This book is part of the CALCE Electronic Packaging Research Center's approach to packaging, design, and reliability. Central to packaging and microelectronics design is the complete understanding of failure mechanisms. The physics of failure approach will enable us to optimize circuit and package design, material selection, and system implementation. Finally, this book concludes with a discussion of the current status and future plans for electromigration-resistant advanced metallization systems for VLSI.

This book was made possible by grants from the National Science Foundation and from DARPA.

ARIS CHRISTOU

College Park, Maryland

CONTENTS

ELECTROMIGRATION AND ELECTRONIC DEVICE DEGRADATION

1

RELIABILITY AND ELECTROMIGRATION DEGRADATION OF GaAs MICROWAVE MONOLITHIC INTEGRATED CIRCUITS

ARIS CHRISTOU
CALCE Electronic Packaging Research Center, University of Maryland, College Park

1.1 INTRODUCTION

The performance and reliability levels of microwave monolithic integrated circuits are reviewed with emphasis on low-noise amplifiers and power amplifiers. The low-noise amplifiers typically have MTBFs in excess of 10^6 hours for MMIC structures operating up to 26 GHz. In the frequency range of 26–94 GHz, MMICs typically degrade from surface-related effects in the high electron mobility transistor. Data for heterojunction bipolar transistors and InAlAs/InGaAs/InP HEMTs are limited at the present time but the data analyzed allow an MTBF projection of 10^4–10^5 hours to be made. Power amplifiers up to 18 GHz typically suffer from electromigration and gate diffusion effects. A finite element analysis program for simulated electrothermal effects in MMICs is also presented.

Electromigration and Electronic Device Degradation, Edited by Aris Christou.
ISBN 0-471-58489-4

1.2 MMIC STATE-OF-THE-ART

Microwave monolithic integrated circuits (MMICs) are used in many systems requiring high levels of reliability. With the advent of new high-frequency applications, there is a commercial and military requirement for both receiver and transmitter modules to operate at frequencies far into the millimeter-wave region. This chapter examines the reliability of the circuits capable of meeting these requirements and assesses the reliability of the building blocks necessary for understanding the reliability of the entire microwave module. The ultimate goal of ongoing research in the MMIC field has been to technologically enable industry to fabricate the transmit and receive modules as well as power amplifiers in the microwave and millimeter-wave region and to understand their reliability. Here, further exploratory effort on more advanced technologies is imperative with the reliability goals established in the design phase. The reliability problems to be reviewed include those of

- Low-noise amplifier (LNA)
- Mixer
- Synthesized local oscillator (LO)
- A/D converters
- Power amplifiers (SSPAs)

While at frequencies around 1 GHz the specification for most microwave systems could be met by a silicon-based technology, at higher frequencies compound semiconductor technologies are the only ones capable of integrating the analog and digital circuitry required for a cost effective totally integrated solution. The advantages over a silicon-based technology are therefore twofold:

1. The operation frequency at the device level is beyond the scope of current silicon technology.
2. At the proposed frequencies, the monolithic integration of many analog functions cannot be achieved by silicon.

The application of MMICs is very widespread. There are, at present, system design work, performance assessment, and reliability investigations being carried out in

- Mobile communications, 900 MHz and higher

- Small direct line links at 23 GHz and higher, using the HEMT technology
- The automobile industry (small data links between car and stationary beacons), where such applications require GaAs plastic integrated circuits
- Civil radar systems for transportation (car, railway) for safety control and measurements
- Other small systems that benefit from subassemblies, such as automatic payment of tolls (public transport, supermarkets, motorways) or private keying-paging
- Direct broadcasting from satellites (DBS) and geographical location satellites (GPS, Locstar), using MMICs at L-band and X-band frequencies

1.3 REVIEW OF CIRCUIT PERFORMANCE

In this section, present research results are briefly reviewed in order to assess their current reliability levels in comparison to performance specifications.

It can be stated that the compound semiconductor industry is a competitive electronics industry, in particular with MESFET technology products, which were the subject of intensive industrial research prior to 1991. There are respectable, but limited, results in HEMT and HBT technologies so that continuous further development activities are necessary, especially in the area of monolithic integration and reliability.

1.3.1 MESFET Integrated Front-Ends

It is only relatively recent (1985–1991) that the design and fabrication expertise has been sufficiently advanced to make the concept of producing many circuit functions on a single GaAs chip a realistic possibility. The first integrated 12-GHz receivers: one used a mixer diode input and a Gunn diode local oscillator; the other, based on MESFETs, exhibited a 4-dB noise figure [1]. Since that time, many laboratories across the world have utilized improvements in the technology to produce higher performance receiver blocks. However, none have so far integrated the local oscillator with the preamplifier mixer and IF amplifier or demonstrated reliability in all three components at elevated frequency and temperature. A X-Ku band upconverter consisting of a mixer, LO buffer amplifier, and RF amplifier having an 8–16 GHz output frequency has been reported recently [2], able to meet military standard 883 specification. The X-Ku band circuit was contained in a very small chip measuring 96 × 48 mils and again reliability in the 8–16 GHz band was not demonstrated through elevated temperature testing. A monolithic front-end (X-K band) consisting of a RF amplifier, mixer, and IF amplifier has also been reported [3]. The noise figure of this circuit was between 5 and 6 dB, with unpublished reliability levels in excess of 10^7 hours at 110°C. Texas Instruments with GEC Marconi have recently published work on a Tx/Rx circuit (transmit–receive) with a much greater level of integration using a technology whose components have a demonstrated reliability in excess of 10^7 hours. These included receiver and transmit elements together with the necessary switching circuitry. The module [4] measured 4.1 × 2.6 mm^2 and operated at an input frequency of 5–6 GHz. The Texas Instruments circuit [6] also operated at X band but was considerably larger (5 × 3 mm^2). The MMIC design utilized a number of novel circuit design concepts—moving away for the conventional ideas of microstrip matching [4] without sacrificing reliability. At higher frequencies, the possibility of producing a monolithically integrated mixer at 94 GHz was demonstrated [5], showing 6–8 dB of single sideband noise. Field effect transistor mixers monolithically integrated based on dual gate FETs have been demonstrated for applications at frequencies [5] above 100 GHz. The failure mode of these mixers was always reported to be related to surface states in the channel due to damage from the passivation.

1.3.2 HFET Devices and MMICs

HEMT Low-Noise Amplifiers HEMT research in most laboratories, as well as component manufacturers, is concentrating on low-noise (LN) applications [5]. The trend is to go to shorter gate lengths in order to extend the upper frequency limit and to improve the material toward better carrier transport properties. This approach, however, has introduced a number of potential reliability problems related to the disappearing gate and channel passivation. Commercially available HEMTs have gate lengths between 0.15 and 0.5 μm and noise figures down to 0.8 dB at 12 GHz. Research work at the present time is concentrating on gate lengths below or at 0.25 μm. Wisseman et al. [6] have reported on the processing of a HEMT with a dry-etched recess-process and WSi_x-gate on MBE-grown AlGaAs/GaAs layers. It resulted in a minimum noise figure of 0.54 dB and 1.3 dB at 12 GHz and 30 GHz, respectively, with gains of 12.1 dB and 7.2 dB. The WSi_x-gate has achieved high reliability, in excess of 10^7 hours, as has been demonstrated on test structures. The test structures, however, were only subjected to storage tests and not to a fully biased test condition.

HEMT research has progressed to gate lengths below 0.12 μm through novel processing and has presently reached 0.1 μm [7–9]. Using material heterostructures with spike doped AlGaAs/GaAs layers, a transit frequency of 113 GHz has been demonstrated. The devices were wet-etched and fabricated with conventional Ti/Pt/Au-gate metallization. The pseudomorphic AlGaAs/InGaAs/GaAs layers for HEMTs resulted in a further improvement of the minimum noise figure to 3 dB and attained a gain of 5.1 dB at 94 GHz. However, pseudomorphic HEMTs are susceptible to noise figure degradation at elevated temperatures. The pseudomorphic structure with a f_{max} of 350 GHz was demonstrated, with no data available on reliability. The reliability problems associated with pseudomorphic layers are related to the interfacial strain that exists between InGaAs and GaAlAs [59].

MMIC investigations are concentrating presently on LN amplifiers that incorporate the HEMT as the basic building block. The aim is to either increase the operating frequency or manufacture broadband amplifiers with a very broad frequency band and nearly constant characteristics. The fabrication of a three-stage LN preamplifier exhibiting a noise figure of less than 4.1 dB and a gain of more than 20 dB in the range of 100 MHz to 18 GHz has been reported [10]. Research [11,12] on 0.5-μm HEMTs resulted in a LNA that exhibited excellent performance between 12 and 16 GHz with a noise figure of 2.5 dB and a gain of 9 db [11]. Another design [13] for a broadband amplifier between 2 and 42 GHz exhibited a flat gain characteristic of 6 $\pm$ 1 dB with a minimum noise figure of 5.2 dB at 18 GHz. The discrete HEMTs showed a NF_{min} of 1.1 dB at a gain of 9 dB at 12 GHz, prior to integration as part of the LNA. Recently, a two-stage V-band LNA, based on 0.25-μm HEMTs and MBE-grown AlGaAs/GaAs layers, exhibited a 3.2-dB noise figure and 12.7-dB gain at 61 GHz, while storage tests at an elevated temperature up to 250°C indicated no degradation in noise figure [14].

Projections of median life for low-noise MMICs and power MMICs are shown in Figure 1.1. The MMIC reliability levels reported to date are only slightly inferior to those for discrete devices. Passive components gain a significant advantage in lifetime over FETs because they typically operate 25°C cooler than FETs in a given MMIC circuit. Therefore the decrease in reliability for MMICs is probably related to matching problems between the stages when degradation is initiated.

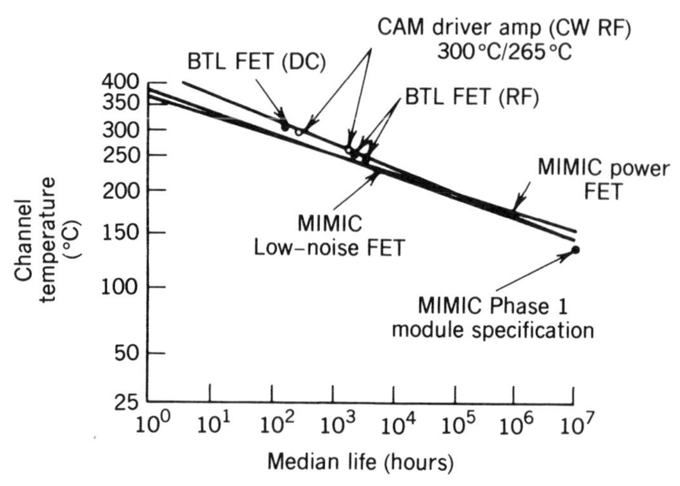

FIGURE 1.1. Projected median life for MMICs and FETs.

DMT (Doped Channel MIS-like Transistors) Based on a doped channel MIS-like transistor (DMT), power densities of 0.83 W/mm with 3-dB gain have been reported at 60 GHz [15]. These results show that with such a device, monolithic integration at high frequency is now possible and opens the way to a future generation of a new type of monolithic integrated circuit. Storage reliability tests reported on DMTs indicate less than 10% degradation in gain at 60 GHz, with projected MTTFs of 10^6 hours.

Materials The above mentioned work was performed on conventional AlGaAs/InGaAs/GaAs layers. Pseudomorphic layers offer better transport properties, especially a higher mobility and a higher sheet carrier concentration. To overcome the limits of this material system, which is related to stress, InAlAs is combined with an InGaAs channel on an InP substrate. This material combination may be achieved with lattice matched [16,17] or strained [18] epitaxial layers. The InP material system has resulted in values for transconductance of more than 1 S/mm (1160 mS/mm Hughes, 1080 mS/mm GE). As ultra-low-noise device with a gate length of 0.25 μm that showed 0.5-dB noise and 15.2-dB gain at 18 GHz and 1.2-dB noise and 8.5-dB gain at 58 GHz has also been demonstrated [14]. An amplifier with 1.5-dB noise and 8-dB gain at 60 GHz and an f_T of 205 GHz has been reported [18]. At the present time, there is no reported reliability data at elevated temperatures for the InP-based structures; however, excess generation-recombination noise at or below 100 Hz has been reported due to trap activation in the InAlAs layers [19].

1.3.3 HBT-Analog-to-Digital Converters (ADCs)

The fastest HBT digital circuit to date is a static divided by four operating at 22 GHz as reported by NTT [20, 21]. A divider operating at 35 GHz has recently been demonstrated [22, 23]. Dynamic dividers operating at 26 GHz have also been reported by NTT. The record for complexity lies with an HBT digital circuit for a 32-bit CPU [24] containing 12,400 gates, pushing HBT circuits to VLSI complexity. The GaAs HBT ADC circuits published to date are designed in the conventional and well-known parallel flash architecture either with or without a sample and hold circuit. Applications of a feedback or a feed-forward ADC design are less common and therefore do not require a stable feedback loop in order to attain reliability.

Research in data converters with HBTs has emphasized a sample and hold circuit using Schottky diode bridges and operation at 2 Gb/s was demonstrated. A transistor performance of $f_T = 53$ GHz and $f_{\max} = 70$ GHz was obtained with small size emitter areas (2.2 μm $\times$ 2.2 μm), InGaAs emitter contact layers, and a non-self-aligned base contact formation. However, degradation of the base characteristics due to enhanced zinc diffusion has been reported for HBTs. ADC circuits, again based on HBTs, have resulted in the fabrication of a 4-bit flash ADC with current mode logic (CML) operating at more than 1 Gb/s. The results on the ADC circuit were based on a non-self-aligned mesa HBT-process and Schottky diodes [25]. A standard HBT with 3 μm $\times$ 10 μm emitter area showed a transit frequency of f_T = 15–20 GHz. A self-aligned base HBT-process to fabricate a 4-bit ADC circuit has also been implemented in (CML) [26]. Operation at more than 1 Gb/s was demonstrated with HBTs of 2 μm $\times$ 3.5 μm size and InGaAs emitter contact layer.

A new processing technology based on non-self-aligned HBTs was reported recently [26]. The process provides for 1.4-THz Schottky diodes, nichrome resistors, MIM capacitors, and air bridge inductors. A pulser circuit using Schottky diodes has demonstrated an 8.6-ps rise time. In addition, the static divide by eight circuits showed no significant degradation during parametric testing. Testing of the HBT circuits, however, showed emitter–base temperature in excess of 260°C and degradation of the base characteristics due to base–dopant diffusion.

The highest speed of a commercially available ADC was obtained by a combination of a Si ADC and GaAs MESFET track and hold circuit as the key element for obtaining high-speed operation [27–29]. The best results for such a circuit were recently published by Wakimoto et al. [30], showing operation of 2 Gb/s. A breakthrough in performance was reported with the fabrication of a 4-bit, 3-Gb/s based on a GaAs MES-

FET technology using a flip-flop as load in the comparator. The same technology was recently extended to obtain 5-bit resolution at 2.2 Gb/s [31, 32]. The potential reliability problems for such circuits are based on the degradation mechanisms of the GaAs MESFET, and an extension of the hybrid circuit law stating that the overall circuit failure rate is the sum of the failure rates of each individual circuit component [33].

1.3.4 HBT Power Amplifiers

The GaAs MESFET has for a long time been used as the basic component for power devices. A MESFET power amplifier with 12 W output power at 20 GHz has been demonstrated; however, with fairly low power gain (5 dB) and low power added efficiency (PAE = 15.5%) [34]. Only very recently, interest has focused on the power applications of HBTs. Compared to the mature GaAs MESFET technologies, the HBT power device development is still in its infancy. HBTs exhibit features making them very attractive for power applications, for example, high PAE and high output power per unit emitter length; however, reliability problems related to the thin base and shallow emitter contact have not been solved.

Japanese and American companies lead the field of discrete power HBTs with the best performance of discrete power HBTs being 1 W and 2.4 W output power with 49% and 30% PAE, respectively, achieved by Toshiba [34] and Texas Instruments [35]. A collector–base breakdown voltage of 23 V was reported for HBTs fabricated by Rockwell [32], and 10-dB power gain at 12 GHz and 12-mW output power was demonstrated [34]. The output power has been raised by a factor of 2 in pulsed operation [34] in comparison with CW operation. In recent work, a 75-GHz current-gain bandwidth and a 175-GHz maximum frequency of oscillation of a HBT device have been reported [35]. At 10 GHz, 4 W/mm CW output power per unit emitter periphery has been measured and the results obtained show an associated PAE of 48% [36–39]. Two-stage amplifiers built with an input stage and an emitter–follower output stage showed an amplification higher than 12 GHz and may be applied to wideband operation [35, 40]. These amplifiers required almost no off chip matching and, therefore, the amplifier reliability was determined by the discrete HBT.

These results show that a rapid improvement in the performance of HBT power devices has been achieved. HBT power devices will find applications in areas where their potential advantages over MESFET are essential, for example, applications with low overall power consumption (high PAE), with high output power in pulsed mode operation, and integration with HBT-based signal processing modules. The reliability problems, however, are significant enough so as to prevent the rapid integration of HBTs into commercial circuits. Due to the diffusion of zinc and carbon in the base, a reliability of only 10^4 hours at 100°C can be projected [41, 42].

Although GaAs HBTs are attractive for power applications, the temperature elevation during large signal operation may result in reliability problems. Maintaining safe operating temperatures is more difficult in GaAs HBTs than in silicon because of the lower thermal conductivity in GaAs. The speed and gain of GaAs HBTs are best with high current densities and with short and wide emitters, both of which increase the peak device junction temperatures. Recent thermal simulations [62] have indicated that thermal effects may be addressed through circuit redesign, without sacrificing die area. The HBT large signal simulation [62] has also indicated that degradation effects through self-heating may be avoided if the intrinsic HBT temperature increase is maintained at less than 100°C or a power dissipation of 13 mW for the standard transistor.

1.4 RELIABILITY CONCERNS OF GaAs INTEGRATED CIRCUITS

There are various aspects of concern in GaAs reliability such as the relationship of reliability and device performance, temperature stability of the circuits, interconnections and wire bonding, and packaging [41–43]. The reliability of the circuits' active and passive elements continues to be of primary importance. Typical active MMIC device reliability has been demonstrated through life tests under DC or RF bias at elevated temperatures [41–48]. The commonly reported degradations under DC bias are the decreases of (a) drain-to-source current I_{DSS} and (b) pinch-off voltage V_p and increase of gate leakage current. For HBTs, a common degradation mechanism has been the increase in the base–collector current. When RF biased, the devices tend to degrade by a decrease in gain [41, 42] accompanied by an increase in noise figure. In addition, formation of interelectrode metallic paths between metal gate and contacts has also been found [44], resulting in device failure via short circuit formation. In HBTs, base punch-through and emitter contact resistivity increase usually occurs. From a properly defined failure criterion, usually a certain percentage of decrease in I_{DSS} or gain, lifetime can be determined and has been reported in the literature. The failure distribution of active three-terminal devices has been assumed to be lognormal and the mean time to failure has been obtained to be in the range of 10^7–10^8 hours at the moderate channel temperature range (130°C–200°C) [41, 43, 45, 46]. Various experimental approaches, including scanning optical and electron microscopy, Auger, energy dispersive x-ray analysis (EDXA), and SIMS analysis, have been employed to identify the cause of failures described above. It is now believed that the primary failure mechanism of a GaAs FET is the metal–semiconductor interface electromigration. Specifically, atoms from the metallizations are driven into the GaAs active layer, in a diffusion-controlled process, resulting in a reduction of channel thickness, compensation of active carriers, increase of contact resistivity, and, consequently, the observed degradation of I_{DSS}, V_p and gain. Surface quality is another factor reported to affect device reliability. The increase in leakage current and the formation of interelectrode paths are directly related to the quality of the GaAs surface. In addition, device reliability is also process and materials dependent, which has been shown by Esfandiari et al. [47]. Improvement of surface quality can be achieved by several approaches including initial polishing of the substrate or by depositing passivation layers, after device processing [48, 49]. However, little work has been reported on techniques for suppressing metal–GaAs interface electromigration. The mechanism for HBT degradation has not been investigated in-depth and is, therefore, not understood, although the initial reports concentrated on base diffusion and surface problems.

Passive devices in GaAs MMICs include MIM capacitors, resistors, and inductors. The failure of passive devices has not received much attention since it has been believed that the active devices control the failure characteristics of GaAs MMICs, and that the failure rate of passive devices is insignificant compared to active devices [47, 50–52]. However, such conclusions cannot be generalized since the failure rate of passive devices with respect to active devices depends on the specific circuit layout and bias condition. Recent investigations have indicated that the failure of passive devices may be very significant in the sense that they are responsible for the early failure of the circuits.

Previous reliability investigations have utilized elevated temperature life tests. The mean time to failure of the circuits at low temperature was then predicted by extrapolating the high-temperature result. However, the validity of such an extrapolation must be justified. By testing identical samples at both high and low temperatures, Esfandiari and coworkers [47] showed correlation between the two results and therefore provided the confidence for such an extrapolation method. The mean time to failure of a typical two-stage feedback amplifier has been determined to be 8×10^7 hours at 125°C for Cr-doped substrates. It has also been suggested that increasing the level of integration of the circuits does not necessarily result in a corresponding decrease in reliability [53].

Other important concerns in the study of MMIC reliability include the accurate determination of channel temperature and thermal impedance, since they provide the basis for the reliability prediction. Several approaches have been suggested [54–59]. The predominant methods are by liquid crystal, electrical diode measurements, or infrared microscopy. The failure of a MMIC has been assumed to be described by a single distribution function. This again may not be a valid assumption and recent investigations [60] have indicated that a dual failure distribution function may be necessary to describe the failure contributions from both active and passive device failures.

Early reports concerning MMIC reliability have indicated that mean time-to-failure (MTF) levels of longer than 10^7 hours at 110°C have been obtained. However, before the appropriate extrapolation may be made, it is imperative that the shape of the failure distribution function be obtained and analyzed. Statistically, one may treat the MMIC as a distribution of active components capable of power dissipation, connected to a second distribution of totally passive components through an appropriate correlation function. Therefore previous assumptions concerning single failure distribution functions for MMICs may not be valid. To date, most investigations have only concentrated on the active components due to the low transistor density of the MMIC structure, which allows for the treatment of each transistor as an independent device. It has been assumed that the active device failure distribution would dominate and may be used to predict total circuit reliability.

In order to accurately predict temperature variations in MMICs during RF reliability testing, a finite element analysis computer calculation has been developed [58] that can simulate both thermal and microwave responses of the circuit. At the device level, a 0.5-W power device is shown (Figure 1.2) to reach a peak channel temperature of 159°C. The active FET is then used to measure the entire thermal response of a broadband 6–18-GHz MMIC shown in Figure 1.3. A total difference of approximately 12°C is shown to exist between passive and active devices. Likewise, the same technique can be extended to examine the actual deformation of a MMIC package during RF testing as shown in Figure 1.4. Pulse microwave testing results in a physical deformation of approximately 0.038%, which may induce wafer cracking and microcracks at package discontinuities.

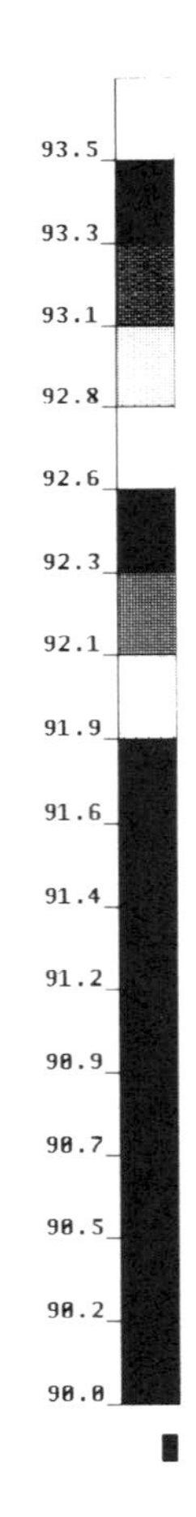

FIGURE 1.2. Finite element analysis of a power FET during RF bias showing a peak temperature of 93.5°C.

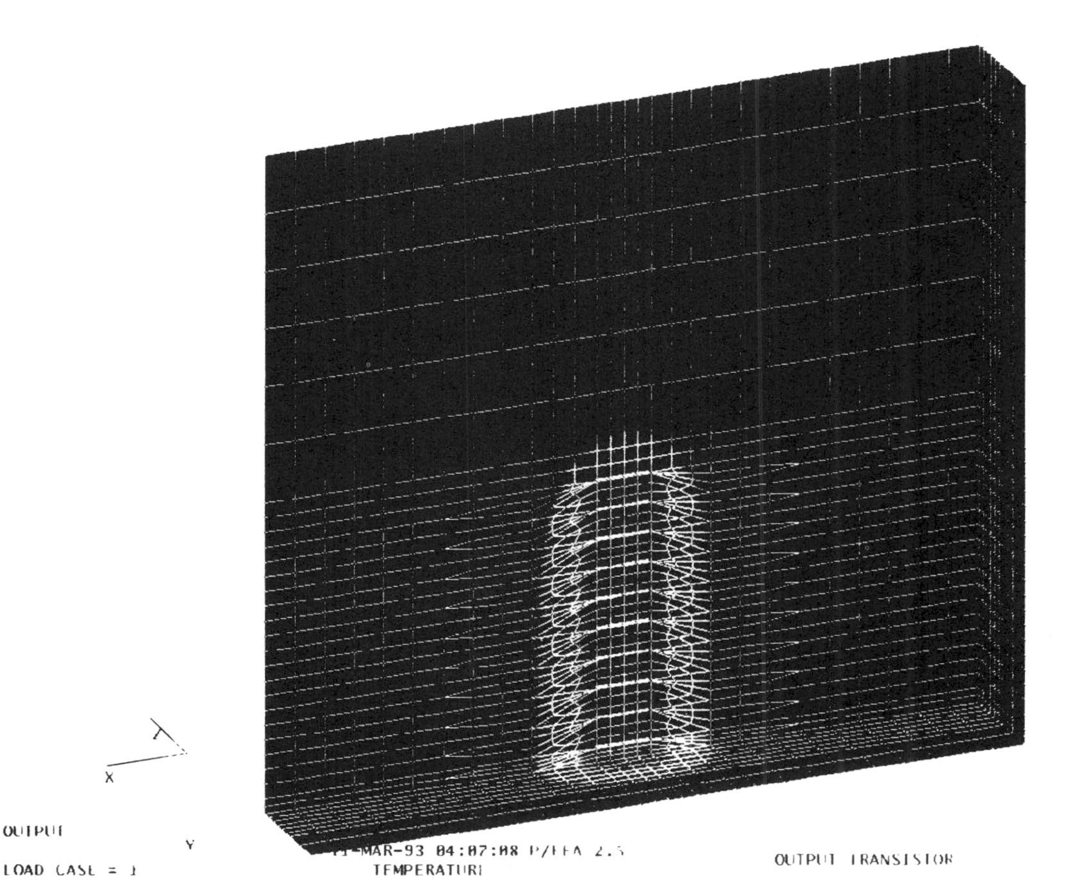

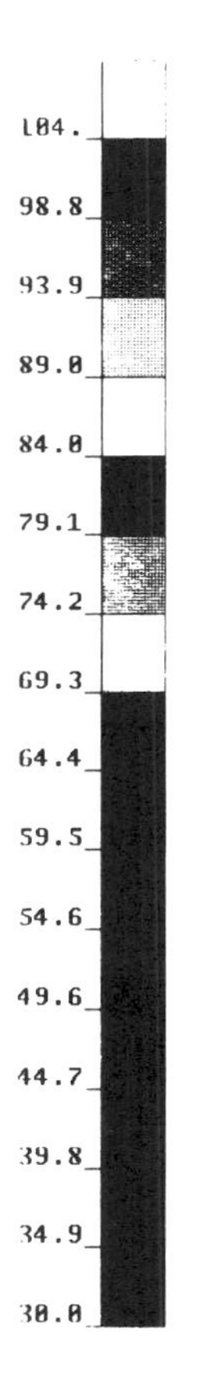

FIGURE 1.3. Finite element analysis of a MMIC showing the three-dimensional thermal distributions.

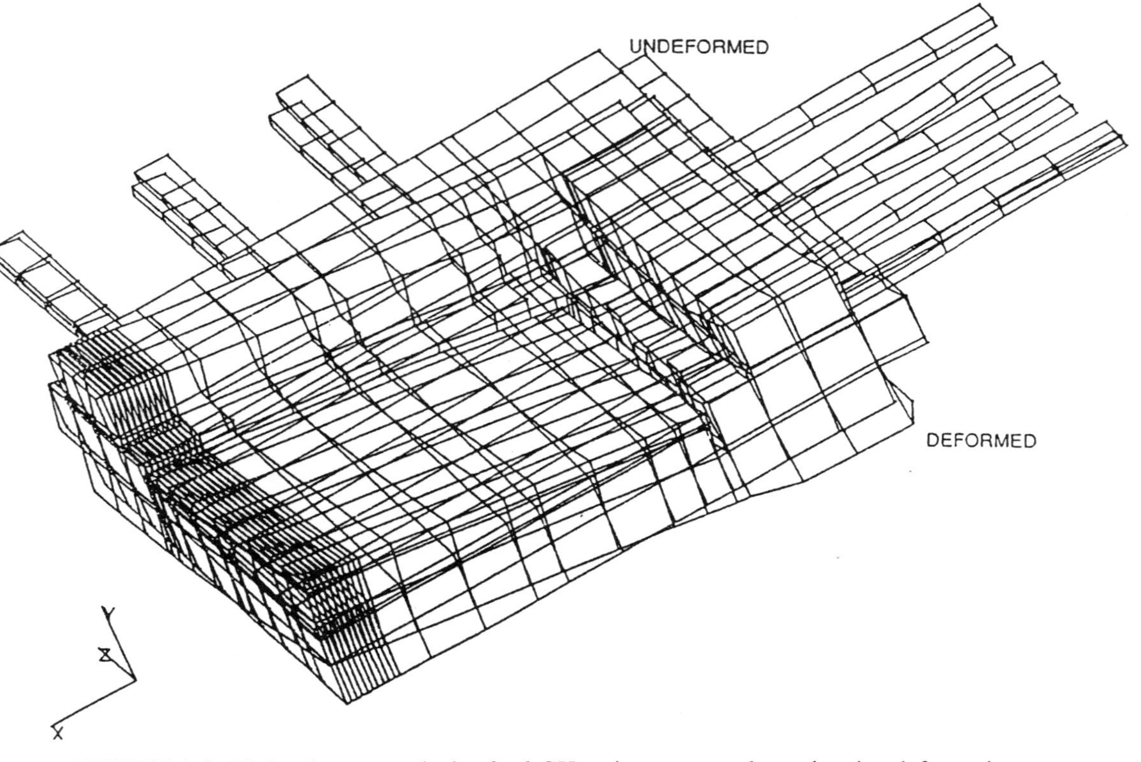

FIGURE 1.4. Finite element analysis of a 6-GHz microwave package showing deformations during RF testing.

1.5 RELIABILITY STUDY OF HIGH-POWER MMICs

In order to begin to understand the reliability problems in high-power broadband MMICs, a limited study was made of the MMIC type, 1-W, 6–18-GHz amplifier at bias conditions into the power saturation region. Life test results for a 5-W MMIC amplifier again biased into saturation are shown in Figure 1.5. The devices (14 total sample size) were held to within 3°C of the base plate temperature of 170°C or 215°C. Approximately 50% of the devices failed by burnout while under high-temperature RF life testing and the remaining circuits failed by gradual degradation of I_D or Po by 20% within the military specification temperature range of −55°C to 125°C. Median time before failure (MTBF) was approximately 600 hours at 170°C and 220 hours at 215°C, assuming a lognormal distribution. Extrapolation results in the prediction of MTBF of 3×10^4 hours at 50°C under RF drive. Failure analysis shows that electromigration of AuGe ohmic contact metallization occurred in the case of MMICs, which degraded by a 20% power loss.

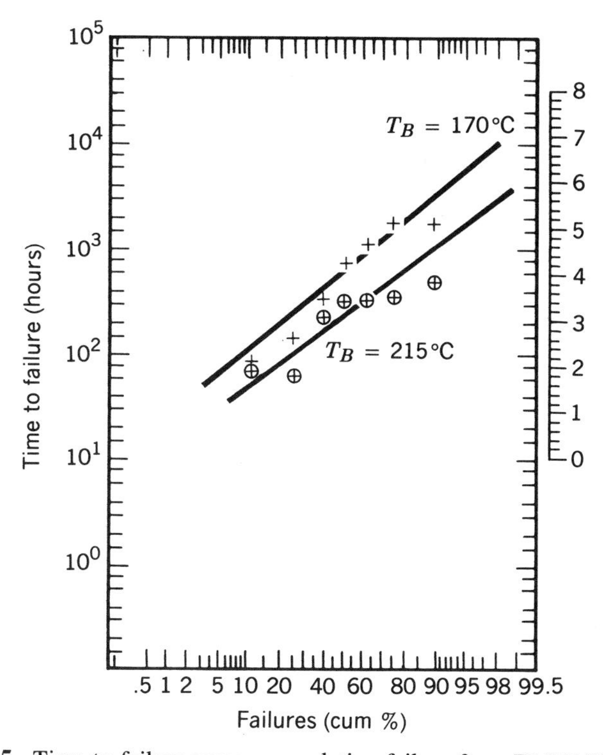

FIGURE 1.5. Time to failure versus cumulative failure for a TI EG 8014 MMIC.

These results, although based on a small population and only two temperatures, follow the classical failure mechanisms, which were observed for power field effect transistors, and illustrate the need for accelerated life testing under actual RF conditions. The results of Figure 1.5 indicate the existence of a low activation energy and a small MTBF under RF drive. These investigations are summarized in this section in order to indicate the present reliability concerns for high-power MMIC circuits, under RF accelerated life tests. Investigations at the Naval Research Laboratory have correctly emphasized accelerated life testing under RF drive.

In contrast with the above investigation, a somewhat lower power output MMIC was accelerated life tested at base plate temperatures of 170°C and 125°C. In this group of MMICs, the observed noncatastrophically failed circuits failed by a wearout mechanism related to leakage currents through the passivation [61]. Therefore, the lower power MMICs may initially wear out by surface effects rather than ohmic contact or electromigration related problems. The role of surface phenomena in MMICs is presently being investigated, both for MESFETs and for HBTs [62].

Therefore, the reliability concerns for high power MMICs fabricated with the standard 1 micron gate length, ion implantation technology are related to the drain current degradation which accompanies output power degradation and may be indicative of ohmic contact interdiffusion. The problems with ohmic contact degradation have also been analyzed for low power MMICs and may represent the typical mode of failure where trapping phenomenon and surface degradation is not present. Apparently, in low power MMICs, surface phenomena may be the dominant wearout mechanism.

The extensive reliability investigations carried out by Texas Instruments concentrated on understanding the degradation mechanisms in GaAs MMICs operated at higher than normal drain voltages and high levels of RF compression. The TI experiments [63] were also reexamined by Anderson et al. [61] and indicated device median life of approximately 10^7–10^8 hours at 140°C channel temperature, for the low- to medium-power amplifiers. The RF compression tests were carried out with a two-stage, 1-W, wideband power amplifier with 5 mm of gate width, operating at 7 V, 700 mA, and 2-dB compression. The base plate temperature was set to 50°C, corresponding to a 140°C channel temperature. Having observed no failures, the authors concluded that MMICs designed for high-compression operation will be highly reliable when operated within the designed limits.

The investigations of Anderson et al. [61] consisting of life tests of 1-dB compression, predicted a MTBF value at 125°C of 2×10^4 hours and 9×10^3 hours for two different power amplifier MMICs. The principal failure mechanism was reported to be excessive leakage current in the channel region. AuGe ohmic contact electromigration was also observed although such a mechanism may not have been the principal cause for failure. Surface degradation appears to be the dominant mechanism. Physically, the coupling between passive circuit degradation and the active device performance exists through the amplifier matching requirements. The effect of the circuit matching on the failure distribution functions must be determined, as well as surface thermal distributions resulting in surface damage leakage currents.

1.6 ELECTROMIGRATION DEGRADATION

Power amplifiers and power field effect transistors have been assessed as alternatives to TWTAs in a variety of space communications satellite systems with a design criterion of 10 years of life. As an example of these investigations, the results obtained with the Fujitsu FLC-30, 3.0-W and FLC-50, 5.0-W transistors will be reviewed [64]. These circuits, in principle, meet the system design requirements of 44% power added efficiency at 2.5-W power output. The Fujitsu devices use AuGe/Au source and drain contacts and Al gates. For RF testing, the following electrical parameters are employed:

$$I_{DS} = 1/2\ I_{DSS}$$

$$\text{RF input} = 1\ \text{W}$$

$$V_{DS} = 9\ \text{volts}$$

V_{GS} is adjusted to provide $I_{DS} = 500$ mA at testing temperatures. Channel temperatures up to 225°C were used. The FLC-30 devices degraded gracefully over 1150 hours, exhibiting increases in gate current from a few microamperes at the start of the test to approximately 1 mA at the end of the test. The latter value corresponds to 2×10^5 A/cm^2. All the FLC-50 devices tested at 225°C failed within a few minutes. Devices tested at 200°C resulted in an MTTF of 2200 hours with the source–drain electromigration as the primary failure mode. Based on a 1.1-eV activation energy, the extrapolated MTTF at 100°C is 3×10^6 hours [65]. Failed circuits shown in Figures 1.6–1.8 show electromi-

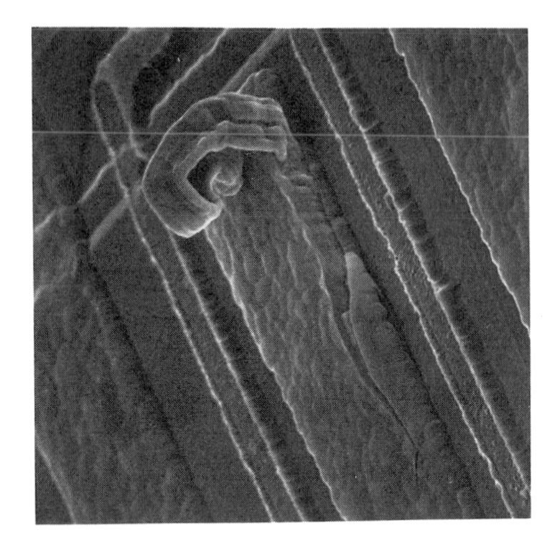

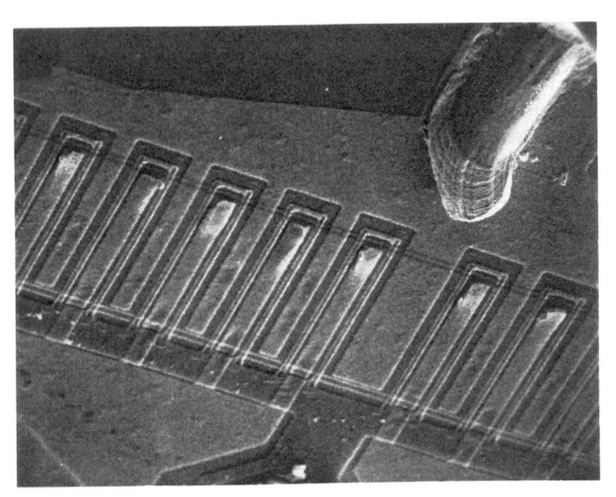

FIGURE 1.6. Typical electromigration effects in GaAs power FETs showing Ga whisker growth.

gration, Ga interdiffusion in the contact system with subsequent Ga whisker growth, and gate void formation due to the AuAl couple formation in the gate.

Electromigration effects in FETs readily occur in devices and circuits operated at microwave saturation. Microwave saturation can drive the gate current (or the rms of gate current) well above the electromigration limits of 10^6 A/cm^2. This effect can be shown with Texas Instruments X-band parts, where the devices were tested with power input of 320 mW, for which the channel temperature was estimated to be between

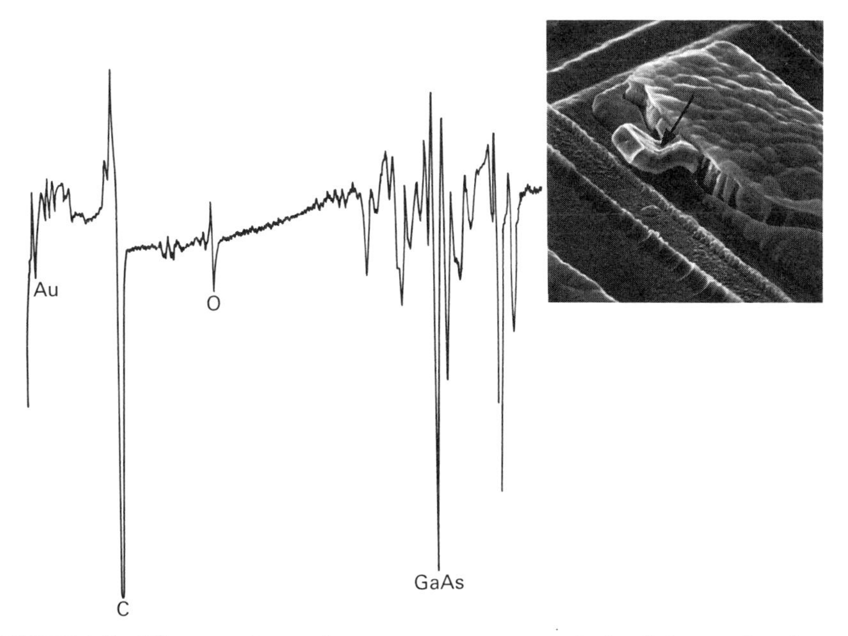

FIGURE 1.7. Microspot Auger electron spectroscopy analysis of electromigration region.

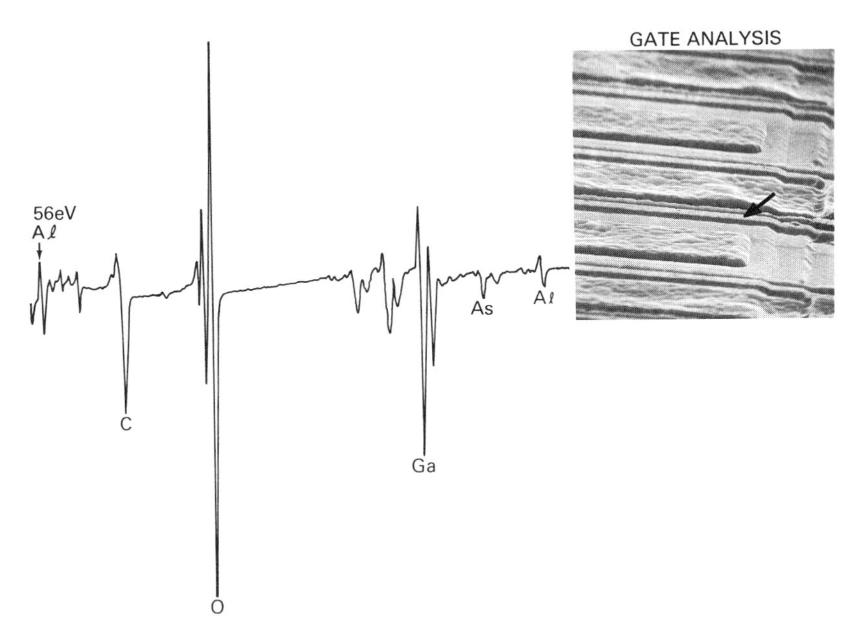

FIGURE 1.8. Microspot Auger electron spectroscopy analysis of aluminum gate region.

120°C and 130°C. The devices were characterized by one or more sharp drops in the strip recording. The devices also failed to pinch off after return to DC bias testing. The devices were examined with electron beam-induced current mode (EBIC) in the SEM, which showed a disappearing gate region. Figure 1.9 displays the EBIC and SEM superimposed, showing voids in the gates, and white strips being good or conducting gates. It can clearly be seen that only half of the fifth gate from the left is operative and that a gate void has occurred under one of the source bond wires. Figure 1.10 is a highly magnified photograph of a void located by EBIC. All these circuits failed by gate void formation and were subsequently analyzed by micro-spot Auger electron spectroscopy (AES) and energy dispersive x-ray (EDAX) techniques. The analysis of the gates revealed the presence of a surface compound of AuAl extending from the gate pad to the edge of the void, as well as the presence of Ga in the same region. These failures are directly related to the gate current flow during RF operation. The RF current can be estimated as

$$I_{\mathrm{rms}} \approx \frac{(P_I)^{1/2}}{R_I}$$

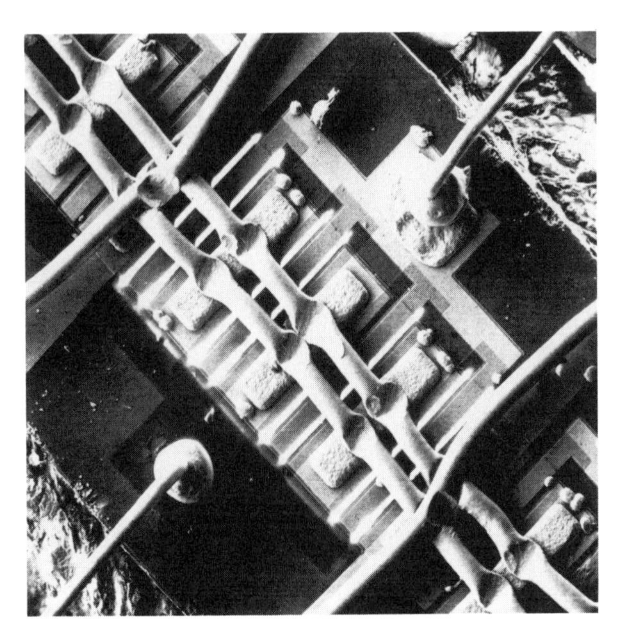

FIGURE 1.9. Superposition of a SEM and EBIC image; active gates show up as white strips.

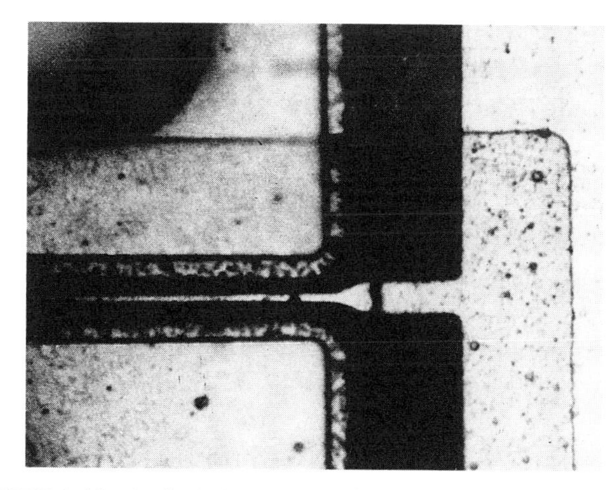

FIGURE 1.10. Optical photograph of a gate void located by EBIC.

where R_I is the device input resistance. For the devices tested, R_I is a few ohms (10–20 ohms) and at a P_I of 500 mW the rms microwave gate current can be as high as 10^7 A/cm^2 [66]. The failed devices indicated that failure occurred by AuAl reactions resulting in voids followed by electromigration in the gate due to the presence of a concentration gradient.

1.7 CONCLUSIONS AND SUMMARY

The performance levels of MMICs has reached the level where rapid insertion is occurring in systems that require such circuits. The MMIC technologies are based on MESFETs, HEMTs, and HBTs. The failure physics investigations summarized in this chapter for power MMICs and HEMT ICs indicate the presence of unique failure mechanisms, which must be further discussed.

This chapter has also discussed the failure distribution for commercial MMICs. It is shown that one must carefully distinguish between the mean time to first failure (passive components) and the general mean time to failure for the entire circuit. Including both types of distributions in a single cumulative distribution function results in a more optimistic MTTF prediction than previously reported. The coupling between the early failures and the main failure distribution must be further assessed through an appropriate failure of physics model. Physically, the coupling between passive circuit degradation and the active device performance exists through the amplifier matching requirements. The effect of the circuit matching on the failure distribution functions must be determined. The classical metallization related problems and surface effects continue to affect power MMIC reliability levels.

REFERENCES

1. P. Harrop, P. Lesartre, and T. Vlek, "Low Noise 12 GHz Front End Designs for Direct Satellite Television Reception," *Philips Tech. Rev.*, Vol. 39, No. 10, 252–270 (1980).
2. A. Fazal, S. Mogh, and R. Ramachandran, "A Highly Integrated Monolithic X-Ku Band Upconverter," *GaAs IC Symp. Tech. Digest*, 157–160 (1988).
3. D.C. Yang, T.S. Lin, R. Esfandiari, S. Bui, T.J. O'Neil, and D. Chow, "Wideband Fully Monolithic X and K-Band GaAs Receiver," *GaAs IC Symp. Tech. Digest*, 165–167 (1988).
4. A.A. Lane, J.A. Jenkins, C.R. Green, and F.A. Meyers, "Passive Elements for X-Band Amplifiers," *GaAs IC Symp. Tech. Digest*, 259-262 (1989).
5. B. Adelseck, "Monolithically Integrated 94 GHz Balanced Mixer," *Proceedings 1990 MTT-S*, 193–196 (1990).
6. W.R. Wisseman, L.C. Witkowski, and G.E. Brehm, "X-Band GaAs T/R Module," *Microwave J.*, 167–173 (1987).
7. A.N. Lepore, H.M. Levy, R.C. Tiberio, P.J. Tasker, E.D. Wolf, and L. Eastman, *Electron. Lett.*, Vol. 24, 365–366 (1988).
8. L.F. Lester, P.M. Smith, P. Ho, P.C. Chao, R.C. Tiberio, and K.H.G. Duh, *IEDM Tech. Digest*, Vol. 172–175 (1988).
9. P.C. Chao, K.H.G. Duh, P. Ho, P.M. Smith, J.M. Ballingall, and A.A. Jabra, *Electron. Lett.*, Vol. 25, 504–505 (1989).
10. N. Ohkawa, "20 GHz Bandwidth Low Noise HEMT Preamplifier for Optical Receivers," *Electron. Lett.*, Vol. 24, 1061–1062 (1988).
11. R. Pyndiah, A. Desarte, M. Wolney, J.C. Meunier, and P. Chambery, *Electron. Lett.*, Vol. 25, 442–443 (1989).
12. P. Gamand, A. Desware, and J.-C. Meunier, *Electron. Lett.*, Vol. 25, 451–453 (1989).
13. P. Gamand, A. Deswarte, J.-C. Meunier, M. Wolny, and P. Chambery, *10th GaAs IC Symp. Digest*, 109–111 (1988).
14. K.H.G. Duh, P.C. Chao, P.M. Smith, L.F. Lester, B.R. Lee, J.M. Ballingall, and M.Y. Kao, *IEEE MTT-S Digest*, 923–926 (1988).
15. H.Q. Tserng, B. Kim, P. Saunier, H.D. Shih, and M.A. Khatibzabeh, *Microwave J.*, No. 4, 125–135 (1989).
16. P. Ho, P.C. Chao, K.H.G. Huh, A.A. Jabra, J.M. Ballingall, and P.M. Smith, *IEDM Tech. Digest*, 184–186 (1988).

17. U.K. Mishra, A.S. Brown, and S.E. Rosenbaum, *IEDM Tech. Digest*, 180–181, (1988).
18. U.K. Mishra, A.S. Brown, S.E. Rosenbaum, C.E. Hooper, and M.W. Pierce, *IEEE Electron. Device Lett.*, Vol. 9, No. 12, 647–649 (1988).
19. U.K. Mishra, A.S. Brown, L.M. Jelloian, and L.H. Hackett, *IEEE Electron. Device Lett.*, Vol. 9, No. 1, 41–44 (1988).
20. N. Nakata, O. Nakajama, T. Nittono, H. Ito, and T. Ishibashi, *Electron. Lett.*, Vol. 23, 64–65 (1987).
21. T. Ishibashi and Y. Yamauchi, "A Possible Near-Ballistic Collection in an AlGaAs/GaAs HBT," *IEEE Trans.*, Vol. 35, 401–404 (1988).
22. J.A. Higgins, "Heterojunction Bipolar Transistors for High Efficiency Power Amplifiers," *GaAs IC Symp. Tech. Digest*, 33–36 (1988).
23. Y. Yamauchi, O. Nakajima, K. Nagata, H. Ito, T. Nitton, and T. Ishibashi, "High Performance Ka-Band HBT Circuits," *Electron. Lett.*, Vol. 23, 881–882 (1987).
24. Y. Yamauchi, O. Nakajima, K. Nagata, and H. Ito, "A 20–26 Watt HBT Power Amplifier," *GaAs IC Symp. Tech. Digest*, 121–124 (1989).
25. S. Evans, J. Delaney, C. Fuller, D. Boone, C. Dubberly, and J. Hoff, "HBT Digital Circuits for High Density CPU," *GaAs IC Symp. Tech. Digest*, 109–112 (1987).
26. K. Poulton, J.S. Kang, and H.J. Corcoran, "Thermal Response of HBT Power MMICs," *GaAs IC Symp. Tech. Digest*, 199–202 (1988).
27. A.K. Oki, M.E. Kim, J.B. Camou, C.L. Robertson, G.M. Gorman, and K.B. Weber, "Self Aligned Emitter HBT Technology," *GaAs IC Symp. Tech. Digest*, 137–140 (1987).
28. K.C. Wang, P.M. Asbeck, M.F. Chang, G.J. Sullivan, and D.L. Miller, "High Speed ADC Mesfet Circuits," *GaAs IC Symp. Tech. Digest*, 83–86 (1987).
29. K. Poulton, J.J. Corcoran, and T. Hornak, "ADC Circuits for High Data Rate Applications," *IEEE Trans. Solid State Circuits*, Vol. SC-22, 961–970 (1987).
30. T. Wakimoto, Y. Akazaw, and S. Konaca, "10 Gb/s Sample and Hold Circuits," *ISSCC Tech. Digest*, 232–233 (1988).
31. T. Docourant, M. Binet, J.-C. Baelde, C. Rocher, and J.-M. Gibereau, "GaAs Mesfet Comparator With 5 bit Resolution," *GaAs IC Symp. Tech. Digest*, 337–340 (1989).
32. F.S. Auricchio and R.A. Rhodes, "A 12 Watt 20 GHz FET Power Amplifier," *IEEE MTT-S Int. Microwave Symp. Digest*, Vol. III, 933–936 (1989).
33. A. Fraser and D. Obgonnah, "Reliability Investigation of GaAs I.C. Components," *GaAs IC Symp. Tech. Digest*, 161–164 (1985).
34. K. Tsuda, J. Akagiand, and J. Yoshida, "High Breakdown Voltage HBTs

for 12 GHz Applications,'' *19th Conf. Solid State Dev. Mater.*, Tokyo, 271–274 (1987).

35. B. Bayraktaroglu, R.D. Hudgens, M.A. Khatizadeh, and H.Q. Tserng, ''High Power HBTs for Wideband Applications,'' *IEEE MTT-S Int. Microwave Symp. Digest*, Vol. III, 1057–1060 (1989).
36. N.H. Sheng, M.F. Chang, P.M. Asbeck, K.C. Wang, G.J. Sullivan, D.L. Miller, and J.A. Higgins, ''HBTs With High Power Added Efficiency,'' *IEDM Tech. Digest*, 619–622 (1987).
37. M.E. Kim, A.K. Oki, J.B. Camou, P.D. Chow, B.L. Neison, and D.M. Smith, ''12–40 GHz Low Harmonic Distortion and Phase Noise Performance of GaAs Heterojunction Bipolar Transistors,'' *GaAs IC Symp. Tech. Digest*, 117–121 (1988).
38. B. Bayraktaroglu, N. Camilleri, H.D. Shih, and H.Q. Tserng, ''2.5 W CW X-Band HBT,'' *IEEE MTT-S Digest*, 969–972 (1987).
39. N. Hayama, S.R. Lesage, M. Madihian, and K. Honjo, ''Coupled Logic Circuits Using AlGaAs HBTs,'' *IEEE MTT-S Tech. Digest*, 679–682 (1988).
40. Y. Yamauchi and T. Ishibashi, ''Direct Coupled Amplifier Using AlGaAs/GaAs CBTs,'' *GaAs IC Symp. Tech. Digest*, 121–124 (1988).
41. T. Preston, ''Accelerated Testing,'' Eighth Annual Reliability Testing Institute, Tucson, AZ (1982).
42. MIL-HDBK-217, ''Reliability Prediction of Electronic Equipment,'' p. 5.1.3.10 (1982).
43. M.J. Howes and D.V. Morgan, ''*Reliability and Degradation: Semiconductor Devices and Circuits*, Wiley, New York, p. 197 (1987).
44. I.T. Alexanian and D.E. Brodie, ''A Method for Estimating the Reliability of ICs,'' *IEEE Trans. Reliab.*, Vol. R-26, 359–361 (1977).
45. F. Jensen, ''Case Studies in System Burn-In,'' *Reliab. Eng.*, Vol. 3, 13–22 (1982).
46. R. Esfandiari, T.J. O'Neill, T.S. Lin, and R.K. Kono, ''Accelerated Aging and Long-Term Reliability Study on Ion-Implanted GaAs MMIC IF Amplifier,'' *IEEE Trans. Electron Devices*, Vol. 36, 1174 (1990).
47. R. Esfandiari, T. Sato, J. Furuya, L. Pawlowicz, and L.J. Lee, ''Reliability and Failure Analysis of MMIC Amplifier Fabricated on Various GaAs Substrates,'' *IEEE GaAs IC Symp.*, 325 (1990).
48. P. Ersland and J.P. Lanteri, ''GaAs FET MMIC Switch Reliability,'' *IEEE GaAs IC Symp.*, 57 (1988).
49. M.M. Dumas, G. Kervarrec, J.F. Bres, J. Boulaire, M. Gauneau, and D. Lecrosnier, ''Investigation on Interelectrode Metallic 'Paths' Affecting the Operation of IC MESFETs,'' *IEEE GaAs IC Symp.*, 15 (1987).
50. W.J. Roesch and M.F. Peters, ''Depletion Mode GaAs IC Reliability,'' *IEEE GaAs IC Symp.*, 27 (1987).

51. M. Spector and G.A. Dodson, ''Reliability Evaluation of a GaAs IC Preamplifier HIC,'' *IEEE GaAs IC Symp.*, 19 (1987).
52. K. Katsukawa, T. Kimura, K. Ueda, and T. Noguchi, ''Reliability Investigation on S-Band GaAs MMIC,'' *IEEE Microwave Monolithic Circuits Symp.*, 57 (1987).
53. Y. Hosono, H. Sato, Y. Mima, S. Ichikawa, H. Hirayana, K. Katsukawa, K. Ueda, K. Uetate, T. Noguchi, and H. Kohzu, ''Stability and Reliability Investigation of Fully ECL Compatible High Speed GaAs Logic ICs,'' *IEEE Microwave Monolithic Circuits. Symp.*, 49 (1987).
54. W.J. Roesch, ''Thermo-Reliability Relationships of GaAs ICs,'' *IEEE GaAs IC Symp.*, 61 (1988).
55. M. Nishiguchi, J. Fujihira, A. Miki, and H. Nishizawa, ''Precisional Comparison of Surface Temperature Measurement Techniques for GaAs ICs,'' *IEEE SEMI-THERM Symp.*, 34 (1991).
56. J.L. Wright, B.W. Marks, and K.D. Decker, ''Modeling of MMIC Devices for Determining MMIC Channel Temperatures During Life Tests,'' *IEEE SEMI-THERM Symp.*, 131 (1991).
57. W. Seely, ''Determine MMIC Maximum Processing Temperature,'' *Microwave J.*, 170 (Aug. 1989).
58. A. Christou, University of Maryland, MMIC Progress Reports, 1991–92.
59. A. Christou, J.M. Hu, and W.T. Anderson, ''Reliability of InGaAs HEMTs on GaAs Substrates,'' *1991 IEEE/IRPS*, 200–206 (1991).
60. A. Christou, J.M. Hu, and P. Tang, ''Reliability of 2–8 GHz MMIC Amplifier,'' *1991 IEEE/GaAs Man. Tech. Conf.*, 200–204 (1991).
61. W.T. Anderson, J.A. Roussos, and K.A. Christianson, ''GaAs MMIC Reliability Studies,'' *ESREF 91*, Vol. 1, 411–422 (1991).
62. K. Poulton, L. Knud, L. Knudsen, and J. Corcoran, ''Thermal Simulation and Design of a GaAs HBT Sample and Hold Circuit,'' *IEEE GaAs IC Symp.*, 129–132 (1991).
63. K. Decker, ''Reliability of Power MMICs Subjected to 2 dB RF Compression,'' 1991 GaAs Reliability Workshop, EIA-JEDEC JC-50, 35–40 (1991).
64. W. Sluzark, ''GaAs Power FET Reliability,'' Cornell Conference on Active Microwave Devices and Circuits, Aug. 14, 1979, Paper IV-7.
65. A. Christou, ''GaAs MMIC Design and Reliability,'' European GAAS Applications Conference April 6–8, 1990, Rome, Italy, Paper 2-1.
66. A.C. Macpherson, K.R. Gleason, and A. Christou, ''Voids in Power FETs Under Microwave Testing,'' AFTA Symposium, Los Angeles, 1978, Paper 21.

2

SIMULATION AND COMPUTER MODELS FOR ELECTROMIGRATION

PIN FANG TANG
CALCE Electronic Packaging Center
*University of Maryland, College Park**

2.1 ELECTROMIGRATION MODELING

2.1.1 Introduction

Among various approaches developed to study electromigration phenomena, computer simulation is a powerful and efficient one. Through a computer simulation, it is easier to consider multiple mechanisms involved in electromigration to enhance the accuracy of the model. Also, it enables one to observe the macroscopic effects of the microscopic variables to reveal the insights of this phenomenon, which may not be experimentally possible. For example, it would be difficult to analytically or experimentally keep tracking the temperature and stress distributions along a conductor line during electromigration, which are important factors affecting the degradation process, whereas the instantaneous numerical solutions of the temperature and stress field equations provide such information; it would be practically impossible to know the grain boundary misorientation angles and the inclination angles with respect to electron flow for all the grain boundaries in a conductor film, and therefore difficult to know their quantitative effects on the failure rate, while they can easily be found out with the computer-

*Present address: IBM, Essex Junction, Vermont.

Electromigration and Electronic Device Degradation, Edited by Aris Christou.
ISBN 0-471-58489-4

constructed films. Furthermore, only with the aid of computer modeling would it be possible to "conduct" life tests on systems having a large number of conductor lines, especially at normal operating conditions, to get statistically meaningful results on median time-to-failure (MTF).

As electromigration becomes a more and more concerned reliability issue for the integrated circuits, an accurate, self-contained computer simulation model is desired. It is the purpose of this chapter to introduce the concepts and methodologies on establishing such a simulation model. It is the author's hope that this introductory work will provide the necessary background on developing a useful tool for circuit simulation and modeling to predict electromigration reliability.

2.1.2 Failure Mechanisms

It is now understood that electromigration is a mass transport in a diffusion-controlled process under a certain driving force. In the temperature range commonly concerned ($<0.5T_{\text{melt}}$), this diffusion is mainly through grain boundaries. Such mass transport therefore obeys the well established diffusion equation [1–3]. However, the driving force here is more complicated than what is involved in a pure diffusion process, in which the concentration gradient of the moving species is the only component. The electrical driving force for electromigration consists of the "electron wind force" and the direct field force [2–6]. The electron wind force refers to the effect of momentum exchange between the moving electrons and the ionic atoms when an electrical current is applied to a conductor. When current density, which is proportional to the electron flux density, is high enough, this momentum exchange effect becomes significant, resulting in a noticeable mass transport—electromigration. Nevertheless, while the ions tend to move in the direction of the impulse during the momentum transfer, which is in the direction opposite to the electrical field, they also tend to move in the direction of the applied field since they are positively ionized. The balance of these two forces determines the movement of the ions. For gold and aluminum, electron wind force dominates over the direct field force; the net electrical driving force is therefore in the direction of the electron wind [7–9]. For simplicity, the term "electron wind force" often refers to the net effect of these two electrical forces. This simplification will also be used throughout the following discussion. Besides the electron wind force, other components of the electromigration driving force basically all result from the inhomogeneities caused by the electromigration-induced damage itself [10–17]. Most of the analytical models developed to date include some components of these damage-induced driving forces and few models take all of them into consideration due to the increased complexity. In this chapter, the components that are of major effect on electromigration will be discussed separately and may be integrated into one complete model for further development.

The three predominant mechanisms in electromigration failure process discussed in this chapter include those associated with (a) the metallurgical-statistical properties of the conducting film, (b) the thermal accelerating process, and (c) the healing effects.

The metallurgical-statistical properties of a conductor film refer to the microstructure parameters of the conductor material, such as the grain size distribution, the distribution of grain boundary misorientation angles, and the inclinations of grain boundaries with respect to electron flow. These metallurgical parameters can only be dealt with statistically because they usually appear to be random. As illustrated in Figure 2.1, the misorientation angle, θ, between the two grains defining the grain boundary determines the mobility of the atoms in that boundary; the grain boundary inclination with respect to the electron flow, ϕ, partially determined by grain size variation, determines the effectiveness of the applied field in that grain boundary; and the grain size variation determines the change in the number of the atomic paths across a cross section of the conductor line. The variations of all these microstructural parameters over a film cause a nonuniform distribution of atomic flow rate. Therefore nonzero atomic flux divergence exists at the places where the number of atoms flowing into the area is not equal to the number of atoms flowing out per unit time [18–27]. With the nonzero atomic flux

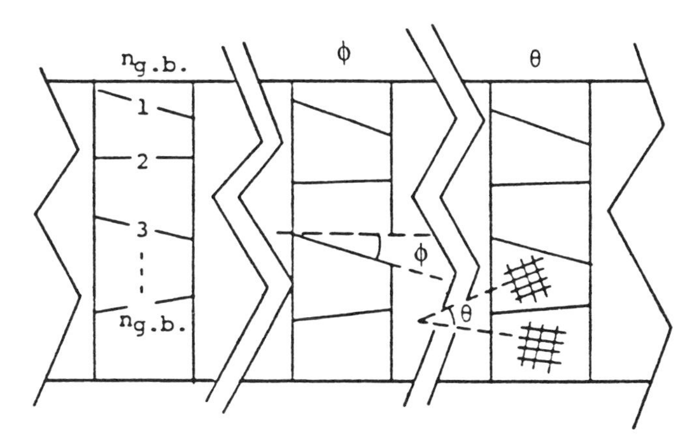

FIGURE 2.1. Schematic illustration of grains, grain boundaries, grain boudnary misorientation angles, θ's, and inclinations angles, ϕ's. Aftger Nikawa Kiyoshi, *IEEE Int. Reliab. Phys. Symp.*, CH1619-6, 175 (1981). © IEEE 1981. Reprinted with permission.

divergence, there will be either a mass depletion (divergence > 0) or accumulation (divergence < 0), leading to formation of voids or hillocks. Microstructural inhomogeneities alone can cause nonzero flux divergence, in spite of the temperature distribution, concentration gradient, and so on. It is therefore the basic mechanism that should be considered.

The thermal accelerating process refers to the acceleration process of electromigration damage due to the local temperature rising. A uniform temperature distribution along a conductor film is possible only before any electromigration damage occurs. Once a void is initiated, it causes the current density to increase in the vicinity around itself because it reduces the cross-sectional area of the conductor. The increase of the local current density is referred as the current crowding. Since joule heating is proportional to the square of current density, the current crowding effect leads to a local temperature rise around the void, which in turn further accelerates the void growth. The whole process continues till the void is large enough to break the line [21, 28–29]. Such a self-accelerating, or thermal run-away, process can be illustrated by the positive feedback cycle in Figure 2.2. When the heat dissipation through the film substrate is poor, or under certain extreme biasing conditions, the thermal run-away process could be the dominant electromigration failure mechanism.

In addition to the aforementioned thermal acceleration process, temperature gradient may exist along a conductor line even before any dam-

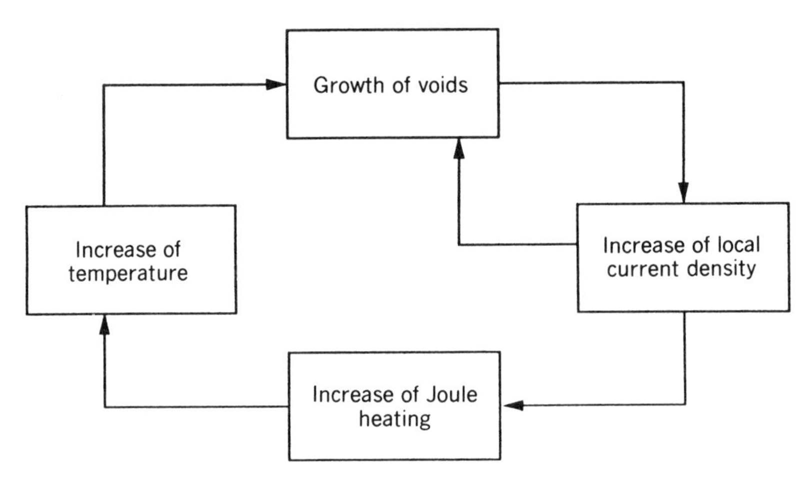

FIGURE 2.2. Thermal acceleration loop during electromigration. After Kiyoshi Nikawa, *IEEE Int. Rel. Phys. Symp.*, CH1619-6, 175 (1981), © IEEE 1981. Reprinted with permission.

age formation due to the limited controllability on the temperature uniformity. Temperature gradient itself could also lead to nonuniform flux divergence since the atomic mobility depends on it in a sensitive way. Thus an accurate temperature determination along the conductor film is critical in modeling the electromigration failure process.

The healing effects refer to those caused by the atomic flow in the direction opposite to the electron wind force, the backflow, during or after electromigration. This backflow of mass begins to take place once a redistribution of mass has begun to form. It tends to reduce the failure rate during electromigration and partially heals the damage after the current is removed. The cause of this backflow of mass is the inhomogeneities, such as temperature and/or concentration gradients, resulting from electromigration damage. These inhomogeneities move the system away from its equilibrium state. When the damage is below certain critical level the disturbed system then tends to relax back to its original state to minimize its total system energy by generating the backflow flux [10–17, 30]. For example, one of the components in this mass backflow is that of the stress-induced discovered by Blech [10, 11]. It was found that the mass accumulation and depletion caused by electromigration at the two ends of a conductor line produce a stress gradient across it. A mass backflow is thus formed to relax this stress gradient. Other examples of mass backflow include those driven by atomic concentration gradient [22, 23, 31] and increase of capillarity [22, 32–34]. Readers are referred to the original works for further information on modeling of the healing effects.

As a result of this mass backflow, there exists a threshold current density for electromigration to become effective [10–17]. The value of this threshold current density corresponds to a minimum energy barrier that the atoms have to overcome to balance off the backflow driving forces. The healing effect is important because a thorough understanding of such a phenomenon would make it possible for one to utilize it in reducing the electromigration effect with the application of AC or pulsed DC current [30, 33].

2.1.3 Modeling Approaches

Several types of computer simulation models for electromigration have been developed [20–21, 28, 31–32, 34–39]. One of the simplest is the probability model. With a probability model, the statistical measures of metallization reliability such as MTF distribution, the standard deviation, and their dependence on line geometry are the main concerns and outputs [37, 40]. The model is elaborated with a probability calculation.

It assumes that a conductor line is composed of a number of unit elements connected in a certain manner (series, parallel, or combination of both) and the failure probability of each unit element is an independent random variable following a lognormal distribution. The failure probability of the conductor as a whole is calculated based on the failure probability of each element and the way they are connected to each other. With a predefined failure criterion, the failure time of each conductor film can then be obtained. The probability model yields qualitative agreements on the line length and width dependence of MTF with experimental results and shows certain statistical features of electromigration failure in a simple way. However, it provides little understanding on the failure mechanisms and the underlying physics. Also, the assumption of a specific type of failure probability, lognormal, for example, for each unit element has to be made as the basis of the calculations, which is sometimes arbitrary or less justified. Therefore the probability modeling method will not be discussed in this chapter. Readers are referred to the original works for more details.

The computer simulation approaches presented in this chapter are based on either the metallurgical-statistical or the thermal mechanism, or a combination of both. Healing effects are also considered although in practice they are sometimes neglected for simplicity.

In the following sections, the physics governing the failure mechanisms will first be explained. The driving forces and the above-mentioned failure mechanisms, which form the basis of a computer simulation, will be analyzed in detail. Monte Carlo approaches will then be introduced to realize the models with statistical considerations. As a conclusion, some typical results will be shown with brief discussions.

2.2 PHYSICAL MODELS

The physical models discussed in this section are two-dimensional. In most applications, such an assumption is physically valid since the thicknesses of metallization lines are usually small (~2000–5000 Å) compared with the linewidth and the size of the voids that have become significant in degrading the line structures (<1 μm). The only reason to make such an approximation is to simplify the mathematical analysis and also to reduce the computing time. For the situations where the two-dimensional approximation does not apply, it is possible to extend a two-dimensional model to a three-dimensional model since the physical analysis at the metal plane would essentially be the same. The diffusion perpendicular to the plane needs to be considered then. Furthermore, the mass transport during electromigration will be assumed to be along grain boundaries only since, as mentioned in the earlier section, they are the main paths of the atomic movement in the temperature range of interest. Therefore the grain boundary diffusion will be the central concern.

2.2.1 Failure Sites and Flux Divergence

A two-dimensional conductor film can be considered as an ensemble of grain boundaries and their intersections as illustrated in Figure 2.3. Experimental observations have indicated that in most cases, mass depletion and accumulation initiate at grain boundary intersections [22, 25, 27] such as triple junctions. The former would eventually lead to the formation of voids or cracks and the latter to hillocks or whiskers. The reason that the grain boundary intersections are likely to be the failure sites is that they often represent the spots where the mass flux would diverge the most. At the grain boundary intersections, there could be an abrupt change in grain size, which produces a change in the number of paths for mass movement; there also could be a change in atomic diffusivity due to the change in grain boundary microstructure [22, 24, 40–41]. Thus the failure sites, meaning the nucleation sites of mass accumulation or depletion, in the following discussion will be considered to be at grain boundary intersections only.

The atomic or ion flux **J** is related to the total driving force **F** by the diffusion Equation (2.1)–(2.3):

$$\mathbf{J} = \frac{ND}{kT}\mathbf{F} \tag{2.1}$$

where N is the atomic concentration, D is the diffusion coefficient, k is the Boltzmann constant, and T is the absolute temperature. In general, the atomic flux in the ith grain boundary is

$$J_i = \frac{N_i D_i}{kT_i} F_i \tag{2.2}$$

where the subscript i denotes the ith grain boundary.

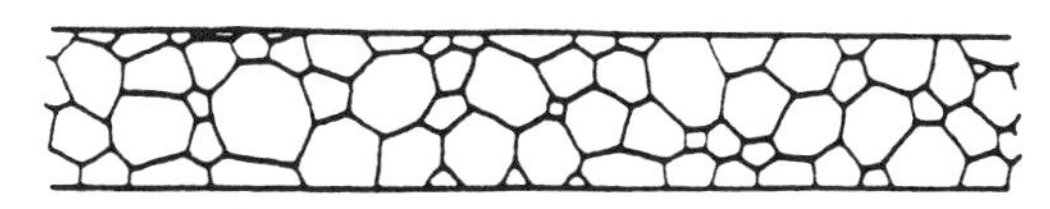

FIGURE 2.3. A two-dimensional grain texture. After Cho J., and Thompson, C. V. *Appl. Phys. Lett.*, Vol. 54, No. 25, 2577 (1989). Reprinted with permission.

In the following discussion, the atomic concentration is assumed to be constant and is equal to its grain boundary value, $N_i = N_{gb}$; the temperature is assumed to depend only on time and the position in a conductor line, namely, $T_i = T(x, y, t)$. Here, x and y specify a position on the metal plane in a Cartesian coordinate system. For a given system and temperature, the diffusion coefficient in a grain boundary is determined by the grain boundary misorientation angle, θ, which has been described previously. A commonly accepted expression for D_i is

$$D_i = D_0 \sin \frac{\theta_i}{2} e^{-Q_0/kT} \qquad \theta_i < 37^\circ \tag{2.3}$$

$$D_i = D_0 \sin 37^\circ e^{-Q_0/kT} \qquad (37^\circ < \theta_i < 60^\circ) \tag{2.4}$$

where D_0 and Q_0 are the prefactor and the grain boundary activation energy of the film, respectively [19, 20, 42, 43]. The activation energy in the above two equations is assumed to be constant over a film surface. Alternatively, the diffusion coefficient D_i can also be expressed by

$$D_i = D_0' e^{-(Q_0 + \Delta Q_i(\theta_i))/kT} \tag{2.5}$$

where D_0' is the prefactor, $Q_i = Q_0 + \Delta Q_i(\theta_i)$ is the total activation energy of the ith grain boundary, and Q_0 here is considered as the average grain boundary activation energy [32, 34]. The purpose of introducing Equation (2.5) is to represent the variation of θ's through the variation of activation energy ΔQ's over the entire ensemble of grain boundaries. The prefactors D_0 and D_0' are assumed to be constant in both approaches. The diffusion coefficient and its microstructure dependence can be understood in the following way. Since θ varies from one boundary to another, the bonding force between ions varies, and consequently the energy required to move an ion along each boundary would also vary, resulting in the variation of diffusion coefficient. The effect of the microstructure variation can thus be equivalently expressed in either approach described above. In both cases, the θ dependence can be represented by Θ,

$$\Theta = \sin \frac{\theta}{2} \qquad (\theta < 37^\circ) \tag{2.6}$$

$$\Theta = \frac{\sin 37^\circ}{2} \qquad (37^\circ < \theta < 60^\circ) \tag{2.7}$$

or

$$\Theta = e^{-\Delta Q_i(\theta_i)/kT} \tag{2.8}$$

and the diffusion coefficient D_i becomes

$$D_i = D_0 \Theta_i e^{-Q_0/kT} \tag{2.9}$$

where Θ_i is defined by Equations (2.6), (2.7), or (2.8), and D_0 is a constant determined by the material of the conductor film. Equation (2.9) defines the microstructural dependence of grain boundary diffusion coefficient solely through Θ.

The driving force on the ions due to the applied electrical field, **E,** in the ith grain boundary, is

$$F_i = Z^* qE \cos \phi_i = Z^* q\rho_0 j [1 + \alpha(T - T_0)] \cos \phi_i \tag{2.10}$$

where Z^*q is the effective charge of the ions. Z^* counts for the balance between the electron wind force and the direct electrical field force, **j** is the applied current density, ϕ_i is the inclination angle of the ith grain boundary with respect to the electron flow, T is the temperature at time t, and T_0 is the temperature at a reference temperature. ρ_0 is the resistivity of the film at the reference temperature and α is the temperature coefficient of the resistivity. Apparently, in Equation (2.10), Ohm's law, $\mathbf{E} = \rho \mathbf{j}$, and $\rho = \rho_0[1 + \alpha(T - T_0)]$ have been applied.

Combining Equations (2.2), (2.9), and (2.10), the atomic flux in the ith grain boundary can be expressed by

$$J_i = \frac{N_{gb} D_0}{kT} Z^* q\rho_0 [1 + \alpha(T - T_0)] \Theta j \cos \phi_i e^{-Q_0/kT} \tag{2.11}$$

Defining

$$\Delta Y = \sum_{i=1}^{n_{gb}} \Theta_i \cos \phi_i \tag{2.12}$$

where n_{gb} is the number of grain boundaries defining a intersection, the flux divergence at this intersection then becomes

$$\nabla \cdot \mathbf{J} = \sum_{i=1}^{n_{gb}} J_i = \frac{N_{gb} D_0}{kT} Z^* q\rho_0 [1 + \alpha \Delta T] \, \Delta Y \, j e^{-Q_0/kT} \tag{2.13}$$

where $\Delta T = T - T_0$. Considering the healing effects by adding in a component of counteracting current density, the threshold current density j_c [21, 29, 35], Equation (2.13) can be modified to

$$\nabla \cdot \mathbf{J} = \frac{N_{gb} D_0}{kT} Z^* q \rho_0 [1 + \alpha \Delta T][j - j_c] \Delta Y e^{-Q_0/kT} \tag{2.14}$$

The number of ions, N, flowing into an arbitrary intersection over a time period Δt is

$$\Delta N = \delta h \, \nabla \cdot \mathbf{J} \, \Delta t \tag{2.15}$$

where δ is the grain boundary width, h is the film thickness, and $\nabla \cdot \mathbf{J}$ is the flux divergence at the intersection. Thus the growth rate of the volume (V) of the mass depleted (or accumulated) at the grain boundary intersection under consideration is

$$\frac{\partial V}{\partial t} = \Omega_0 \frac{\partial (\Delta N)}{\partial t} = \delta h \Omega_0 \, \nabla \cdot \mathbf{J} \tag{2.16}$$

where the Ω_0 is the atomic volume.

2.2.2 Current Crowding Effect

The current density j in Equation (2.14) is usually not a constant. As soon as the electromigration-induced damage begins to form, current density becomes nonuniform due to the current crowding effect introduced in the earlier section. As the voids or cracks grow, the nonuniformity of the current density over a conductor line increases. Since the Joule heating is proportional to the square of the current density, the local temperature will also increase rapidly. The current crowding effect therefore plays double roles here: both the elevated local current density and temperature accelerate the electromigration process [Equation (2.14)]. Thus obtaining an accurate current density distribution is necessary in determining the flux divergence.

The instantaneous current density field can be obtained by solving the two-dimensional Laplace equation for electrical potential [35] $u = u(x, y)$ at each time step:

$$\frac{\partial}{\partial x}\left(k \frac{\partial u}{\partial x}\right) + \frac{\partial}{\partial y}\left(k \frac{\partial u}{\partial x}\right) = 0 \tag{2.17}$$

where $k = k(x, y)$ is the electrical conductivity. In solving the above equation, the existence of the voids can be considered by assigning $k = 0$ to those regions where material has been depleted away and no current is being conducted. The current density is then obtained through the electrical potential according to

$$\mathbf{j}(x, y) = -k \nabla \cdot u(x, y) \tag{2.18}$$

Since the line structure changes with time as the voids or cracks grow, Equation (2.17) needs to be solved at every time step, which may be time consuming. Therefore current crowding effect is sometimes described by simplified approaches. For example, when the voids are assumed to be small cylinders perpendicular to the metal plane, the current density in a particular cross section of the conductor line, as illustrated in Figure 2.4*a*, is

$$j_x = \frac{j_0}{1 - \sum_{i=1}^{n_c} \frac{d_i}{W}} \tag{2.19}$$

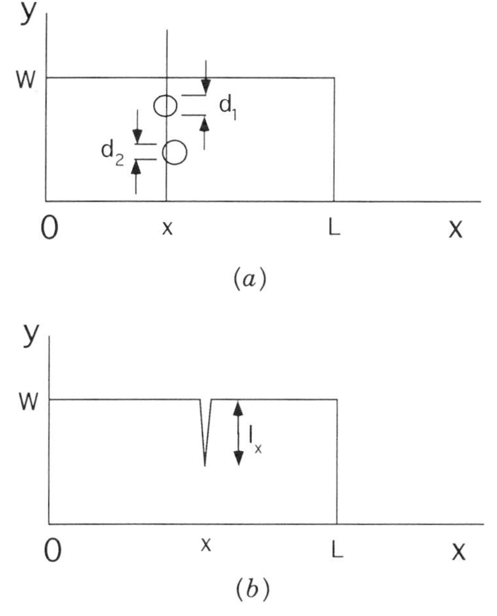

FIGURE 2.4. Schematic illustration of (*a*) cylindrical voids, and (*b*) crack in a conductor line of *W* wide and *L* long.

where j_0 is the initial current density, j_x is the current density at the cross section at x, n_c is the number of voids that intercept the cross section, d_i is the chord of the ith void on the cross section, and W is the linewidth [32, 34]. When the voids are mainly cracks growing from the edges to the inner region as shown in Figure 2.4*b*, the current density in the cross section where the crack is formed becomes

$$j_x = \frac{j_0}{1 - \dfrac{l_x}{W}} \tag{2.20}$$

where l_x is the crack length at x [28]. The simplification implied in both Equations (2.19) and (2.20) is that the current is assumed to be uniform within the undamaged regions.

2.2.3 Threshold Current Density

Threshold current density results from the mass backflow introduced as the healing effect in the previous section. Among all the components of the mass backflow, the stress-induced one has been more intensively studied [10–17]. In a drift-velocity experiment, Blech [10] studied the stress gradient across a conductor stripe as a result of electromigration-induced mass transport. He found that the mass accumulation and depletion formed at the two ends of the thin film stripe produce a stress gradient across it, resulting in a force that generates an ion flow in the direction opposite to the stress gradient, and also opposite to the electron wind force. In a simple one-dimensional case, this stress-generated force is

$$F_s = \Omega_0 \frac{\partial \sigma}{\partial x} \tag{2.21}$$

where x is the coordinate along the conductor length and σ is the stress along x. A direct consequence of this effect is the so-called electromigration threshold current density j_{th}. The nonelectromigration condition below the threshold current density requires that the electrical driving force F be less than F_s. Assuming that the maximum stress $\sigma_{\max}$ is reached at the anode side and neglecting the small stress at the cathode side, the partial derivative of σ with respect to x can be approximated as $\sigma_{\max}/L$, where L is the stripe length. When the electrical driving force becomes the same as F_s, the following threshold equation is obtained by simply combining Equations (2.10) and (2.21):

$$(jL)_{\text{th}} = \left| \frac{\Omega_0 \sigma_{\max}}{Z^* q'} \right| \tag{2.22}$$

In the above equation, $(jL)_{\text{th}}$ defines the threshold value for the product of line length and current density. Such a threshold product is useful in circuit design.

The concept of stress-induced mass backflow can be extended to be more precise. For a polycrystalline metal line, whose length is large compared with its width, the mass accumulation and depletion do not appear only at the ends of the lines; rather, they occur at every vacancy source and sink pair [44]. These vacancy sources and sinks are usually grain boundary intersections. Thus the length l in Equation (2.22) should be modified to be the "dipole length" of each pair of mass accumulation–depletion sites and it therefore is a variable along the conductor line. In a conductor line of many flux divergence dipoles, a stress field is generated around each dipole as the result of mass accumulation/depletion at the two sites of the dipole. This stress field is also superposed by the stress field of adjacent dipoles. The superposition of stress fields gives the total stress at any point in the line. The concept of "flux divergence dipole" has made it possible to explain the fact that not every grain boundary intersection in a conductor line—although they may all be nonzero flux divergence sites—would result in damage. The explanation is that each intersection has its own value of j_c. The applied electrical current may be high enough to overcome some of the j_c's but not all, leading to a different response in damage formation at every intersection. However, since the number of grain boundary intersections is usually quite large, sometimes it could be a good approximation to work with a single averaged threshold current density for the whole conductor stripe.

2.2.4 Thermal Modeling

The temperature variation on a conductor line can be the most important factor in accelerating electromigration degradation because of the exponential dependence of the atomic diffusivity on temperature. Temperature gradient may exist before any electromigration effect takes place. For instance, places closer to the bonding pads are colder than the center region of a line, and places closer to a heat source (such as a gate junction) or resistors are hotter than the other areas. As a result, flux divergence may occur solely due to the initial temperature gradient [2, 3, 28]. As the damage starts to form, the additional Joule heat generated by current crowding further increases the temperature gradient. The temperature gradient in turn accelerates the growth rate of the voids/hillocks following the positive feedback loop in Figure 2.2.

One of the approaches to obtain the temperature distribution is to solve the thermal equation assuming constant boundary conditions, that is, constant ambient temperature at the ends of the lines [35]:

$$\frac{\partial}{\partial x}\left(\tau \frac{\partial T}{\partial x}\right) + \frac{\partial}{\partial y}\left(\tau \frac{\partial T}{\partial y}\right) + j^2 \rho_0(1 + \alpha\,\Delta T) = 0 \tag{2.23}$$

where $\tau = \tau(x, y)$ is the thermal conductivity coefficient. This assumption may not be valid in the cases where the two ends of the conductors are not kept at a fixed temperature. In most experiments, the substrate of the conductors (semiconductors or sample holders) are kept either in a hot stage or in a constant-temperature chamber. It is therefore more reasonable to assume a constant temperature at the substrates. Thus the thermal equation becomes [21, 31]

$$-\tau \frac{\partial^2 T}{\partial x^2} - \tau \frac{\partial^2 T}{\partial x^2} = Q - \frac{\lambda}{h}(T - T_s) \tag{2.24}$$

where λ is the heat transfer coefficient between the film and the substrate and Q is the Joule heat generated in unit volume per unit time. T_s is the substrate temperature and is obtained by

$$T_s = T_a + P_d R_{sa} \tag{2.25}$$

where T_a is the ambient temperature, P_d is the power dissipated in the entire line, and R_{sa} is the thermal resistance between the substrate and the ambient. The Joule heat Q can be obtained from

$$Q = j^2 \frac{Wh}{\lambda_0} R_{\text{th}} \tag{2.26}$$

where λ_0 is the average grain size and R_{th} is the thermal resistance per unit volume. The modification from Equation (2.23) to (2.24) can be further generalized. Sometimes only the heat sink or package is held at ambient temperature. The heat dissipation through the adhesive materials between the semiconductor and the package and between the package and the heat sink also need to be considered. The substrate temperature can now be obtained from [39]

$$T_s = T_a + P_d \sum_{i=1}^{n_l} \theta_{R_i} \tag{2.27}$$

where n_l is the number of layers between the semiconductor substrate and the heat sink, and θ_{R_i} is the thermal resistance of the ith layer.

Thermal modeling is critical for an accurate determination of temperature distribution. Interested readers are referred to more detailed works [28, 29, 39, 45].

2.2.5 Structural Factor

The structural factor ΔY in Equation (2.14) reflects the effects of grain boundary microstructure on electromigration and plays an important role in forming the flux divergence. As defined by Equation (2.12), it is clear that the larger the *variation* in θ's and ϕ's, the greater the value of ΔY will be. Thus ΔY characterizes the grain boundary structural inhomogeneity.

To analyze the structural factor in greater detail, an intersection of three grain boundaries—a triple junction—is considered in the following. A triple junction shown in Figure 2.5*a* can be equivalently described as shown in either Figure 2.5*b* or 2.5*c*. In Figure 2.5*b*, the triple junction is characterized by the three relative angles (ϕ_{ij}'s) between each pair of adjacent grain boundaries, an orientation angle (θ_0) between a representative boundary (boundary 1) with respect to electron flow, and the grain boundary misorientation angles (θ_i's). In Figure 2.5*c*, the same triple junction can also be characterized by the orientation angles of individual grain boundaries with respect to the electron flow (ϕ_i's) and θ_i's. With Figure 2.5*b* as a reference, the structural factor can be reexpressed as

$$\Delta Y = \cos\theta_0\, e^{-\Delta Q_1/kT} + \cos(\theta_0 + \phi_{12}) e^{-\Delta Q_2/kT} + \cos(\theta_0 + \phi_{12} + \phi_{23}) e^{-\Delta Q_3/kT} \tag{2.28}$$

where ϕ_{ij} is the relative angle between the ith and the jth grain boundary and $i, j = 1, 2, 3$. It can easily be shown that ΔY, and therefore the flux divergence, vanishes for any value of θ_0 if the following conditions are met:

$$\Delta Q_1 = \Delta Q_2 = \Delta Q_3 = 0 \tag{2.29}$$

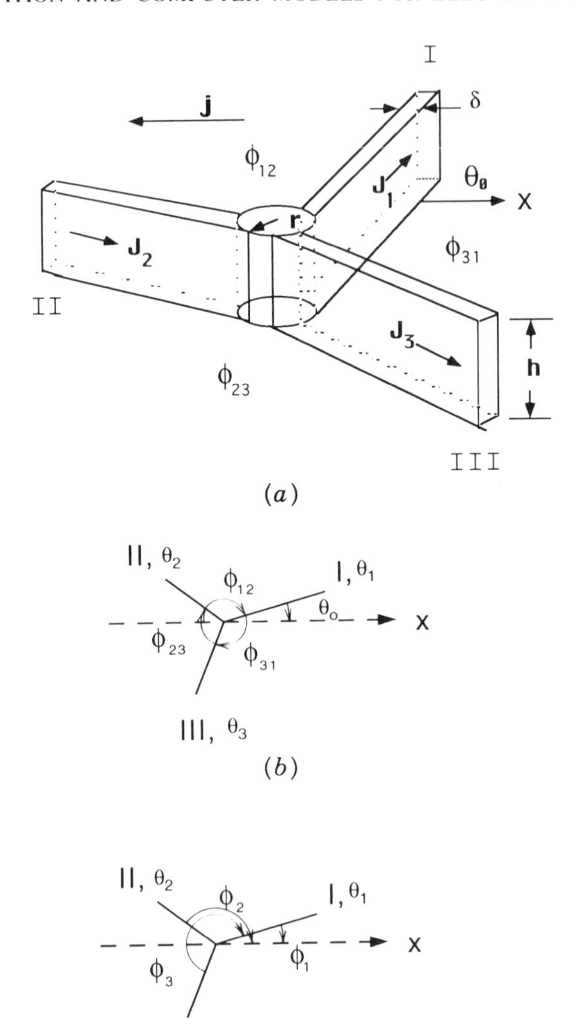

FIGURE 2.5. Schematic illustration of (*a*) a triple junction defined by grain boundaries I, II, and III, (*b*) and (*c*) the top view of the triple junction. x represents the direction of electron flow.

and

$$\phi_{12} = \phi_{23} = \phi_{31} = 120^\circ \tag{2.30}$$

This states that ΔY would be zero regardless of the orientation of the whole triple junction with respect to the direction of electron flow when the above conditions are satisfied. Equation (2.29) represents a case of uniform diffusion coefficient for all the grain boundaries, while Equation (2.30) requires a perfect hexagonal grain texture. However, due to the variations in crystal orientation in a polycrystalline film, the activation energy and therefore the diffusion coefficient vary from one grain boundary to another. Also, the relative angles between two adjacent grain boundaries are normally not 120° due to the variations of grain sizes [20, 25, 27, 40]. Flux divergence is therefore likely to exist at triple junctions as the result of the structural inhomogeneities. A mass depletion ($\Delta Y < 0$) or accumulation ($\Delta Y > 0$) occurs depending on the sign of ΔY. The knowledge of ΔY therefore is helpful for one to improve electromigration reliability by reducing the value of the structural factor. For example, high-temperature annealing of the metallization lines results in a more uniform grain size and grain boundary misorientation. Consequently, the flux divergences are reduced.

The properties of ΔY can also be viewed graphically [32, 34, 46]. Rearranging Equation (2.28), it becomes

$$\Delta Y = A \cos \theta_0 + B \sin \theta_0 \tag{2.31}$$

where

$$\begin{aligned} A &= e^{-\Delta Q_1/kT} + \cos \phi_{12}\, e^{-\Delta Q_2/kT} + \cos \phi_{31}\, e^{-\Delta Q_3/kT} \\ B &= \sin \phi_{31}\, e^{-\Delta Q_3/kT} - \sin \phi_{12}\, e^{-\Delta Q_2/kT} \end{aligned} \tag{2.32}$$

Note that θ_0 is the angle between the electron current direction and boundary 1. In a polar coordinate system with the axis representing the direction of electron current, Equation (2.31) defines two circles tangential to the origin and also to each other, as shown in Figure 2.6. The blank circle represents the negative values of ΔY while the hatched cir-

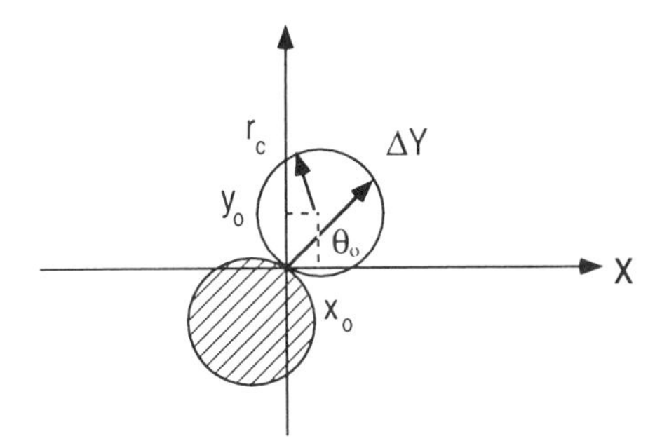

FIGURE 2.6. Graphical illustration of structural factor, ΔY.

cle is for the positive values. Thus the hatched circle corresponds to triple junctions of hillocks, whereas the blank one corresponds to triple junctions of voids. The position of the circle center (x_0, y_0) and the radius, r, are determined by ϕ's and ΔQ's, as it can easily be shown that $x_0 = A/2$, $y_0 = B/2$, and

$$r_c = \frac{\sqrt{A^2 + B^2}}{2} \tag{2.33}$$

With the orientation of a triple junction defined by θ_0, the graphical properties of ΔY can be summarized as follows. Each circle, characterized by its center position and radius, represents a particular configuration of triple junctions characterized by ϕ's and ΔQ's. The angle defined by the axis and a vector from the origin to a point on the circle, θ_0 specifies a particular orientation among these triple junctions. The magnitude of the vector is proportional to the value of the flux divergence. By rotating the end point of a vector along the circle, the orientation is rotated with respect to the electron flow, and the flux divergence is changed as reflected by the change in vector length.

It is interesting to note that for a particular triple junction there always exists a direction (perpendicular to the line connecting the centers of the two circles) along which the flux divergence is zero in spite of the possible variations of ϕ's and θ's. Ideally, if every triple junction in a conductor line could be aligned to such a direction, it would be possible to eliminate the flux divergence.

2.2.6 Concentration Gradient

Concentration gradient in grain boundaries results in another component of the total force felt by each migrating ion. Before the condensation of vacancies occurs, which leads to the formation of voids, the atomic concentration changes as a result of mass transport [22–24, 26]. The concentration gradient exerts a force on the ions in a way to oppose the force that causes this inhomogeneity. Such a force induces a backflow against the electromigration. The general expression of the concentration gradient-induced force is [31]

$$F_c = -\frac{kT}{N_{\text{gb}}} \nabla N_{\text{gb}} \tag{2.34}$$

where N_{gb} denotes the grain boundary atomic concentration. This force component is sometimes expressed in a form similar to that of the electrical driving force [31]:

$$F_c = Z^*qE_{\text{th}} = \frac{kT}{L} \ln \left(\frac{1 - F_g}{1 - F_l} \right) \tag{2.35}$$

where L is the length of the conductor line and F_g and F_l are the average fractional mass accumulation and depletion along the line, respectively. Their values are determined by the mass distribution along the conductor line, which needs to be obtained from solving the mass transport equation. The mass distribution, in turn, is determined by the total diffusion driving force including F_c. With both Equations (2.34) and (2.35), taking F_c into consideration may make it necessary to solve the mass transport equation (2.14) iteratively. An alternative approach is to consider the effect of the concentration gradient together with that of the mechanical stress, and to build the total effect of the two into the threshold current density since they both induce backflow of mass and oppose the electron wind force.

It has been pointed out [23] that the vacancy concentration reaches its supersaturation in a very short time period compared to the damaging process, usually within a few minutes. After the supersaturation, vacancy condensation takes place. If voids nucleate, the atomic concentration is roughly a step function of the position, that is, zero or constant. As it has been estimated, the mass flux due to concentration gradient is small compared to those induced by electrical current and stress gradient [22]. The concentration gradient then becomes less significant. The existence of the voids is then taken care of by the zero electrical conductivity, current crowding, and the associated temperature rising. Thus, as an approximation, one may neglect the effect of concentration gradient especially if the time frame involved is well beyond the transient period.

2.2.7 Pulsed DC Current

In many applications, pulsed DC or AC current is employed. Computer simulation has not been exercised on this subject extensively. A simple approach utilizing the DC-stress modeling for pulsed DC stress is introduced in this section.

A commonly employed design rule for the application of pulsed current is to translate the DC design guidelines to that of the pulsed by substituting the *average* current density of the pulsed current for the DC value [31, 47–48]. Thus if the fraction of time during which the pulsed current is "on" is defined as the duty factor r, and the current density during the "on" time is j_p, the equivalent DC current density, j_{DC}, is

$$j_{DC} = rj_p \tag{2.36}$$

For a periodic DC pulsed current, if the current-on time and the period are δ and Γ, respectively, the Joule heat per volume per second generated during one period is [31]

$$\frac{dQ}{dV} = \frac{\delta \rho j_p^2}{\Gamma} = r\rho j_p^2 \tag{2.37}$$

Thus the simulation model described in the previous sections may be extended to the pulsed current situation with the current density and Joule heat in Equations (2.14) and (2.24) modified by the above two equations. At a frequency higher than about 1 MHz, IC metallization line temperature cannot follow individual pulses but, instead, responds to the time-averaged Joule heating due to these pulses. This would provide a justification for the application of Equation (2.37). On the other hand, when the frequency is low (i.e., in the range of a few hertz), the healing effect during the current "off" period should be considered.

The validity of Equations (2.36) and (2.37) still remains open although it has been used for practical purposes. To be more general, an exponent has been introduced into the above equation [31, 33, 49–51]:

$$\frac{dQ}{dV} = r\rho j_p^2 = r\rho \left(\frac{j_{DC}}{r^m}\right)^2 = \rho \frac{j_{DC}^2}{r^{2m-1}} \tag{2.38}$$

and the value of m could be a fitting parameter in a computer model and has been suggested to be smaller than unity.

2.3 MONTE CARLO SIMULATION

2.3.1 Generation of Grain Boundary Texture

In this section, methods that have mostly been used to simulate grain boundary textures are introduced. The first thing to do in a computer simulation of electromigration is to generate a geometrical pattern that simulates the grain texture of the polycrystalline metal thin film under consideration. In this two-dimensional modeling, only grain boundary diffusion parallel to the metal plane and flux divergence at the grain boundary triple junctions are of concern. The problem is then to generate an appropriate two-dimensional grain boundary network.

Voronoi Polygon Method The most commonly accepted method for generating such a grain boundary network is the "Voronoi polygon" approach [36]. In this approach, polygons are generated in a random fashion to represent the grains in the metal film. First, the conductor stripe is visually discretized into a gridlike network with each cell being rectangular in shape. All the cells are equal in size, which should be about the desired average grain size. Then the method begins with laying down randomly distributed points on the discretized stripe surface at a prescribed cell density (number of points per cell). These points can be considered as the crystal seed points, that is, nucleating centers. The edges of the polygons are formed by constructing the perpendicular bisectors of rays connecting a given seed point and its neighboring seed points. The points enclosing a polygon generated this way form a set of points that are an equal distance from the cell's seed as well as from the seeds of neighboring cells [35]. The way these polygons are generated simulates the nucleating-growing-equilibrating process of the grains during the film deposition. These polygons thus represent a carpet of metal grains. Figure 2.7 shows a typical grain network generated from this approach, and Figure 2.8 shows a set of typical results of grain network based on random Poisson-distributed seeds [35]: (a) the area distribution of the grain size, (b) the diameter distribution of the grains, (c) the distribution of segment length, and (d) the number of vertices. These distributions simulate the characteristics of the grain structure of a deposited film. The model is thus useful in studying the correlation between electromigration failure and the microstructure parameters of polycrystalline films.

Triple-Junction-Lattice Method In the following discussion, triple junctions will be the only type of intersections considered since they represent the majority of grain boundary intersections in most applications [25, 27, 40].

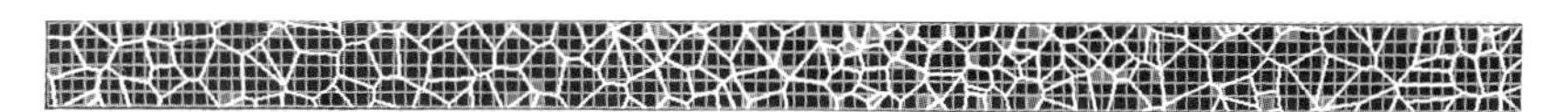

FIGURE 2.7. A typical two-dimensional grain texture generated from Voronoi polygon approach. After Marcoux Paul J., Paul P. Merchant, Vladimir Naroditsky, and Wulf D. Rehder, *Hewlett-Packard J.*, June 1989, 79 (1989). © Copyright Hewlett-Packard Company. Reprinted with permission.

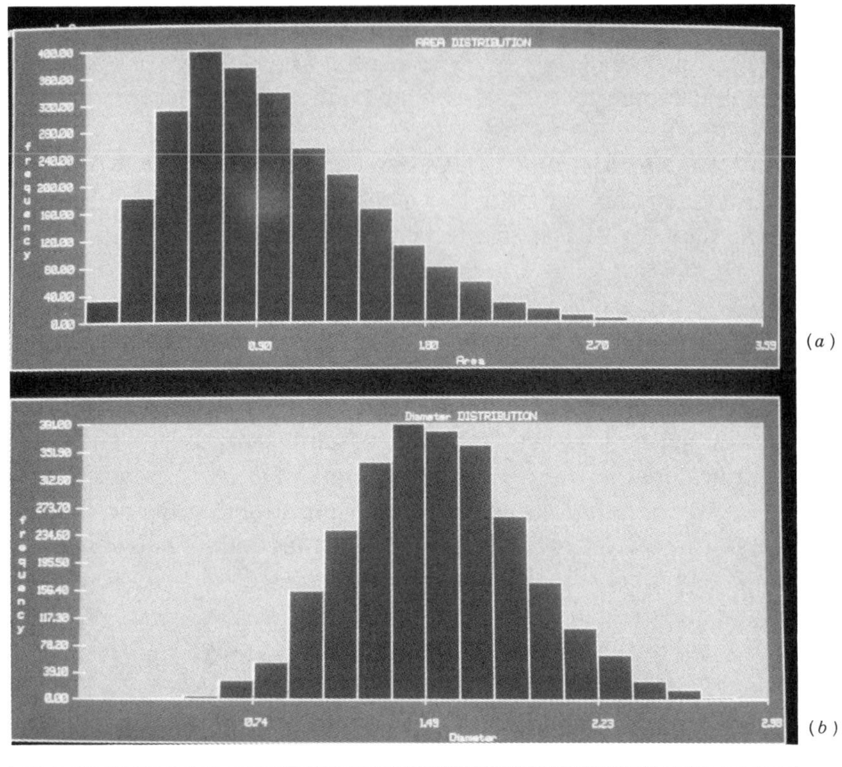
AREA DISTRIBUTION
400.00
360.00
320.00
280.00
240.00
200.00
160.00
120.00
80.00
40.00
0.00
frequency
0.90
1.80
2.70
3.59
Area
(a)
Diameter DISTRIBUTION
391.00
351.90
312.80
273.70
234.60
195.50
156.40
117.30
78.20
39.10
0.00
frequency
0.74
1.49
2.23
2.98
Diameter
(b)

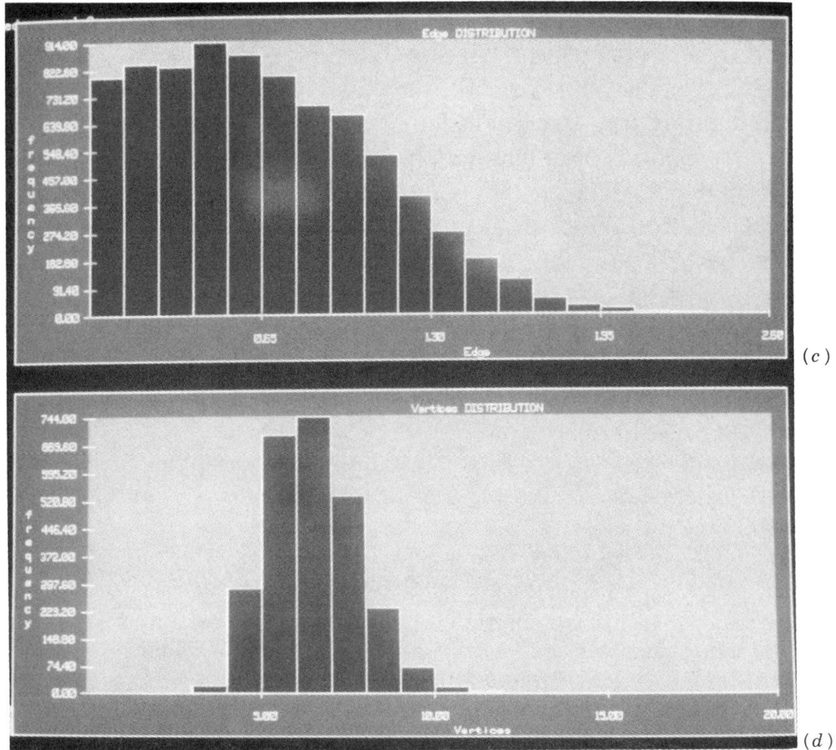
Edge DISTRIBUTION
914.00
822.60
731.20
639.80
548.40
457.00
365.60
274.20
182.80
91.40
0.00
frequency
0.65
1.30
1.95
2.60
Edge
(c)
Vertices DISTRIBUTION
744.00
669.60
595.20
520.80
446.40
372.00
297.60
223.20
148.80
74.40
0.00
frequency
5.00
10.00
15.00
20.00
Vertices
(d)

When only the grain boundary intersections are of interest, for example, if the flux divergence is assumed to exist at grain boundary intersections only, and the linkage between the grains is less important, the procedure of grain texture generation may be simplified so that only the triple junctions are generated [32, 34]. In this approach, the conductor line is first defined and discretized into rectangular cells. Seeds are then randomly distributed into the cells according to a prescribed cell density. The seeds now represent the triple junctions. Each triple junction is characterized by the microstructural parameters of the three grain boundaries defining the junction, which were discussed earlier. The values of the parameters (ΔQ's and $\Delta\theta$'s) are then assigned to each grain boundary randomly. The random assignment of the microstructural parameters is consistent with the randomness of the grain distribution generated by the Voronoi polygon method. The randomization of the grain boundary structures will be discussed in detail in the next section. When the number of grains and therefore the triple junctions are large, such a simplified approach may significantly reduce the computing time while the results produced can still be statistically accurate.

Stacking-Grains Method Another approach in generating grain texture, the stacking-grain method, can be illustrated by Figure 2.1 [20, 21]. In this method, the conductor stripe is first hypothetically discretized into segments along its length. All the segments are nonoverlapping and equal in size. Their length is equal to the average grain size, d_{50}. The segment boundaries provide the grain boundaries perpendicular to the line length. Then each segment is further discretized by stacking grains into it until the linewidth is filled. The size of each stacking grain is random, following a certain distribution population. The intersection of any two adjacent grains within a segment forms a grain boundary nonperpendicular to the length. The number of grains stacked into each segment, n_{gb}, the inclination angle of each grain boundary, ϕ, and the orientation mismatch angle, θ, between two adjacent grains defining the boundary are therefore also randomly distributed over the entire line. Comparing with Voronoi polygon approach, the stacking-grain method requires much less computation on geometrical aspects. On the other hand, it may be oversimplified when partitioning the conductor plane into grains.

FIGURE 2.8. Results of electromigration simulation on the distributions of (*a*) grain areas, (*b*) average diameters of grains, (*c*) length of grain boundary segments, and (*d*) number of vertices per grain. After Marcoux Paul J., Paul P. Merchant, Vladimir Naroditsky, and Wulf D. Rehder, *Hewlett-Packard J.*, Vol. 79, June (1989). © Copyright 1989 Hewlett-Packard Company. Reprinted with permission.

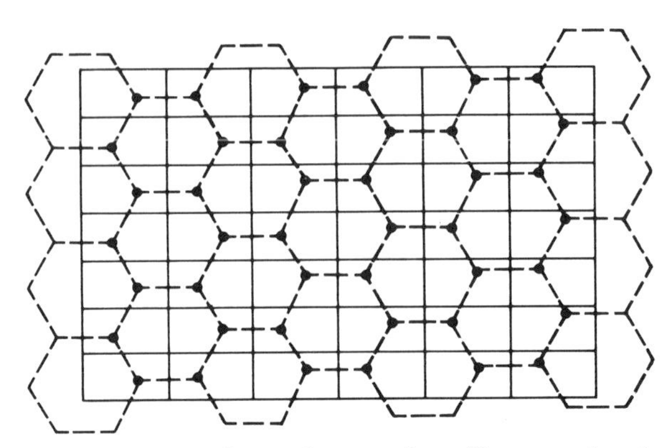

FIGURE 2.9. Partition of a conductor stripe of hexagonal grain texture.

A Perturbed Hexagonal Grain Structure A special grain structure produced by the triple-junction-lattice method is introduced next as the "perturbed hexagonal grain structure."

For a ⟨111⟩ oriented metallization system, the ideal two-dimensional grain texture is hexagonal. The relative angles between any two adjacent boundaries, ϕ_{ij}'s, are close to 120° [20]. The dashed lines in Figure 2.9 represent a perfect hexagonal grain boundary network, and the solid dots correspond to the triple junctions. Let the solid lines be a rectangularly discretized conductor stripe; it is clear that each cell contains one and only one triple junction. For this ideal metal film, the size of each cell and the position of the triple junction within each cell can both be precisely determined knowing the number of the cells and the dimensions of the line. This is also true if the line is rotated 60° with respect to the grain boundary network.

For a nonideal, but practically obtainable, metal film, the grain boundaries can be considered to form a perturbed hexagonal network. Such a perturbed hexagonal network may be achieved by moving each triple junction around inside its cell. With this perturbed hexagonal grain texture, the relative angles between two adjacent grain boundaries will no longer always be 120°; the diffusivity along each grain boundary will also be varying due to the variation in spatial orientations of the two grains defining the boundary. The network thus simulates a more realistic grain texture of the metal film. This perturbed hexagonal network can be realized by applying the triple-junction-lattice method with the cell density of seeds being unity [32, 34].

2.3.2 Structure-Induced Resistance Change

Electrical resistance change of a conductor provides one way to monitor electromigration degradation since the value of resistance is sensitive to the structural change. The resistance calculation to be presented is based on the aforementioned perturbed hexagonal grain structure model depicted in Figure 2.9. Also, in the resistance calculation, it is assumed that voids increase the resistance of the conductor whereas the hillocks have little effect on the resistance change since very little current is likely to be diverted into a hillock. Therefore only the negative flux divergence sites are of interest. Strictly speaking, the resistance of the conductor line should be calculated by integrating the elemental resistance of a small volume over the entire line, which may be irregular in shape with the electromigration-induced damage. The simplified approach employed here is to calculate the elemental resistance of each cell in the perturbed hexagonal model and then add up the resistances of all the cells as if they are discrete resistors connected together. Once the cells are defined, the fractional resistance change of each single cell is determined by the amount of mass that has been transported and the shape of the void inside the cell, if the triple junction in the cell is a negative flux divergence site. For the simplest case where the void is cylindrical in shape and perpendicular to the conductor plane, the elemental fractional resistance change of the cell on the ith row and jth column, $(\Delta R/R_0)_{ij}$, can easily be shown to be [32]

$$\left(\frac{\Delta R}{R_0}\right)_{ij} = \frac{w}{l}\left[\frac{2}{\sqrt{1-x^2}}\tan^{-1}\left(\frac{\sqrt{1+x}}{\sqrt{1-x}}\right)\right] - x - \frac{\pi}{2} \quad (2.39)$$

where w and l are the width and length of the cell and x is the diameter of the cylinder, d, normalized by the cell width: $x = d/w$. The diameter of a cylindrical hole can be obtained from Equation (2.16) since the volume V is related to d by $V = \pi d^2 h/4$. When x is small, it is straightforward [32] to see that $(\Delta R/R_0)_{ij} \approx \pi r^2/wl$, where $r = d/2$ is the radius of the cylindrical void. For simplicity, the subscripts i and j are omitted for x, r, and d, although they vary from cell to cell. In the above calculation, it has been assumed that the electrical current is always in the direction of the conductor line length. This obviously is an approximation since the current has to be deflected in the vicinity around the void. However, in the early stages of electromigration, when x is much small than unity, it has been found to be a good approximation [32].

To obtain the total resistance of the line, it has to be first determined how the cell resistors are connected together. There are two possibilities of making the simplest electric circuit analogy: the "parallel of series" and the "series of parallel" (PS and SP hereafter) models. In the SP model, the resistance of each column (perpendicular to the length) is first calculated as if the cells in the column are resistors connected in parallel. The total resistance of the conductor line is then obtained by adding up the resistances of all the columns (from left to right). For the PS model, however, the two steps are reversed; the resistance along each row (parallel to the length) is first calculated as if the cells in the row are resistors connected in series, and the total resistance is then obtained by treating the rows as being connected in parallel. Neither approach is strictly correct since a metal film is a continuum material. When the number of the grains is large and the partition is fine enough compared with the size of the conductor line, both represent good approximations. In the case where the length of the conductor line is much larger than the width, the SP circuit analogy should be employed. The total line resistance now is

$$R_T(t) = \frac{n_w R_T(0)}{n_l} \sum_{i=1}^{n_l} \left\{ \sum_{j=1}^{n_w} \left[1 + \left(\frac{\Delta R}{R_0} \right)_{ij} \right]^{-1} \right\}^{-1} \tag{2.40}$$

where $R_T(0)$ is the initial value of the total resistance, n_l and n_w denote the number of cells along the length and across the width, respectively, and $(\Delta R/R_0)_{ij}$ is given by Equation (2.39). The method presented above provides a convenient way to calculate resistance for electromigration-induced structural damage. It, however, is based on a number of assumptions. Basically, it is good for early stage electromigration only. Other methods of resistance calculation are also available. For example, to avoid dealing explicitly with the exact shape of voids, a concept of *porosity* has been introduced. Interested readers are referred to the original works [18, 31].

2.3.3 Failure Criterion

The most important results expected from a computer simulation include MTF, its standard deviation σ, and their geometrical dependence. These quantities are of central concern in the prediction of electromigration reliability. In order to calculate MTF, a properly defined failure criterion is necessary. The following discussion on failure criterion is again based on the model of perturbed hexagonal grain structure.

A traditional criterion defining a failure is a complete opening of the conductor line. For example, when solving the electrical potential field in Equation (2.17), if the potential gradient is zero across two adjacent vertical (perpendicular to current flow) partitioning lines, it signifies that no current is flowing and therefore an open circuit has occurred [35].

Alternatively, the failure can be defined as when the maximum crack length reaches the width of the line or the maximum film temperature rises to the melting point of the conductor [18, 28].

In most applications, the early stage electromigration is of more interest. A complete open circuit is beyond our concern and requires extensive computation time. The failure criterion may then be defined to identify early failures such as a certain percentage increase in resistance or a certain percentage loss of mass along the conductor line. The mass-loss criterion for the perturbed hexagonal grain network shown in Figure 2.9 is described as follows [32, 34]. Define a "column diameter" that is equal to the sum of all the hole diameters along a column. The maximum column diameter is monitored among all the columns during the whole degradation process. The failure is then defined to have happened when the maximum column diameter reaches a certain percentage of the conductor linewidth:

$$\text{Max}\left(\sum_{i=1}^{n_w} d_{ij}\right) = \frac{W}{f} \qquad j = 1, 2, \cdots, n_l \tag{2.41}$$

where d_{ij} is the diameter of the void in the cell at the ith row and jth column, if it exists. $1/f$ is related to the fractional loss of the cross-sectional area in the column. In the work by Tang et al., f has been chosen to be 2 [32, 34].

The resistance criterion is straightforward and is more frequently used since it makes the comparison with experimental results easier. Readers are referred to the original works for this subject [31, 32, 34, 52–56].

2.3.4 Random Number Distribution

When generating the seeds randomly to construct a grain texture, the distribution population from which the random numbers are chosen needs to be considered. In the case where the random number generator of a computer provides a uniform distribution only, a transformation from uniform to the desired distribution is required. For example, it is believed that the nucleation seeds of grains follow a Poisson distribution [20, 35]. Therefore when applying the Voronoi polygon method to generate a film, one needs to transform the uniformly distributed random numbers generated by the computer so that they follow a Poisson distribution. Such a transformation is briefly summarized below for the reader's convenience. More detailed discussion can be found elsewhere.

The probability density function, ρ, of a uniform distribution is a constant and is equal to unity when normalized. Thus $\rho(x) = 1$, while

x, ranging from 0 to 1, is the random number generated by a computer. Denoting the transformation function as $y = f(x)$, the ensemble of the uniformly distributed random number x will yield an ensemble of normally distributed random number y. The probability density function of a normal distribution is

$$\rho(y) = \frac{1}{\sqrt{2\pi}\sigma} e^{-(y-\mu)^2/2\sigma^2} \tag{2.42}$$

where μ and σ are the median value of y and the standard deviation of the distribution, respectively. Figure 2.10 shows the probability density function of (*a*) a uniform distribution, (*b*) a normal distribution, and (*c*) the schematic showing the correlation between y and x. With the aid of Figure 2.13*c*, it is easy to understand the conservation of probability between the two distributions; that is, the probability for x to occur in the interval dx, $\rho(x)\, dx$, is equal to the probability for y to occur in the corresponding interval dy, $\rho(y)\, dy$. Thus

$$\rho(y)\, dy = \rho(x)\, dx = dx \qquad [0 \le x \le 1] \tag{2.43}$$

or

$$\int_{y_{\min}}^{y} \rho(y)\, dy = \int_{x_{\min}}^{x} \rho(x)\, dx \tag{2.44}$$

where $x_{\min}$ and $y_{\min}$ are the minimum values of x and y, respectively. For the present case, Equation 2.44 simplifies to

$$\frac{1}{\sqrt{2\pi}\sigma} \int_{-\infty}^{y} e^{-(y-\mu)^2/2\sigma^2}\, dy = \int_{0}^{x} dx = x \tag{2.45}$$

The left side of Equation (2.45) can be further simplified knowing that the total probability of y is equal to unity:

$$\frac{1}{\sqrt{2\pi}\sigma} \int_{-\infty}^{\infty} e^{-(y-\mu)^2/2\sigma^2}\, dy = \frac{2}{\sqrt{2\pi}\sigma} \int_{-\infty}^{\mu} e^{-(y-\mu)^2/2\sigma^2}\, dy = 1 \tag{2.46}$$

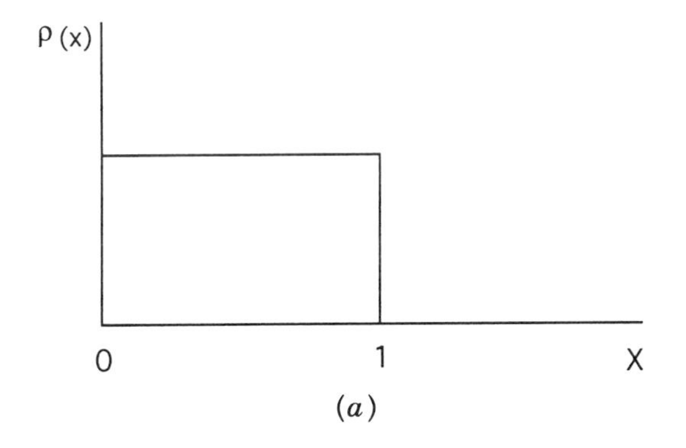

(*a*)

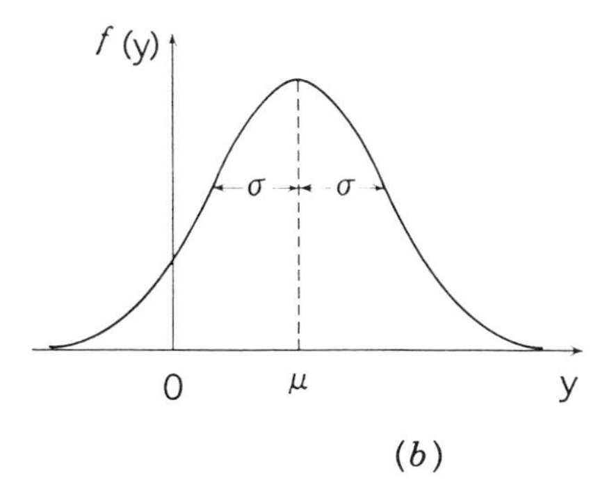

(*b*)

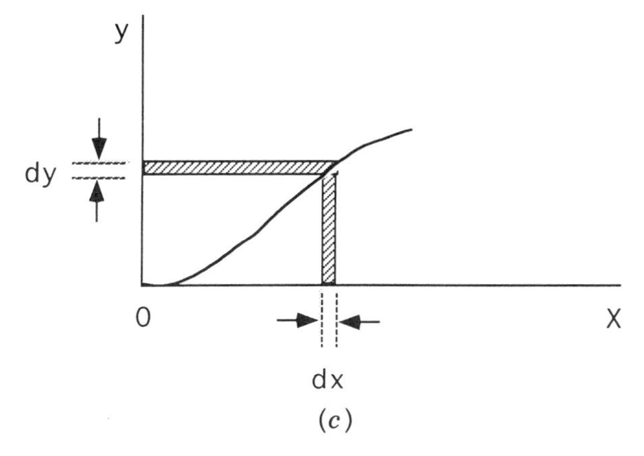

(*c*)

FIGURE 2.10. Schematic illustration of (*a*) a uniform distribution, (*b*) a normal distribution, and (*c*) correlation between $y(x)$ and x.

and

$$\frac{1}{\sqrt{2\pi}\sigma}\int_{-\infty}^{y} e^{-(y-\mu)^2/2\sigma^2}\,dy$$
$$= \frac{1}{\sqrt{2\pi}\sigma}\left[\int_{-\infty}^{\mu} e^{-(y-\mu)^2/2\sigma^2}\,dy + \int_{\mu}^{y} e^{-(y-\mu)^2/2\sigma^2}\,dy\right] \quad (2.47)$$

Combining Equations (2.46) and (2.47), Equation (2.45) then becomes

$$x = \frac{1}{2} + \frac{1}{\sqrt{2\pi}\sigma}\int_{\mu}^{y} e^{-(y-\mu)^2/2\sigma^2}\,dy \quad (2.48)$$

Equation (2.48) completely defines the functional relationship between x and y; that is, $y = f(x)$. Thus a set of normally distributed random numbers y is obtained from the set of uniformly distributed random numbers x.

The derivation for Equation (2.48) can be extended to obtain transformation functions between other distributions following a similar procedure, which will not be discussed here.

2.3.5 Randomization of Grain Boundary Microstructure

When the triple-junction-lattice method is used to generate a film, grain boundary microstructure needs to be specified after the triple-junction seeds are distributed. To be consistent with the randomness of grain size and grain orientation, the grain boundary microstructure is also randomized. This section focuses on the distribution and the constraints of the randomized quantities in generating such a metal grain structure. For convenience, a triple junction will be characterized by the three relative angles ϕ_{ij}'s, the orientation angle, θ_0, and the activation energy variations, ΔQ_i's, as depicted in Figure 2.5*b*.

The distribution of ϕ_{ij}'s is related to that of the grain size. It has been reported that the grain size of thin aluminum films [20] follows a lognormal distribution. Therefore ϕ_{ij}'s may be assigned randomly following a lognormal distribution. However, to know precisely the distribution of ϕ_{ij}'s, the functional relationship between ϕ_{ij}'s and grain sizes needs to be determined first. As preliminary models, ϕ_{ij}'s are sometimes assumed to be uniformly distributed [20, 21, 32, 34]. The first constraint on the ϕ_{ij}'s is that they sum to 360°:

$$\sum_{i,j=1}^{3} \phi_{ij} = 360^\circ \qquad (\phi_{ij} = \phi_{ji}) \quad (2.49)$$

Second, the value of the ϕ_{ij}'s should not be too far away from their equilibrium value: 120° because the conductor lines are assumed to be sufficiently annealed and metallurgically stable. Denoting the maximum deviation of ϕ_{ij} from its equilibrium to be $\Delta\phi_{max}$, the value of ϕ_{ij} should be confined in the following range:

$$120^\circ - \Delta\phi_{max} \leq \phi_{12}, \phi_{23}, \phi_{31} \leq 120^\circ + \Delta\phi_{max} \tag{2.50}$$

The value of $\Delta\phi_{max}$ varies depending on film properties. In the work by Tang et al. [32, 34], it is limited below 20°. Lastly, at each grain boundary triple junction, the force balance has to be maintained. Denoting grain boundary surface tension for the three boundaries as γ_1, γ_2, and γ_3, the following condition represents the force balance between the three grain boundaries:

$$\frac{\sin \phi_{12}}{\gamma_1} = \frac{\sin \phi_{23}}{\gamma_2} = \frac{\sin \phi_{31}}{\gamma_3} \tag{2.51}$$

The value of γ has been reported mostly to be in the range of 300–500 dyn/cm [3], which can be used as the limits for the random number assignment. Thus, among the six variables in Equation (2.51), only three of them are independent since there are three constraints [Equations (2.49) and (2.51)] relating them to each other. For example, if ϕ_{12}, ϕ_{23}, and γ_1 are chosen to be independent, ϕ_{31} is determined by Equation (2.49), and γ_2, and γ_3 can be obtained by solving Equation (2.51):

$$300 \text{ dyn/cm} \leq \gamma_1 \leq 500 \text{ dyn/cm} \tag{2.52}$$

$$\gamma_2 = \gamma_1 \frac{\sin \phi_{23}}{\sin \phi_{12}} \tag{2.53}$$

$$\gamma_3 = \gamma_1 \frac{\sin \phi_{31}}{\sin \phi_{12}} \tag{2.54}$$

The variation in activation energy is zero for a perfect metallization system: $\Delta Q = 0$. In a Monte Carlo simulation, they should be confined within a small range about zero:

$$-\Delta Q_{max} \leq \Delta Q_i \leq \Delta Q_{max} \qquad (i = 1, 2, 3) \tag{2.55}$$

where ΔQ_{max} denotes the maximum variation range of ΔQ's and is limited below 0.02 eV in the work by Tang et al. [32, 34].

Finally, the orientation angle, θ_0, is merely the orientation of a particular triple junction with respect to the electron flow. When the number of the triple junctions is large, those with the same ϕ_{ij}'s may orient themselves in all possible directions. Thus the value of θ_0 can be allowed to vary in its full range:

$$0° \leq \theta_0 \leq 360° \tag{2.56}$$

As mentioned in the earlier sections, a triple junction can be characterized equivalently by the orientation angles of individual grain boundaries with respect to electron flow and their misorientation angles (ϕ_i's and θ_i's in Figure 2.5*c*). In this case, θ_i is usually confined between 0 and 60° [20, 21]. The variation range of ϕ_i's can be obtained from that of ϕ_{ij}'s and θ_0 through the conversion between them employed in Equation (2.28). The distributions for ϕ_i's and θ_i's are sometimes assumed to be uniform [20, 21].

2.3.6 Model Implementation

With the physical models built, the failure criterion defined, and the randomization characteristics (distribution, variation range, etc.) determined, a computer model of electromigration based on a Monte Carlo simulation can now be developed.

The model implementation is summarized in the flowchart shown in Figure 2.11. The grain structure of a conductor line is first constructed, based on the randomized-grain-boundary approaches discussed as the Voronoi polygon or triple-junction-lattice method. The structure factor ΔY's are then calculated for all the grain boundary triple junctions over the entire line. At the beginning of each time interval Δt, the current density distribution is first calculated using equations for current and current crowding effect. The temperature distribution is obtained by solving the thermal equations. The flux divergence and therefore the mass that has been transported into or out of each triple junction are then determined. The volume of all the voids and the line resistance are monitored at every time step. The failure criterion is checked each time when the void sizes are calculated. If the line fails, the failure time is recorded. Otherwise, the above steps are repeated for the next time interval, $t + \Delta t$, until the line fails. The same calculation procedures will then be applied toward the next line.

The above algorithm can also be interpreted from the physical process point of view. The electrical field, grain boundary diffusivity, and grain boundary structural factors determine the atomic flux and the flux di-

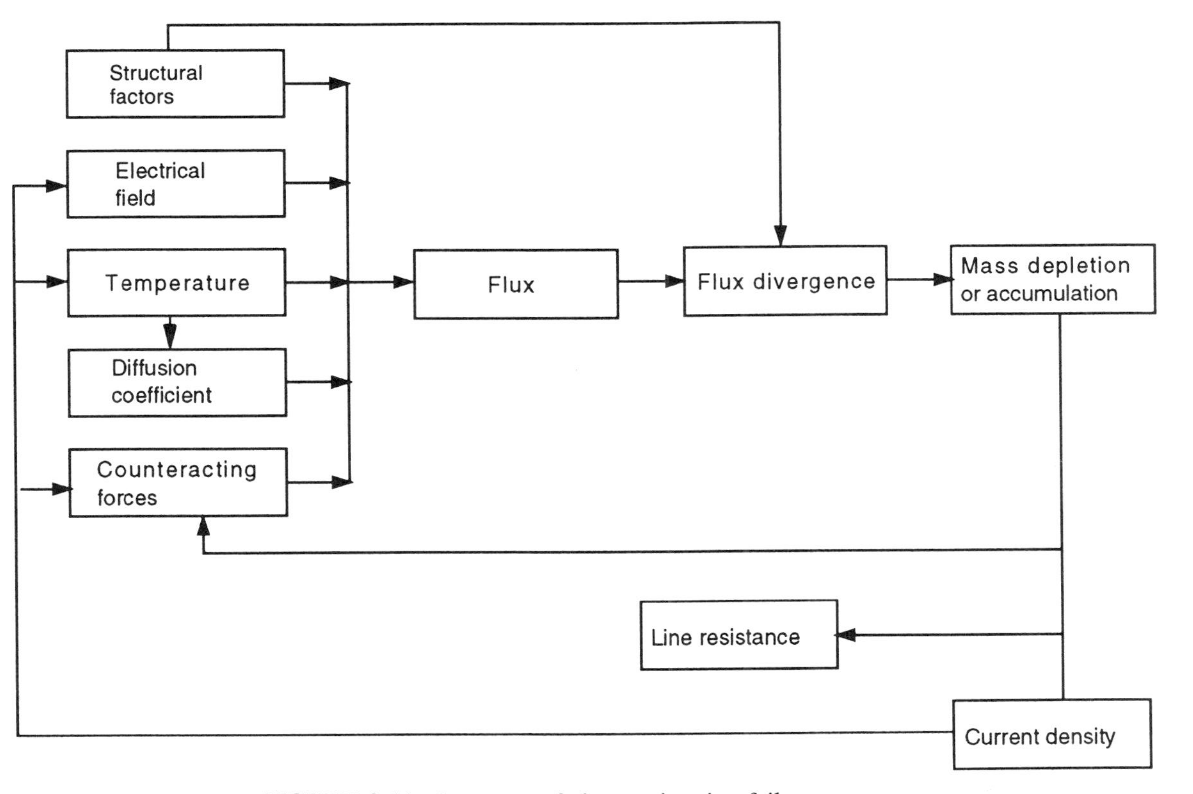

FIGURE 2.11. Summary of electromigration failure process.

vergence. The mass accumulation and depletion caused by the nonzero flux divergence directly influence the mass backflow, the line resistance, and the current density distribution. The change in current density further changes the temperature distribution, which in turn changes the atomic diffusivities and the backflow rate. The flux is therefore changed again. The entire process continues until the mass depletion or accumulation becomes severe enough to destroy the conductor line.

To obtain the median lifetime to failure, MTF, for a particular type of conductor film, failure time should be calculated for a large number of lines in order to obtain statistically meaningful results. These lines should be macroscopically identical; that is, they should have the same dimensions, the same microstructural parameters, and should be stressed under the same conditions, such as the initial current density and temperature. The statistical characteristics should also be the same, meaning that the randomly generated quantities should follow the same distribution with the same constraints. The number of lines calculated to obtain the MTF is 3000–5000 in the work by Tang et al. but can be varied depending on the individual case. Statistical analysis should be carried out once the distribution of failure time has been obtained. The major interest includes MTF and σ, and their dependence on line geometry.

2.4 TYPICAL RESULTS

Typical results have been selected from some representative works on electromigration modeling for illustration purposes. They are presented in this section with brief discussions. Readers are referred to the original works for more results as referenced in the text.

2.4.1 Distribution of Transported Mass

Figure 2.12 is a result for volume distribution of transported mass [32, 34] in a gold conductor film. The simulation is based on a simplified model for the early stage of electromigration, namely, before current crowding and the local heating effect become significant. The line is constructed by the triple-junction-lattice method. The figure shows the number of triple junctions normalized by the total versus the volume of the transported mass normalized by the cell volume, wl. The negative values correspond to mass accumulation, while the positive values correspond to mass depletion. The result is for t = 259 hours at 250°C, with $\Delta\phi_{max}$ = 10°, ΔQ_{max} = 0.02 eV, and j = 3 MA/cm^2. The line

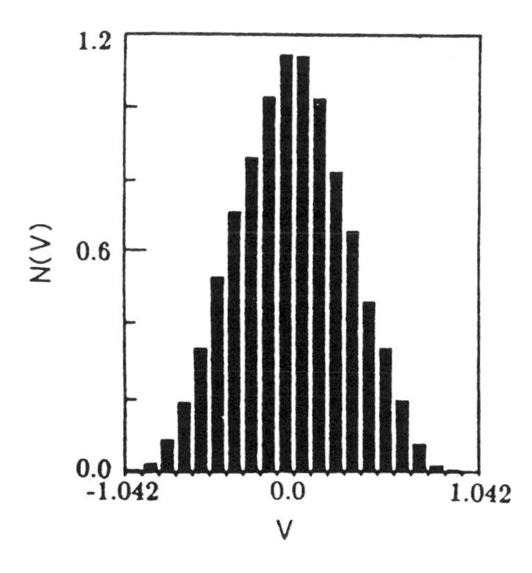

FIGURE 2.12. Normalized number of grain boundary triple junctions, characterized by normalized hole(positive)/hillock(negative) volume V, $N(V)$ versus V for $t = 259$ hours at 250°C, with $\Delta\phi_{max} = 10°$ and $\Delta Q_{max} = 0.02$ eV. After Tang P. F., A. G. Milnes, C. L. Bauer, and S. Mahajan, *Proc. Mater. Res. Soc.*, Vol. 167, 341 (1989).

dimensions are 20 cells across the linewidth, and 1400 cells along the length. With an average grain size of 1 μm, the line is about 8.7 μm wide and 1050 μm long. The figure also shows that the distributions of voids and hillocks are symmetric about zero, as expected from mass conservation. Similar results can be produced from other models. Comparison of volume distribution with experimental results is possible with the aid of microscopic techniques such as the scanning electron microscope (SEM) or transmissional electron microscope (TEM).

2.4.2 Resistance Change

Figure 2.13 shows three simulation results on fractional resistance increase with time (solid lines) compared with experimental data (discrete symbols) [32, 34], based on the same model and with the same system used to obtain Figure 2.12. The total resistance in this simulation is calculated by the SP circuit analogy. The initial current density in this simulation is 3 MA/cm^2, and the temperature is 220°C for (*a*), 240°C for (*b*), and 260°C for (*c*). The results show a fairly linear relationship between the resistance and time at the early stage electromigration up to 0.2% increase, and a good agreement with the corresponding experimental data. The linearity has also been observed by other workers [3, 31, 52, 56, 57]. However, since the model has been simplified for early stage electromigration, the results cannot be extrapolated to higher percentage increase of resistance, due to the neglect of the current crowding and local heating effects.

The effects of grain boundary microstructure on electromigration can

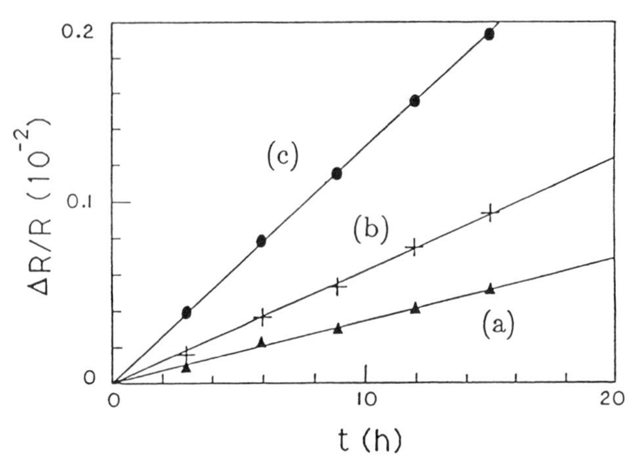

FIGURE 2.13. Typical computer simulation results of fractional resistance versus time (solid lines) compared with experimental data (solid dots) for $j = 3 \times 10^6$ A/cm^2 at (*a*) 220°, (*b*) 240°, and (*c*) 260°C. After Tang P. F., et. al. *Proc. Mater. Res. Soc.*, vol. 167, 341 (1989).

be viewed through Figure 2.14, which shows the fractional resistance change for a different set of microstructural parameters. The model employed, the system simulated, and the stress conditions applied are otherwise identical to that used to produce Figure 2.12 except for the grain boundary parameters. The two groups of curves in Figure 2.14 differ by the range of activation energy variation, $\Delta Q_{\max}$. Group (*a*) is the result for $\Delta Q_{\max} = 0.01$ eV, with the three curves corresponding to $\Delta\phi_{\max} = 20°$, 15°, and 10° from top to bottom; group (*b*) represents the results for $\Delta Q_{\max} = 0.02$ eV, with the same corresponding $\Delta\phi_{\max}$'s as in group (*a*). The time here is normalized by a time constant, t_0, which is determined by the material and stress conditions [32, 34, 46]. The purpose of introducing t_0 is to separate out all other factors and therefore to show the microstructural effects only. It can be seen that the positions of the two groups are mainly determined by the value of $\Delta Q_{\max}$ since the two groups are separated further than the curves within each group. It is therefore suggested that the variation in grain boundary diffusivity plays a more significant role in forming the structure inhomogeneities than the grain size variation does, under the stress conditions applied in this simulation.

It has been reported that a passivation layer on top of the metal film may be beneficial with respect to electromigration reliability [3, 58–60]. The reason is that the passivation layer produces a compressive stress, which suppresses the growth of the hillocks, and consequently the voids [61]. The effect of applying such a passivation layer on electromigration-induced damage is presented in Figure 2.15 [31]. The conductor

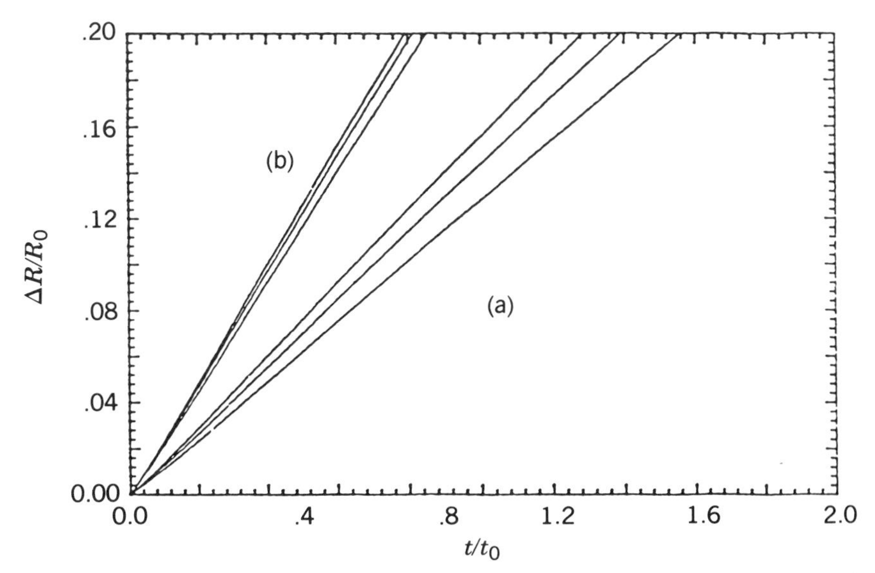

FIGURE 2.14. Effects of grain boundary microstructure parameters: fractional resistance versus normalized time for $j = 3 \times 10^6$ A/cm^2 at 250°C. $\Delta Q_{\max} = 0.01$ eV for group (*a*) and 0.02 eV for group (*b*). In both groups, $\Delta\phi_{\max} = 10°$, 15°, and 20°C from the top to the bottom curve. After Tang P.F., et al, *Proc. Mater. Res. Soc.*, vol. 167, 341 (1989).

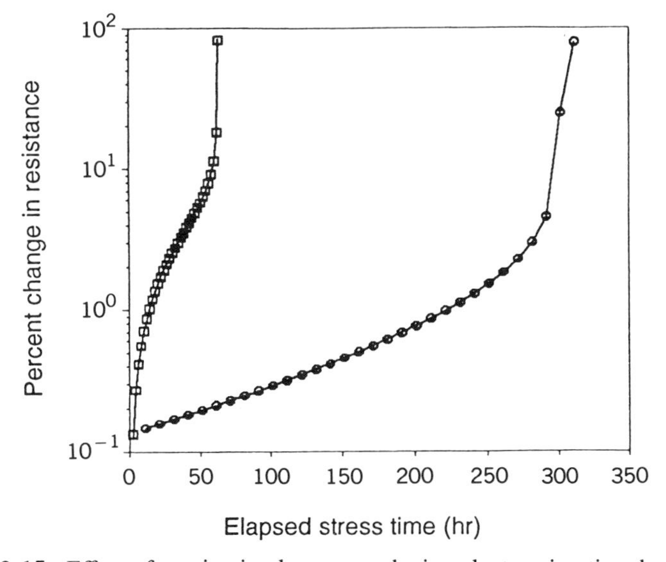

FIGURE 2.15. Effect of passivation layer on reducing electromigration damage: fractional resistance versus time for $j = 1 \times 10^6$ A/cm^2 at 200°C. After Harrison Jr., James W., *IEEE Trans. Electron Devices*, Vol. 35, No. 12, 2170 (1988). © 1988 IEEE Reprinted with permission.

film modeled in this simulation is an Al-Cu alloy line about 50 μm long and 2 μm wide. The mean grain size is 2 μm, and the mechanical stress is $s = 0.5$ N/m^2. The stress conditions employed are $j = 1$ MA/cm^2 and $T = 200°$C. The model includes both current crowding and local heating calculations and assumes an average mechanical stress induced by the passivation layer. The total resistance, however, is calculated by the PS model. The left and right curves are the results without and with a passivation layer, respectively. For this particular system, it can be seen that there is roughly a sixfold increase in the time at which the line resistance begins to increase rapidly as the damage becomes catastrophic with the application of passivation layer. Similar results have been observed experimentally by many workers [3, 58–63]. Moreover, the general trend of resistance change is also shown in this figure: a relatively slow increase at the early stage and a sudden, rapid increase at some time indicating the beginning of a catastrophic failure.

2.4.3 Mean Time-to-Failure

Distribution The distribution of failure time is a critical factor for the prediction of electromigration reliability. The importance of correctly determining the statistical model governing the electromigration failure process arises from semiconductor manufacturing, in which the extrapolation of circuit performance to low failure percentage (commonly 1% or 0.1%) is required. In this range of low percent failure, the choice of different statistical distributions may yield widely varying results. On the other hand, experimental determination at low failure percentages is costly since it requires a very large statistical sample size. An accurate distribution function of the failure time is therefore essential in predicting circuit reliability.

Failure time of electromigration has been reported to be lognormally distributed in some but not all the applications [3, 34, 40, 52, 64–66]. Much work has been done to experimentally determine such a distribution. It has become more clear now that the failure time is neither strictly lognormally nor normally distributed. The lognormal distribution, however, could be a good approximation for many applications. Such a closeness of failure time distribution to lognormal distribution is related to the statistical nature of the distributions of the grain boundary microstructural properties.

Figures 2.16*a* and 2.16*b* [32, 34] show some results based on the same system simulated, the same stress conditions applied, and the same model used to obtain Figure 2.12. Both figures show the normalized failure probability density versus the normalized time, which was intro-

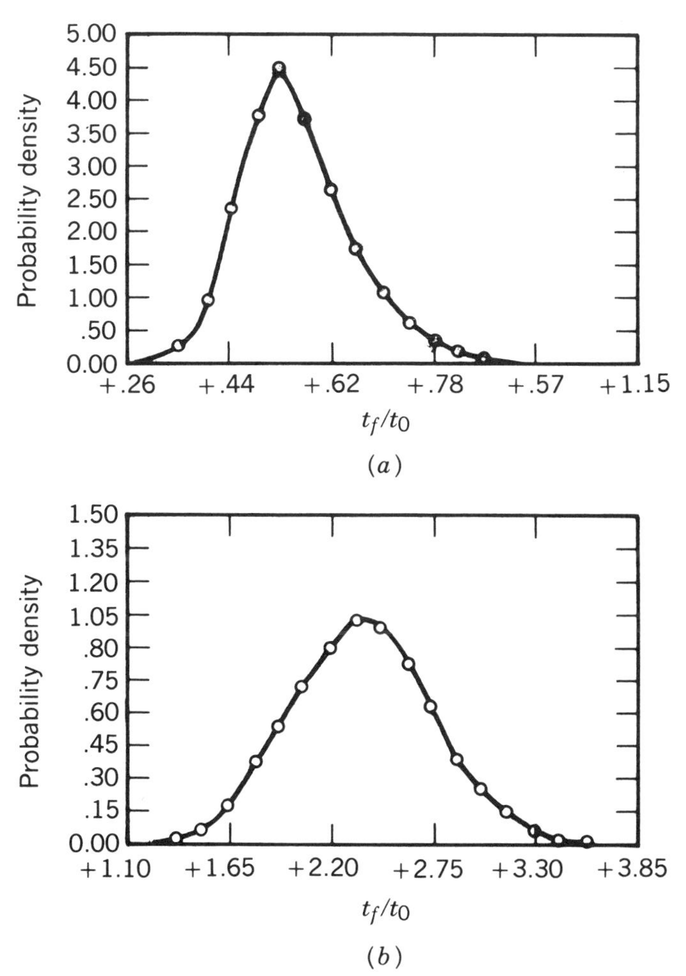

FIGURE 2.16. Simulation result on failure probability distribution of normalized failure time for lines of (*a*) 1 × 50, and (*b*) 15 × 50 cells. After Tang P. F., Modeling of Electromigration with Applications to Au on GaAs, Ph.D thesis, Dept. of ECE, Carnegie-Mellon University, Pittsburgh, PA, April 30, 1990.

duced earlier. The conductor lines simulated in Figure 2.16*a* are 1 cell wide and 50 cells long, in Figure 2.16*b* they are 15 cells wide and 50 cells long. Each distribution is obtained based on 3000 macroscopically identical lines. The probability density function appears to be closer to a normal distribution in Figure 2.16*b* while it is somewhere between lognormal and normal in Figure 2.16*a*. Thus, in this example, the failure time distribution seems to be between normal and lognormal. Figure 2.17 is the cumulative failure probability of the above results for three sets of lines plotted on a lognormal scale. The width × length for the

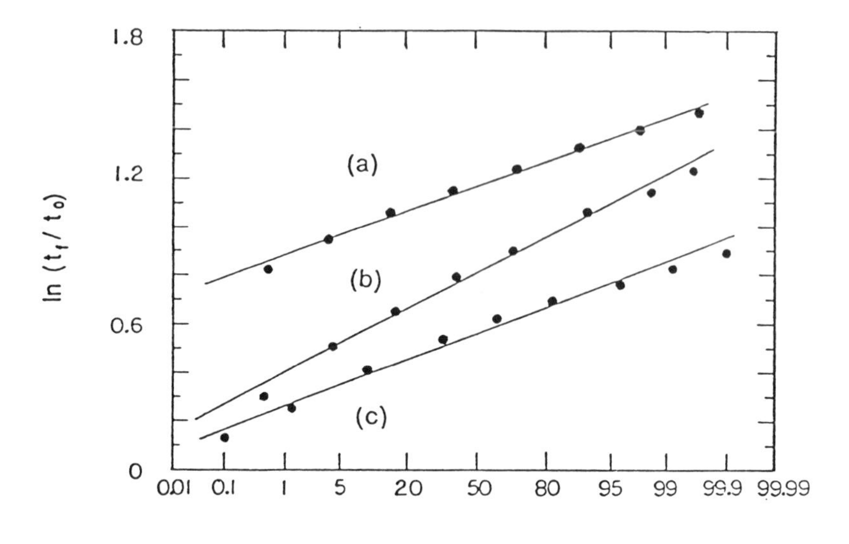

FIGURE 2.17. Simulation result of cumulative failure probability as a function of failure time, shown by dots, on a lognormal scale for lines of (*a*) 40 × 50, (*b*) 15 × 50, and (*c*) 15 × 350 cells. After Tang P. F., et. al. *Proc. Mater. Res. Soc.*, Vol. 167, 341 (1989).

lines are: (*a*) 40 × 50, (*b*) 15 × 50, and (*c*) 15 × 350 cells. The three groups of data fit straight lines fairly well especially in the range of 2–98%. If MTF, which is defined as the 50% failure time, is of major interest, the lognormal distribution may be considered as a good approximation. However, if the early failure is of central importance (i.e., less than 2%), more careful investigations are required and an early stage model may be developed [65, 67].

Geometrical Dependence The cumulative probability plot in Figure 2.17 also indicates the line geometry dependence of MTF and σ. For instance, results (*a*) and (*b*) show that a decrease of linewidth from 40 to 15 cells decreases MTF, but increases σ; while results (*b*) and (*c*) show that an increase of line length from 50 to 350 cells decreases both MTF and σ. The width and length dependencies shown here are consistent with the experimental observations [3, 37, 40, 41, 68–70]. These geometry dependencies of MTF and σ can be explained in the following. Since the microstructure of the grain boundaries is essentially random, the probability for a line to contain a triple junction, which has the largest possible flux divergence, increases as the line length increases; therefore MTF decreases. However, as the linewidth increases, since the microstructure parameters tend to distribute more uniformly when the number of grain boundaries is larger, the probability for all the large flux divergence sites to appear in the same column to form a large column diameter is lowered. Since the failure time depends on the total cumulative damage in the column with the maximum column diameter, MTF then increases. In either case, the larger the number of grains or triple junctions, the lower the statistical standard deviation is.

When the linewidth becomes smaller than the average grain size, both MTF and σ increase rapidly, as shown in Figure 2.18. The lines simulated in Figures 2.18*a* and 2.18*b* are 100 μm long and 0.5 μm thick and are constructed by the stacking-grains method [21]. The stress conditions are $j = 2 \times 10^6$ A/cm^2 and $T = 200°$C. The average grain diameters are 1, 2, and 4 μm, respectively. The model takes both current crowding and local heating effects into consideration, as well as the threshold current density. Figure 2.18*c* is a result from a probability model [37] simulating a Al–2%Cu–0.3%Cr film. Both MTF and σ are calculated as a function of linewidth, normalized by the average grain size. The results from both works show an increase of MTF and σ as

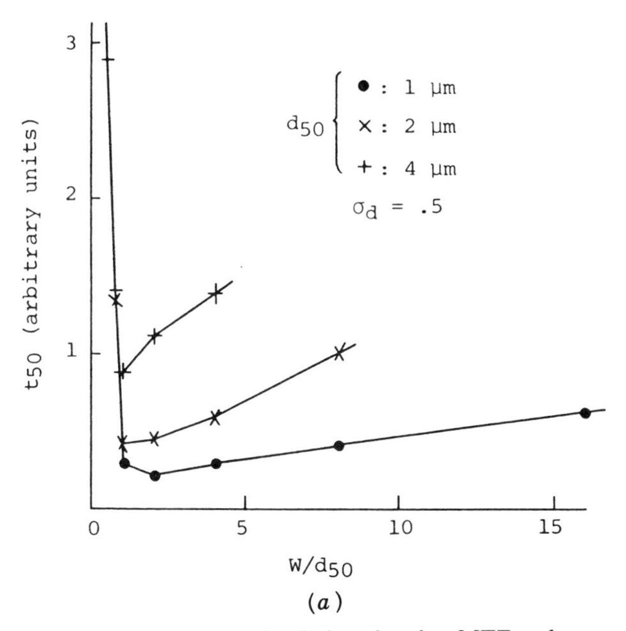

(*a*)

FIGURE 2.18. Result of computer simulation showing MTF and σ versus line width normalized by average grain size. Parts (*a*) and (*b*) after K. Nikawa et. al., *IEEE Int. Reliab. Phys. Symp.*, 175 (1981) © 1981 IEEE; reprinted with permission, and (*c*) after J. Cho, and C. V. Thompson, *Appl. Phys. Lett.*, Vol. 54, No. 25, 2577 (1989). Reprinted with permission.

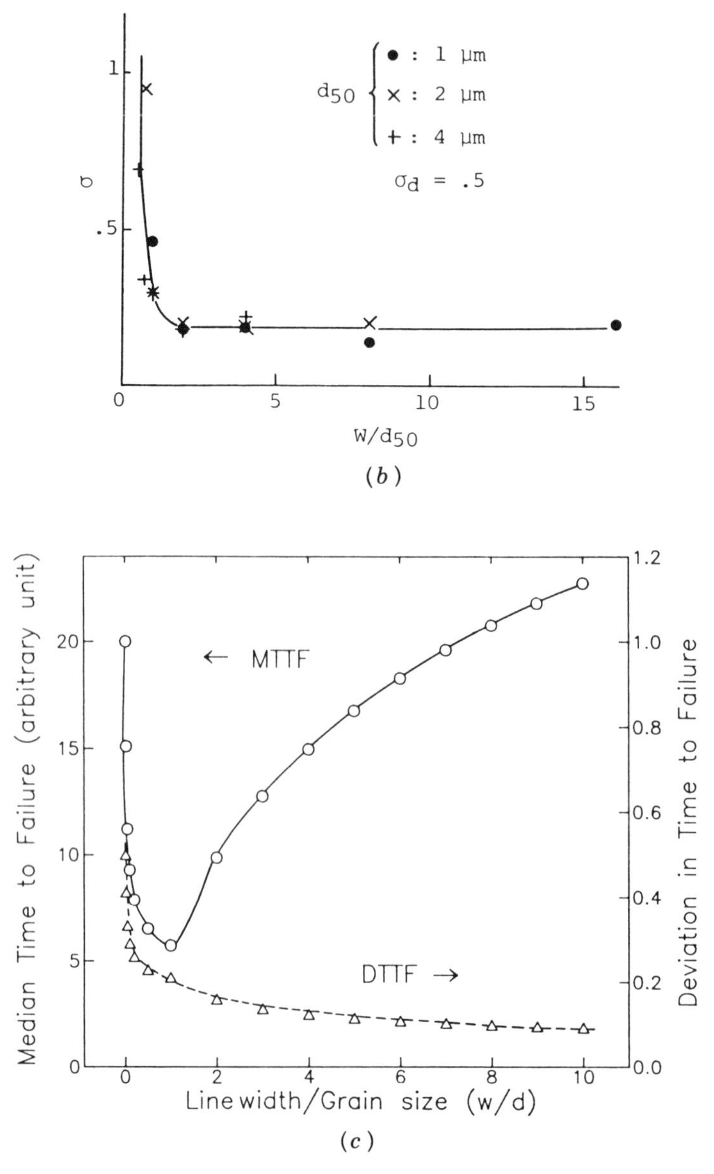

FIGURE 2.18. (*Continued*)

the linewidth becomes smaller than the grain size. This behavior of MTF and σ is due to the formation of the "bamboo structure" [41]. The name comes from the fact that when the linewidth becomes comparable with or smaller than the average grain size, almost no grain boundary is parallel to the line length. The conductor line is sectioned into segments by the nonparallel grain boundaries appearing as a bamboo. The number of paths for atoms to move in the direction of electron wind force is nearly zero in this case. The MTF therefore increases. On the other hand, since the number of grains or grain boundaries is reduced, the statistical error is increased. The dependence of MTF and σ on line geometry is of particular interest since they provide guidelines for metallization design rules [37, 40, 41, 69, 70].

Grain Size The effects of average grain size and grain size variation on MTF are shown in Figure 2.19 [20]. In this simulation, the conductor line is constructed by the stacking-grain method. The model assumes an early stage of electromigration, which allows for the use of a constant current density and temperature distribution. Figure 2.19 shows the mean grain size dependence of MTF for different standard deviations of

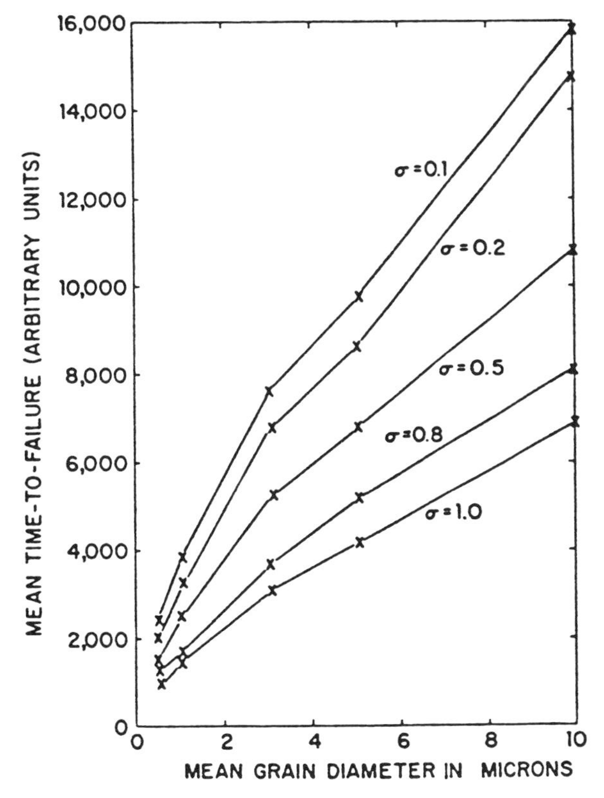

FIGURE 2.19. Computer-simulated dependence of mean time-to-failure upon median grain size. After M. J. Attardo, R. Rutledge, and R. C. Jack, "Statistical metallurgical model for electromigration failure in aluminum thin-film conductors," *J. Appl. Phys.*, Vol. 42, No. 11, 4343 (1971). Printed with permission.

grain size distribution, σ. The result suggests that the MTF be approximately proportional to the mean grain size, d_{50}, and a higher value of σ yields a lower MTF. It can also be seen from this result that the larger the mean grain size is, the more sensitive the MTF on the variations of grain size distribution, or, in other words, an increase in σ of grain size

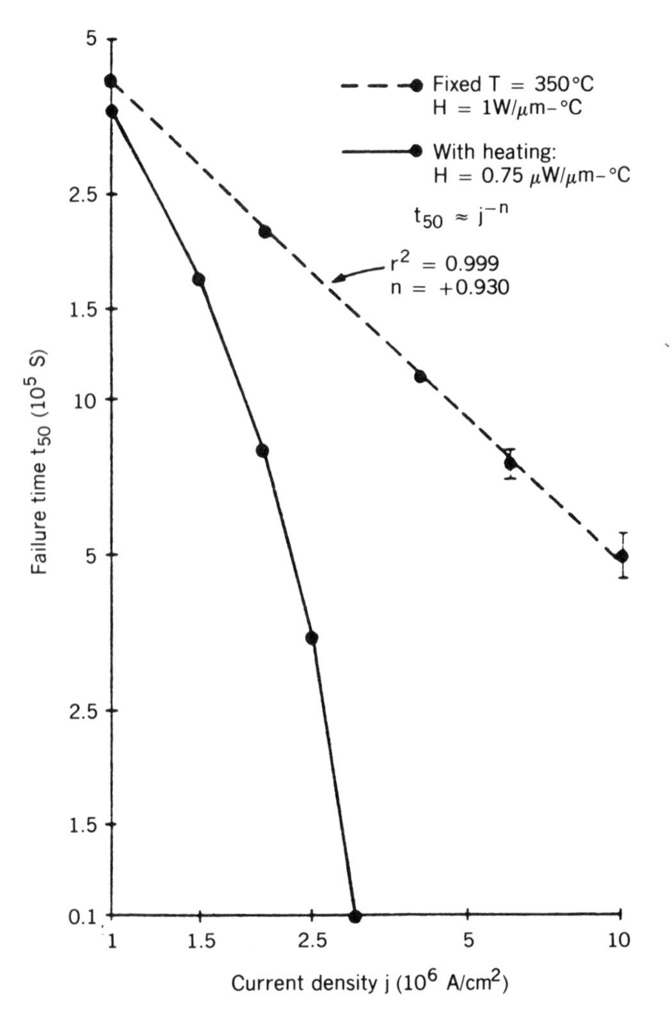

FIGURE 2.20. A result of simulated failure time for various current densities under conditions of normal Joule heating (solid) and fixed temperature (dashed). H is the film-to-substrate heat transfer coefficient. A large value of H ensures a constant temperature. The value of 0.75 $\mu W/\mu M$-°C is typical of real systems. The quantity r^2 is the regression coefficient and n is the slope from the least squares fit. After P. J. Marcoux et. al., *Hewlett-Packard J.*, 79 (1989). © Copyright 1989 Hewlett-Packard Company. Reprinted with permission.

distribution decreases MTF more for larger d_{50} than it does for smaller d_{50}.

2.4.4 Current Density Dependence

Experimental observations have indicated a larger than unity power dependence of MTF on current density. This value of the exponent has been reported to vary largely [3, 71–73]. Such a higher power dependence of MTF on current density has been suggested to be due to the Joule heating rather than a result of some fundamental mechanism. In practice, however, it is difficult to distinguish these two issues experimentally since Joule heating cannot be separated from the electromigration failure process. With computer simulation, such an ''experiment'' is possible. Figure 2.20 shows the result of a Monte Carlo simulation of a conductor line constructed by the Voronoi polygon method. The model includes current crowding, local heating, and healing effects [35]. The results in Figure 2.20 show MTF versus current density for two different ''tests.'' The dashed line represents a situation where the conductor line is kept at a fixed temperature, 350°C. Thus no local Joule heating exists and the temperature is uniform and invariant during the ''test.'' The solid curve represents the result with Joule heating, and only the ambient temperature is fixed. The results show clearly that the exponent of current, n, is close to unity without Joule heating, while the n value is not a constant with Joule heating. Such a simulation result indicates that, fundamentally, the exponent of current density is unity. The occurrence of higher order current density dependence of MTF must be a concomitant effect of Joule heating.

2.5 SUMMARY

Computer modeling of the electromigration failure process based on Monte Carlo simulations has been introduced in the aspects of both concepts and methodologies. The subject has been discussed from the physical modeling and the computer modeling viewpoints, followed by some typical results.

The key factor determining electromigration reliability is the flux divergence of mass transport. As a contributing component, electrical driving force, microstructural properties of the metallization systems, current crowding effect, temperature distribution, damage-induced mass backflow, and atomic concentration gradient have been discussed individually. Each component has been described with its underlying physics and formulated into equations for simulation. Application of pulsed DC current has also been considered, and the concept of equivalent current density has been introduced.

Methods in generating polycrystalline metallization films based on randomized grain microstructure have been described. The randomization of microstructure parameters has been discussed. Evaluation of resistance change and definition of failure criterion have also been discussed based on the computer model and grain microstructure. Statistical analysis can be performed with systems containing a large number of conductor lines.

Finally, results of different simulation works have been presented and briefly discussed in the aspects of mass distribution, resistance change, MTF, and its dependencies on metallization geometry, microstructural parameters, and current density.

Computer simulation of electromigration has a unique capability in simulating a large number of conductor lines, which is required for statistical analysis, and in revealing the effects of microstructural parameters, which is essential for a good understanding of the governing physics. It therefore provides an efficient approach for reliability prediction and for quantitative study of the basic mechanisms involved in electromigration.

REFERENCES

1. H.B. Huntington and A.R. Grone, "Current-Induced Marker Motion in Gold Wires," *J. Phys. Chem. Solids*, Vol. 20, Nos. 1/2, 76 (1961).
2. P.S. Ho, and T. Kwok, "Electromigration in Metals," *Rep. Prog. Phys.*, Vol. 52, 301 (1989).
3. F.M. d'Heurle and P.S. Ho. "Electromigration in Thin Films," *Thin Films—Interdiffusion and Reactions*, Wiley, New York, 1978.
4. R.V. Hesketh, "Electromigration: The Electron Wind," *Phys. Rev. B*, Vol. 19, No. 4, 1727 (1979).
5. Richard S. Sorbello, "Theory of the Direct Force in Electromigration," *Phys. Rev. B*, Vol. 31, No. 2, 798 (1985).
6. R.P. Gupta, "Theory of Electromigration in Noble and Transition Metals," *Phys. Rev. B*, Vol. 25, No. 8, 5188 (1982).
7. I.A. Blech and R. Rosenberg, "On the Direction of Electromigration in Gold Thin Films," *J. Appl. Phys.*, Vol. 46, No. 2, 579 (1975).
8. I.A. Blech and E. Kinsbron, "Electromigration in Thin Gold Films on Molybdenum Surfaces," *Thin Solid Films*, Vol. 25, 327 (1975).

9. K.L. Tai and M. Ohring, "Grain-Boundary Electromigration in Thin Films II, Tracer Measurement in Pure Au," *J. Appl. Phys.*, Vol. 48, No. 1, 36 (1977).
10. I.A. Blech, "Electromigration in Thin Aluminum Films on Titanium Nitride," *J. Appl. Phys.*, Vol. 47, No. 4, 1203 (1976).
11. I.A. Blech and Conyers Herring, "Stress Generation by Electromigration," *Appl. Phys. Lett.*, Vol. 29, 131 (1976).
12. E. Kinsbron, I.A. Blech, and Y. Komem, "The Threshold Current Density and Incubation Time to Electromigration in Gold Films," *Thin Solid Films*, Vol. 46, 139 (1977).
13. H.-U. Schreiber, "Electromigration Threshold in Aluminum Films," *Solid State Electron.*, Vol. 28, No. 6, 617 (1985)
14. C.T. Rosenmayer, F.R. Brotzen, J.W. McPherson, and C.F. Dunn, "Effect of Stresses on Electromigration," *IEEE Int. Reliab. Phys. Symp.*, Vol. CH2974-4/91, 52 (1991).
15. R.S. Hemmert and M. Costa, "Electromigration-Induced Compressive Stresses in Encapsulated Thin-Film Conductors," *IEEE Int. Reliab. Phys. Symp.*, Vol. CH2974-4/91, 64 (1991).
16. C.A. Ross and J.E. Evetts, "A Study of Threshold and Incubation Behavior During Electromigration in Thin Film Metallisation," *Proc. Mater. Res. Soc.*, Vol. 107, 319 (1988).
17. I.A. Blech and K.L. Tai, "Measurement of Stress Gradients Generated by Electromigration," *Appl. Phys. Lett.*, Vol. 30, No. 8, 387 (1977).
18. J.D. Venables and R.G. Lye, "A Statistical Model for Electromigration Induced Failure in Thin Film Conductors," *Proc. 10th IEEE Int. Reliab. Phys. Symp.*, p. 159 (1972).
19. J.M. Schoen, "Monte Carlo Calculations of Structure-Induced Electromigration Failure," *J. Appl. Phys.*, Vol. 51, No. 1, 513 (1980).
20. M.J. Attardo, R. Rutledge, and R.C. Jack, "Statistical Metallurgical Model for Electromigration Failure in Aluminum Thin-Film Conductors," *J. Appl. Phys.*, Vol. 42, No. 11, 4343 (1971).
21. Kiyoshi Nikawa, "Monte Carlo Calculations Based on the Generalized Electromigration Failure Model," *IEEE Int. Reliab. Phys. Symp.*, Vol. CH1619-6, 175 (1981).
22. R. Rosenberg and M. Ohring, "Void Formation and Growth During Electromigration in Thin Films," *J. Appl. Phys.*, Vol. 42, No. 13, 5671 (1971).
23. M. Ohring, "Electromigration Damage in Thin Films Due to Grain Boundary Grooving Processes," *J. Appl. Phys.*, Vol. 42, No. 7, 2653 (1971).
24. M.J. Attardo and R. Rosenberg, "Electromigration Damage in Aluminum Film Conductors," *J. Appl. Phys.*, Vol. 41, No. 6, 2381 (1970).

25. M. Genut, Z. Li, C.L. Bauer, S. Mahajan, P.F. Tang, and A.G. Milnes, "Characterization of Early Stages of Electromigration at Grain Boundary Triple Junctions," *Appl. Phys. Lett.*, Vol. 58, No. 21, 2354 (1991).
26. J.R. Lloyd and S. Nakahara, "Grain Boundary and Vacancy Diffusion Model for Electromigration-Induced Damage in Thin Film Conductors," *Thin Solid Films*, Vol. 72, 451 (1980).
27. L. Berenbaum, "Electromigration Damage of Grain-Boundary Triple Points in Al Thin Films," *J. Appl. Phys.*, Vol. 42, No. 2, p. 880, Feb. (1971).
28. R.A. Sigsbee, "Electromigration and Metallization Lifetimes," *J. Appl. Phys.*, Vol. 44, No. 6, 2533 (1973).
29. A.P. Schwarzenberger, C.A. Ross, J.E. Evetts, and A.L. Greer, "Electromigration in the Presence of a Temperature Gradient: Experimental Study and Modeling," *J. Electron. Mater.*, Vol. 17, No. 5, 473 (1988).
30. Z. Li, C.L. Bauer, S. Mahajan, and A.G. Milnes, "Degradation and Subsequent Healing by Electromigration in Al-1 wt% Si Thin Films," *J. Appl. Phys.*, Vol. 72, No. 5, 1821 (1992).
31. James W. Harrison, Jr., "A Simulation Model for Electromigration in Fine-line Metallization of Integrated Circuits Due to Repetitive Pulsed Currents," *IEEE Trans. Electron Devices* Vol. 35, No. 12, 2170 (1988).
32. P.F. Tang, *Modeling of Electromigration with Applications to Au on GaAs*, Ph.D thesis, Dept. of ECE, Carnegie–Mellon University, Pittsburgh, PA, April 30, 1990.
33. A. Bobbio, A. Ferro, and O. Saracco, "Electromigration Failure in Al Thin Films Under Constant and Reversed DC Powering," *IEEE Trans. Reliab.*, Vol. R-23, No. 3, 194 (1974).
34. P.F. Tang, A.G. Milnes, C.L. Bauer, and S. Mahajan, "Electromigration in Thin Films of Au on GaAs," *Proc. Mater. Res. Soc.*, Vol. 167, 341 (1989).
35. Paul J. Marcoux, Paul P. Merchant, Vladimir Naroditsky, and Wulf D. Rehder, "A New 2D Simulation Model of Electromigration," *Hewlett-Packard J.*, June, p. 79 (1989).
36. P.P. Meng, *Computer Simulation of Electromigration in Thin Films*, M.S. thesis, Rensselaer Polytechnic Institute, Troy, NY, May 1988.
37. J. Cho and C.V. Thompson, "The Grain Size Dependence of Electromigration Induced Failure in Narrow Interconnects," *Appl. Phys. Lett.*, Vol. 54, No. 25, 2577 (1989).
38. Zoltan J. Cendes and Jin-Fa Lee, "The Transfinite Element Method for Modeling MMIC Devices," *IEEE Trans. Microwave Theory and Techniques*, Vol. 36, No. 12, 1639 (1988).
39. M.S. Fan, A. Christou, and M. Pecht, "Two-Dimensional Thermal Modeling of Power Monolithic Microwave Integrated Circuits (MMIC's)," *IEEE Trans. Electron Devices*, Vol. 39, No. 5, 1075 (1992).

40. B.N. Agarwala, M.J. Attardo, and A.P. Ingraham, ''Dependence of Electromigration-Induced Failure Time on Length and Width of Aluminum Thin-Film Conductors,'' *J. Appl. Phys.*, Vol. 41, No. 10, 3954 (1970).

41. S. Vaidya, T.T. Sheng, and A.K. Sinha, ''Linewidth Dependence of Electromigration in Evaporated Al-0.5%Cu,'' *Appl. Phys. Lett.*, Vol. 36, No. 6, 464 (1980).

42. C.M. Li, ''High-Angle Tilt Boundary—A Dislocation Core Model,'' *J. Appl. Phys.*, Vol. 32, No. 3, 525 (1961).

43. D. Turnbull and R.Z. Hoffman, ''The Effect of Relative Crystal and Boundary Orientation on Grain Boundary Diffusion Rates,'' *Acta Metall.*, Vol. 2, 419 (1954).

44. C.A. Ross and J. E. Evetts, ''A Model for Electromigration Behaviour in Terms of Flux Divergences,'' *Scripta Metall.*, Vol. 21, 1077 (1987).

45. C.R. Crowell, C.C. Shih, and V. Tyree, ''Simulation and Testing of Temperature Distribution and Resistance Versus Power for SWEAT and Related Joule-Heated Metal-on-Insulator Structures,'' *Proc. 28th IEEE Int. Reliab. Phys. Symp.*, 37 (1990).

46. C.L. Bauer, P.F. Tang, A.G. Milnes, and S. Mahajan, ''Fundamental Mechanisms for Electromigration in Thin Films,'' *Proceedings of the 4th International Conference of Quality in Electronic Components*, April 25–28, 1989, Bordeaux, France, p. 222.

47. J.M. Towner and E.P. Van den ven, ''Aluminum Electromigration Under Pulsed dc Conditions,'' *21th Annu. Proc. Reliab. Phys.*, 36 (1983).

48. L. Brooke, ''Pulsed Current Electromigration Failure Model,'' *25th Annu. Proc. Reliab. Phys.*, p. 136 (1987).

49. J.S. Suehle and H.A. Schafft, ''Current Density Dependence of Electromigration t_{50} Enhancement Due to Pulsed Operation,'' *IEEE Int. Reliab. Phys. Symp.*, Vol. CH2887-0/90, 106 (1990).

50. R.E. Hummel and H.H. Hoang, ''On the Electromigration Failure Under Pulsed Conditions,'' *J. Appl. Phys.*, Vol. 65, No. 5, 1929 (1989).

51. Laura Brooke, ''Pulsed Current Electromigration Failure Model,'' *IEEE Int. Reliab. Phys. Symp.*, Vol. CH2388-7/87, 136 (1987).

52. A.T. English, K.L. Tai, and P.A. Turner, ''Electromigration of Ti-Au Thin Film Conductors at 180°C,'' *J. Appl. Phys.*, Vol. 45, No. 9, 3757 (1974).

53. Donald J. LaCombe and Earl Parks, ''A Study of Resistance Variation During Electromigration,'' *IEEE Int. Reliab. Phys. Symp.*, Vol. CH2113-9, 74 (1985).

54. R. Rosenburg and L. Berenbaum, ''Resistance Monitoring and Effects of Nonadhesion During Electromigration in Aluminum Films,'' *Appl. Phys. Lett.*, Vol. 12, No. 5, 201 (1968).

55. J.R. Lloyd and R.H. Koch, ''Study of Electromigration-Induced Resis-

tance and Resistance Decay in Al Thin Film Conductors,'' *Appl. Phys. Lett.*, Vol. 52, No. 3, 194 (1988).

56. R.W. Pasco and J.A. Schwartz, ''Temperature-Ramp Resistance Analysis to Characterize Electromigration,'' *Solid-State Electron.*, Vol. 26, No. 5, 445 (1983).
57. R.E. Hummel and H.J. Geier, ''Activation Energy for Electrotransport in Thin Silver and Gold Films,'' *Thin Solid Films*, Vol. 25, 335 (1975).
58. J.R. Black, ''Electromigration Failure Modes in Aluminum Metallization for Semiconductor Devices,'' *Proc. IEEE*, Vol. 57, 1587 (1969).
59. J.R. Black, ''Electromigration—A Brief Survey and Some Recent Results,'' *IEEE Trans. Electron Devices*, Vol. ED-16, 338 (1969).
60. A.J. Learn, ''Effect of Structure and Processing on Electromigration-Induced Failure in Anodized Aluminum,'' *J. Appl. Phys.*, Vol. 44, 1251 (1973).
61. N.G. Ainslie, F.M. d'Heurle, and O.C. Wells, ''Coatings, Mechanical Constraints, and Pressure Effects on Electromigration,'' *Appl. Phys. Lett.*, Vol. 20, 172 (1972).
62. V. Teal, S. Vaidya, and D.B. Fraser, ''Effect of a Contact and Protective Seal on Aluminum Electromigration,'' *Thin Solid Films*, Vol. 136, 21 (1986).
63. Kenji Hinode and Yoshio Homma, ''Improvement of Electromigration Resistance of Layered Aluminum Conductors,'' *IEEE Int. Reliab. Phys. Symp.*, Vol. CH2787-0, 25 (1990).
64. Janet M. Towner, ''Are Electromigration Failures Lognormally Distributed?'' *Proc. IEEE Reliab. Phys. Symp.*, 100 (1990).
65. D.J. LaCombe and E.L. Parks, ''The Distribution of Electromigration Failures,'' *Proc. 24th IEEE Int. Reliab. Phys. Symp.*, p. 1 (1986).
66. J.R. Lloyd, ''On the Log-normal Distribution of Electromigration Lifetimes,'' *J. Appl. Phys.*, Vol. 50, No. 7, 5062 (1979).
67. H.H. Hoang, E.L. Nikkel, J.M. McDavid, and R.B. MacNaughton, ''Electro-migration Early-Failure Distribution,'' *J. Appl. Phys.*, Vol. 65, 1044 (1989).
68. E. Kinsbron, ''A Model for the Width Dependence of Electromigration Lifetimes in Aluminum Thin-Film Stripes,'' *Appl. Phys. Lett.*, Vol. 36, No. 12, 968 (1980).
69. T. Kwok, ''Effect of Metal Line Geometry on Electromigration Lifetime in Al-Cu Submicron Interconnects,'' *Proc. 26th IEEE Int. Reliab. Phys. Symp.*, p. 185 (1988).
70. G.A. Scoggan, B.N. Agarwala, P.P. Peressini, and A. Brouillard, ''Width Dependence of Electromigration Life in Al-Cu, Al-Cu-Si and Ag Conductors,'' *Proc. 13th IEEE Int. Reliab. Phys. Symp.*, p. 151 (1975).
71. T. Kwok, T. Nguyen, P. Ho, and S. Yip, ''Current Density and Tem-

perature Distribution in Multilevel Interconnection with Studs and Vias,'' *IEEE Int. Reliab. Phys. Symp.*, Vol. CH2388-7/87, 130 (1987).

72. H.A. Schafft, T.C. Grant, A.N. Saxena, and C. Kao, ''Electromigration and the Current Dependence,'' *IEEE Int. Reliab. Phys. Symp.*, Vol. CH2113-9/85, 93 (1985).
73. M. Shatzkes and J.R. Lloyd, ''A Model for Conductor Failure Considering Diffusion Concurrently with Electromigration Resulting in a Current Exponent of 2,'' *J. Appl. Phys.*, Vol. 59, No. 11, 3890 (1986).

3

TEMPERATURE DEPENDENCIES ON ELECTROMIGRATION

MICHAEL PECHT AND PRADEEP LALL

CALCE Electronic Packaging Research Center, University of Maryland, College Park

3.1 INTRODUCTION

The phenomenon of electrotransport, known as "electromigration," is the result of high current density (typically of the order of 10^6 amperes/cm^2 in aluminum) in metallization tracks. The result is a net flux of metallization atoms that generally migrate in the same direction as the electron flux. The product is a continuous impact on the metal grains in the metallization, causing the metal to pile up in the direction of the electron flow and produce voids in an upstream direction with respect to the electron flow. The phenomenon has been explained by the momentum exchange between current carriers, that is, the electrons or the holes and metallization atoms.

Temperature is generally considered a key parameter in the design of electronic systems and higher reliability has been associated with lower steady-state temperature, based on Arrhenius and Eyring models. Implicit in these models is the assumption that the failure mechanisms are caused by atomic processes whose reaction rate increases exponentially with temperature. The role of temperature in electromigration is ex-

Electromigration and Electronic Device Degradation, Edited by Aris Christou.
ISBN 0-471-58489-4

tremely complex, especially within normal operating temperatures of less than 125°C. The lifetime due to electromigration is a complex function of temperature and cannot be represented by a simple activation energy. The temperature acceleration can, however, be represented by an apparent activation energy that changes with operating conditions (Figure 3.1). Furthermore, there have been insufficient tests on electromigration at temperatures less than 125°C due to difficulty in conducting such tests. Most of the tests have been performed at elevated temperatures in the neighborhood of 150°C or higher and the results extrapolated to normal operating or room temperatures. Extrapolation of failure rates from stress results at higher temperatures, to provide reliability estimates at lower temperatures, will not be accurate since the physics of failure phenomena changes from grain boundary migration to surface migration.

Electromigration damage forms at sites of atomic flux divergence, of which the three main sources are structural defects, microstructural inhomogeneities, and local temperature gradients. Electromigration failures tend to be localized near sites of maximum temperature gradient [30, 38], even though typical field failures are characterized by structurally induced flux divergences, rather than temperature gradient-in-

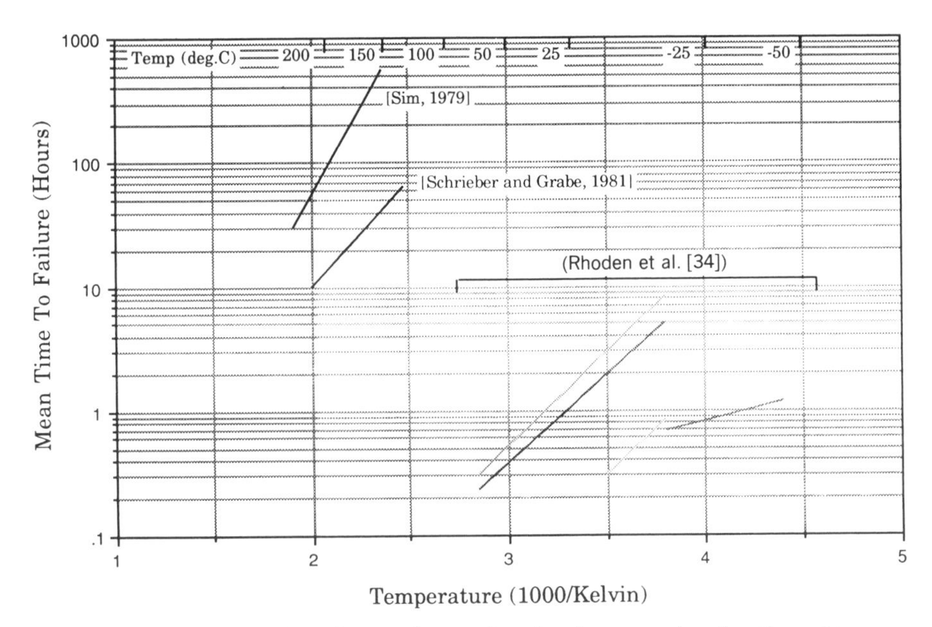

FIGURE 3.1. The lifetime due to electromigration is a complex function of temperature and can be represented by an apparent activation energy that changes with operating temperature.

duced flux divergences [40]. Motivated by the need to explain the nature of temperature dependencies of electromigration, this chapter presents both steady-state temperature dependence and temperature gradient dependence by examining the physics of the failure processes.

3.2 MODELING TEMPERATURE EFFECTS

The phenomenon of electromigration-induced mass transport is attributed to atomic flux during the passage of current through a polycrystalline thin film conductor, as a result of electromigration in the lattice and at the grain boundaries. Diffusivity (D) is the only temperature-dependent term in the flux equation and is exponentially dependent on temperature. The contribution due to electromigration in the lattice is [20]

$$J_1 = \frac{1}{kT} N_1 D_1 j\rho e Z_1^* \tag{3.1}$$

The contribution of electromigration in the grain structure is given by [19]

$$J_b = \frac{1}{kT} \frac{N_b \delta}{d} D_b j\rho e Z_b^* \tag{3.2}$$

where N is the atomic density, D is the diffusivity, j is the current density, ρ is the resistivity, eZ^* is the effective charge, k is the Boltzmann constant, and T is the steady-state temperature. The subscripts "1" and "b" represent lattice terms and grain boundary, respectively. The quantity δ is the effective boundary width (10 Å) for mass transport, and d is the average grain size. In aluminum, the transport is via the grain boundary, so grain boundary parameters such as diffusivity are important.

Electromigration damage can occur only where there is divergence in the electromigration flux, J, caused by variations in any of the parameters on the right-hand side of Equations (3.1) and (3.2). For instance, if the grain size of the metallization changes, perhaps due to change in the substrate, the flux-carrying capacity of the track is altered and there is flux divergence. Cross-section changes in themselves do not lead to flux divergences, because a reduction in the cross-sectional area leads to a higher current density in the remaining section. However, a section change can cause a change in the local self-heating and therefore a temperature-induced divergence.

Temperature changes in the stripe are important sources of flux divergence, because the flux depends exponentially on temperature [38]. Temperature changes may arise from changes in the thermal properties of the substrate, as the metallization passes over other features on the substrate, or by changes in self-heating caused by section changes, for instance, at steps over substrate features. Temperature changes can also lead to thermomigration (transport of material in a temperature gradient), though the magnitude of thermomigration is small compared to electromigration [38].

At moderate temperatures (much lower than $0.5T_m$, where T_m is the melting temperature of the materials), the atomic flux from electromigration in the lattice is vanishingly small as compared with that from the grain boundaries, so the grain boundary electromigration becomes a dominant mode of mass transport for temperatures much lower than $0.5T_m$ [18]. The mass transport in the grain boundaries in most metals is by vacancy diffusion mechanism [27]. The relative contributions of the atomic flux due to lattice diffusion and grain boundary diffusion can be estimated from the ratio of J_1 and J_b, given by (the subscripts 1 and b denote the electromigration parameters of the lattice and grain boundary, respectively) [18]

$$\frac{J_1}{J_b} = \frac{N_1}{N_b} \frac{d}{\delta} \frac{D_1}{D_b} \frac{Z_1^*}{Z_b^*} \tag{3.3}$$

The measured values of Z_1^* and Z_b^* usually do not differ by more than an order of magnitude. For thin films with 1-μm grain size at $0.5T_m$, where T_m is the melting temperature of the materials [18],

$$\frac{N_1}{N_b} = 1, \quad \frac{d}{\delta} \cong 10^3, \quad \frac{D_1}{D_b} \cong 10^{-7}, \quad \frac{J_1}{J_b} \cong 10^{-4} \tag{3.4}$$

Attardo and Rosenberg [2], while conducting experiments on 0.4–0.6 mils wide and 10–12 mils long Al films on silicon wafers having 800 Å of thermally grown SiO_2 and 2000 Å of sputtered quartz, found that mass transport during electromigration shifted from grain boundary diffusion to lattice at threshold temperatures higher than $0.5T_m$. The rate of electromigration was determined by the film's degree of preferred orientation.

Diffusion rates are highly anisotropic. The diffusion parallel to dislocations composing tilt boundaries proceeds at several orders of magnitude greater than that in a direction perpendicular to the dislocation in that boundary. The vacancy flux in the grain boundary with contributions from electromigration and diffusion via grain boundary are given by

$$J_v = -D_v \nabla C_v + J_b \tag{3.5}$$

where D_v is the vacancy diffusivity in the grain boundary [18]. The local variation of the vacancy concentration is given by the equation

$$\frac{dC_v}{dt} = -\nabla \cdot J_v + \frac{C_v - C_v^0}{\tau} \tag{3.6}$$

where the last term expresses the local deviation of the vacancy concentration from thermal equilibrium [18]. The quantity τ is the average lifetime of the vacancy, which is determined by the efficiency of the source and the sinks in creating and annihilating vacancies. Under a steady-state condition, $dC_v/dt = 0$, so

$$C_v - C_v^0 = \tau \nabla \cdot J_v \tag{3.7}$$

The vacancy flux is represented as

$$\frac{dC_v}{dt} = D\left(\frac{d^2C_v}{dx^2} - \frac{dC_v}{dx}\frac{Z^*eE}{kT}\right) - \left(\frac{D}{kT}\frac{dC_v}{dx} - \frac{DC_vZ^*eE}{(kT)^2}\right) \cdot \left(\frac{\Delta H_b}{T}\left(\frac{dT}{dx}\right) + \frac{d(\Delta H_b)}{dx}\right) + \frac{C_v - C_v^0}{\tau_0} \tag{3.8}$$

where ΔH_b is the activation energy for the grain boundary diffusion [2]. The solution of this equation provides the rate of vacancy buildup at the point of divergence and the maximum vacancy concentration that can be achieved during steady-state electromigration. It is evident from Equation (3.8) that the damage from vacancy supersaturation can arise from any number of discontinuities other than temperature, such as structural variations between boundaries or between regions of the stripe consisting of different structures [2]. The actual process of hole formation is a void growth process, and the time required for vacancy buildup to maximum supersaturation is orders of magnitude less than that needed for observation of holes. Rosenberg and Ohring [35] calculated the vacancy supersaturations by considering the case of two boundaries of differing diffusion characteristics and located in the iso-

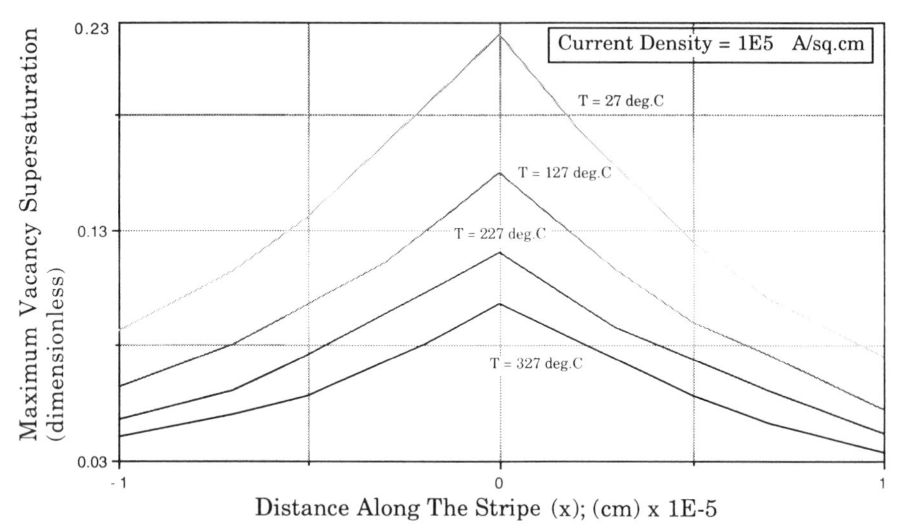

FIGURE 3.2. Temperature dependence of vacancy supersaturation distribution [2].

thermal region ($dT/dx = 0$), joining at $x = 0$, which is the accumulation site. The steady-state solution is represented in Figure 3.2 to demonstrate the effects of temperature on vacancy supersaturation. Assuming a product D_0Z^* for aluminum to be $3 \pm 0.5 \times 10^{-2}$ cm^2/s, for current density of 1×10^6 A/cm^2, Rosenberg and Ohring [35] found the maximum vacancy supersaturation to decrease from 27°C to 327°C. Once the maximum vacancy supersaturation is reached, the driving force responsible for vacancy diffusion due to concentration gradient is lost. Rosenberg and Ohring observed that under current densities of 1×10^6 A/cm^2 and steady-state temperature of 127°C, the maximum supersaturation was between 0.1 and 1.

Generally, electromigration has been shown to have a linear dependence on temperature gradient, an exponential dependence on temperature for temperatures greater than 150°C, and a dependence on current density whose order varies from 2 to 14. The combinations of current density and temperature, which will produce electromigration damage in Ti-Pt-Au metallization, have been characterized in Figure 3.3. The characteristics have been derived for not more than 0.14% cumulative failures (an average of 8 failures/10^9 stripe hours) after 20 years, which is the 3-sigma limit of the lifetime distribution. In Figure 3.3, the lifetimes are assumed to have a fourth dependence on the current density [45].

In an ideal case of structurally uniform conductor with no temperature gradient, there is no flux divergence, so that electromigration damage

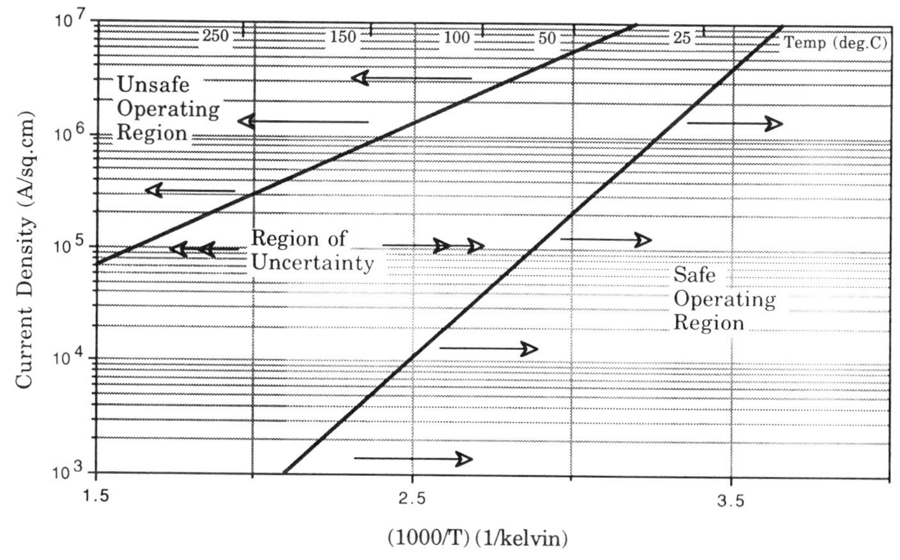

FIGURE 3.3. Combinations of current density and temperature that will produce electromigration damage in Ti-Pt-Au metallization.

will not occur [18, 19]. The extent of deviation of vacancy concentration from equilibrium is proportional to the vacancy flux divergence. Whenever there is a spatial variation in any of the parameters affecting the grain boundary electromigration, which may include structural defects in the form of nonuniform thickness of the conductor, or temperature gradient, a divergence in the atomic flux occurs, giving rise to a local vacancy supersaturation or depletion (void formation).

3.3 TEMPERATURE DEPENDENCE OF ELECTROMIGRATION IN THIN FILM CONDUCTORS

3.3.1 Steady-State Temperature Effects in Presence of Cracks

Effect of cracks in the conductor metallization perpendicular to current flow direction in correlation with temperature has been predicted by Sigsbee [41]. Grain boundary electromigration, internal heat generation, and current crowding at growing voids dominate the rate processes that lead to failure. Assuming that the bottom interface below the substrate is held at a constant temperature T_e, and the conductor experiences a temperature rise ΔT_0 above T_e due to Joule heating, for a crack perpendicular to the current direction, the current crowding was approximated by

$$J_e(X) = \frac{J_{e0}}{\left(1 - \frac{X}{W}\right)} \tag{3.9}$$

Joule heating of the stripe causes an initial temperature rise resulting in instability in the stripe, causing the vacancies in the stripe to migrate along grain boundaries and precipitate on a suitable boundary, forming elongated voids. The crack grows long by accumulation of vacancies flowing along nearby grain boundaries. The atomic flux of vacancies in the crack effected was represented as

$$J_v = J_{v0}\left(\frac{1}{1 - L}\right)\left(\frac{\Delta H\, \Delta T_0[\ln(1 - L) + L]}{kT^2 L}\right) \tag{3.10}$$

where $L = l/W$, l being the crack length and W being the conductor width, T_e is the bottom interface temperature, and ΔT_0 is the temperature rise due to Joule heating in the metallization, J_{v0} is evaluated at initial film temperature $(T_e + \Delta T_0)$, the second term in brackets is due to current crowding, and the third term in brackets is due to self-heating. Assuming that only the grain boundary migration contributed to crack growth, the lifetime was calculated as

$$\tau_f = \frac{cWkT}{\delta N_a q^* \rho J_e D_0 \exp\left(-\Delta \frac{h}{kT}\right)} 2\left(1 + \left(\frac{0.2\Delta H}{kT^2}\right)\Delta T_0\right) \tag{3.11}$$

where c is the crack width, W is the linewidth, k is the Boltzmann constant, T is the steady-state temperature, δ is the effective grain boundary width for transport, N_a is the atomic density, q^* is the effective charge, ρ is the resistivity, J_e is the current density at the crack tip region, D_0 is the diffusion coefficient, ΔH is the activation energy, and ΔT_0 is the initial temperature rise due to Joule heating. The above model neglects the effects due to coefficient of resistance and temperature sensitivity of q^*. Sigsbee showed that these factors had negligible effect on the life prediction estimate. The lifetimes were found to have a J_e^{-n} dependence with n varying from unity at low ΔT_0 levels to 15 for high ΔT_0. The model has been shown to model grain boundary electromigration for temperatures in the neighborhood of 260°C [41]. The above grain boundary grooving mechanism is typically noticed in silver metallizations [43].

3.3.2 Steady-State Temperature Effects in Presence of Voids

The atomic flux is dependent on temperature, therefore local temperature gradients will cause a divergence in atomic flux. The depletion of mass occurs wherever the electron flow is in the direction of increasing temperature. Conversely, the accumulation of mass occurs wherever the electron flow is in the direction of decreasing temperature [43]. Metallized stripes in good thermal contact with their substrates have negligible temperature gradients [43]. Electromigration in thin metallized films is confined mainly to grain boundaries [1, 10, 36]. For this reason the steady-state temperature dependence of electromigration is of the same magnitude as that of grain boundary diffusion [10, 36].

Voids form as a consequence of flux divergences at nonsymmetrical nodes and grow with time, eventually coalescing to form a gap across the conductor. The effect is particularly severe in films with a small grain size because a large number nodes are available to act as nuclei for formation of voids. On the other extreme, if the grain size is comparable with stripe width, the probability that a single grain will cover the entire stripe increases, which introduces an additional source of flux divergence acting as a barrier to the atoms migrating from the negative side, preventing the replacement of atoms that are transported away from connecting boundaries on the other side of the grain [2, 9].

The flow of current through stripes creates voids at grain boundary nodes that are suitably oriented relative to the current flow direction and longitudinal temperature gradient. The resulting porosity increases as a function of density of grain boundary nodes, current density, resistivity, and mobility of metal ions along grain boundaries and is represented by [43]

$$\frac{dp}{dt} = Cnj\rho\mu \tag{3.12}$$

where C is a constant of proportionality, p is the porosity, n is the grain boundary node density, j is the current density, ρ is the resistivity, and μ is the mobility of metal ions along grain boundaries. Pore formation reduces the cross-sectional area of metal stripe available for carrying the current, thereby increasing the local current density within the remaining section [43]:

$$j = \frac{j_0}{(1 - p)} \tag{3.13}$$

where j_0 is the initial current density in the pore-free stripe. The increased current density caused an increase in the current-enhanced motion of the metal atoms in the stripe, at the same time increasing the Joule heating within the remaining conducting portions of the stripe. The local temperature in the stripe increases above the ambient temperature, T_0, by an amount that is proportional to Joule heating given by [43]

$$\Delta T = T - T_0 = \frac{j^2 \rho}{h} \tag{3.14}$$

The temperature rise leads to a corresponding increase in mobility, μm, of the metal atoms along grain boundaries [43]:

$$\mu = \frac{D}{kT} = \left(\frac{D_0}{kT}\right) e^{-Q/kT} \tag{3.15}$$

where D is the diffusion coefficient for motion along grain boundaries, and Q is the activation energy for this process. The increase in temperature leads to a change in the resistivity, ρ, of the metal [43]:

$$\rho = \rho_0(1 + \alpha(T - T_0)) \tag{3.16}$$

where α is the temperature coefficient of resistance. At a constant total current the increase in resistivity causes an additional increase in the local rate of Joule heating and the effective electric field ($j\rho$) experienced by the atoms. The electromigration failure occurs where the grain boundary migration and temperature gradient combine to create suitable conditions for porosity to develop until it exceeds the critical value resulting in the melting of the stripe. Venables and Lye [43] combined the above equations to give the time to failure as

$$T_F = \frac{1}{2Cn}\left(\frac{\tau_0 k T_0}{j_0 \rho_0 D_0 e^{-Q/kT}}\right) \int_{x_0}^{x_1} \frac{e^{-Qx/kT_0}}{x^2(1 - x + x\alpha T_0)}\, dx \tag{3.17}$$

where

$$\tau_0 = \frac{\Delta T_0}{T_0} = \frac{j_0 \rho_0}{h T_0} \tag{3.18}$$

and

$$\begin{aligned} x &= \frac{\tau_0}{(1 - p)^2 + \tau_0(1 - \alpha T_0)} \\ &= 1 - \left(\frac{T_0}{T}\right) \end{aligned} \tag{3.19}$$

$$x_0 = \frac{\tau_0}{1 - \tau_0(\alpha T_0 - 1)} \qquad \text{for } t = 0$$

$$x_1 = 1 - \left(\frac{T_0}{T_m}\right) \qquad \text{for } t = T_F \tag{3.20}$$

where T_m is the melting temperature of the metal. Venables and Lye [43] showed that time to failure versus current density varied in a complex manner, and that a simple power law dependence (as shown by Black) was inadequate to describe the experimental conditions over more than a small range of current densities. The results of Attardo and Rosenberg [2], Black [5], and Blair et al. [9] were tangential to the results from the Venables–Lye model, at temperature of 210°C, and current densities ranging from 1×10^4 to 2×10^6 A/cm^2 (Figure 3.4). Venables and Lye [43] showed that when the temperature dependence of times to failure due to electromigration was represented as an Arrhenius plot, although the curves appeared as accurate straight lines, the slopes yielded only apparent activation energy, which varied with test conditions, indicating that the time to failure was a complex function of baseline temperature and could not be represented by an Arrhenius plot to give an activation energy (Figures 3.5 and 3.6).

3.3.3 Steady-State Temperatures Effects Without Assumption of Structural Defect Magnitudes

A general, but not universal, expression for the mean time-to-failure MTF (or t_{50}, which is the time to reach failure of 50% of a group of

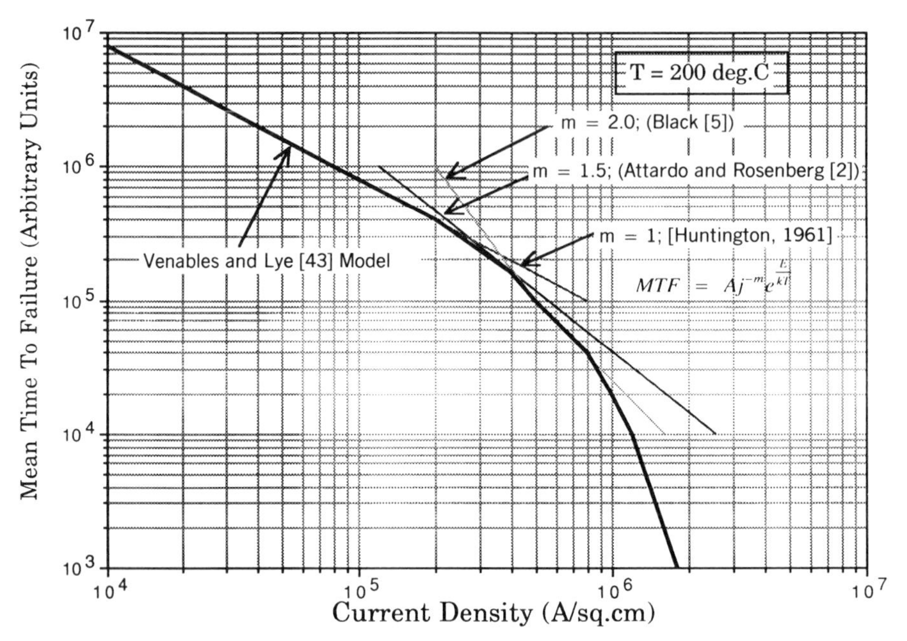

FIGURE 3.4. Mean time to failure varies with current density and temperature in a complex manner. Simple power laws can represent experimental observations over a small range of current densities [43].

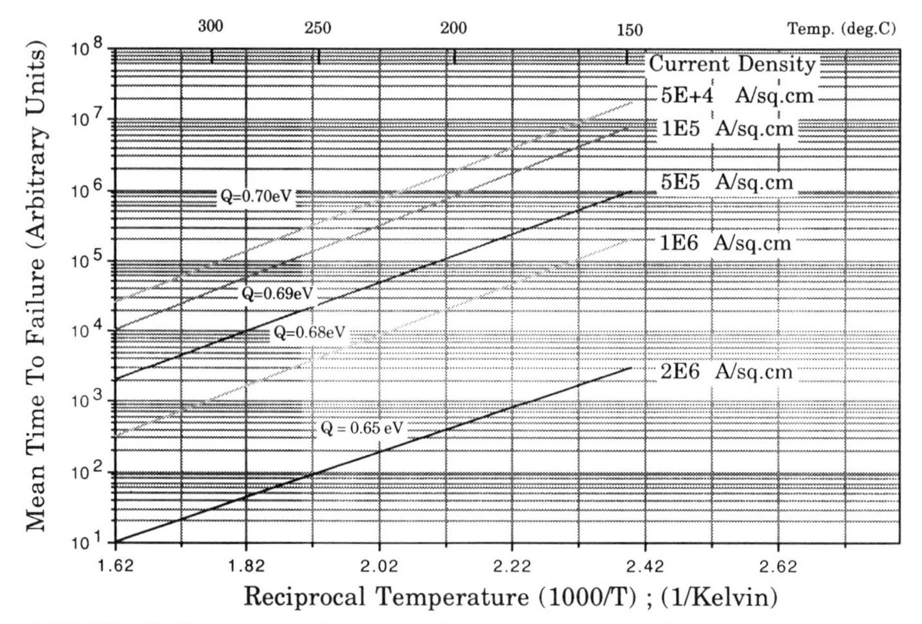

FIGURE 3.5. Temperature dependence has been represented by an apparent activation that changes with test conditions [43].

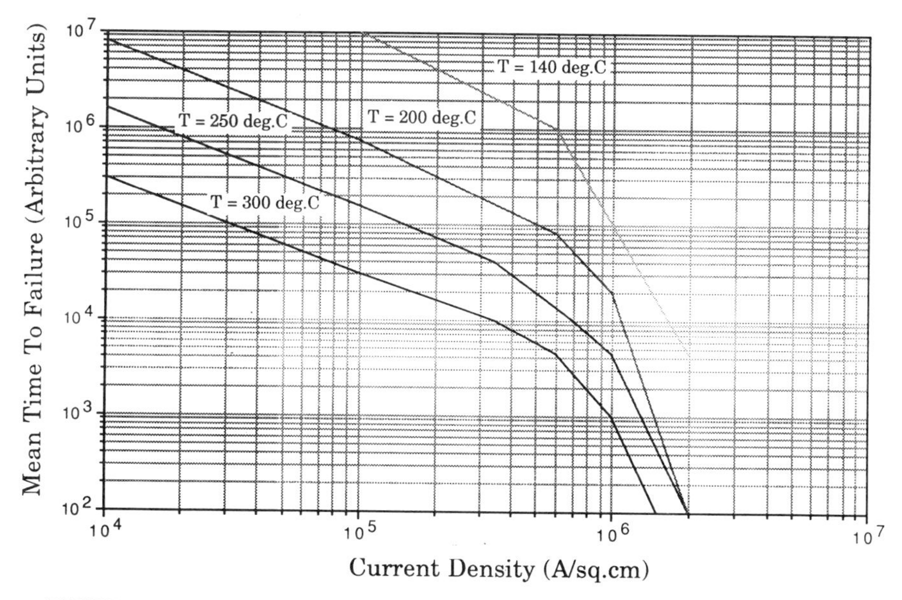

FIGURE 3.6. Influence of baseline temperature on mean time to failure [43].

identical conductor lines) is

$$\text{MTF} = Aj^{-n}e^{E/kT} \tag{3.21}$$

where A is a parameter depending on sample geometry, physical characteristics of the film and substrate, and protection coating, j is the current density (A/cm^2), and n is an experimentally determined exponent (Equation (3.21) is not a failure rate expression, as the failure times are typically observed to follow a lognormal distribution). This model is applicable only to conductor films, which are wider than the average grain size of the aluminum film from which it is constructed. As the conductor width is reduced and approaches or becomes less than the average grain size, the structure begins to "bamboo," that is, most of the grain boundaries become normal to the electron flow. Black's relationship does not apply when bambooing starts to occur [3].

The electromigration lifetime test is carried out under a set of accelerated test conditions at elevated temperatures and with high current density stressing. The data are then extrapolated to device operating conditions, with current density stressing below 5×10^5 A/cm^2, by the Arrhenius-like empirical equation cited above (originally formulated by Black [4, 6, 7]). Values of n that have been reported are

$n = 1\text{–}3$ Chabra and Ainslie [11]
$n = 1.5$ Attardo [2]
$n = 1.7$ Danso and Tullos [12]
$n = 2$ Black [8]
$n = 6\text{–}7$ Blair et al. [9]

Black characterized his data in the range $0.5 \times 10^6 < j_0 < 2.8 \times 10^6$ with an exponent of $n = 2$. Attardo reported $n = 1.5$ in the range $10^5 < j_0 < 10^6$ A/cm^2. Blair et al. reported a value of $n = 4\text{–}5$ in the range of $10^6 < j_0 < 2 \times 10^6$ A/cm^2. Venables and Lye found that the simple power law dependence of time to failure on current density could be used to describe experimental observations in a small range of current densities [43].

Shatzkes and Lloyd [40] treated electromigration failures by superimposing Fickian diffusion and mass transport due to electromigration force, and they derived a modification of Black's equation

$$t_f = \left(\frac{2C_f}{D_0}\right)\left(\frac{k}{Z^* e\rho}\right)^2 T^2 j^{-2} e^{\Delta H/kT} \tag{3.22}$$

where C_f is the critical value of vacancy concentration at which failure occurs, D_0 is the preexponential factor for grain boundary self-diffusivity, k is the Boltzmann constant, Z^* is the effective charge, e is the electronic charge, ρ is the resistivity, T is the steady-state temperature, j is the current density, and ΔH is the activation energy. Equation (3.22) differs from Black's equation in that it has a T^2 preexponential term, but it fits Black's data equally well.

Table 3.1 shows the estimated MTF of Ti:W/Al and Ti:W/Al + Cu film conductors at 85°C for 5×10^5 and 2×10^5 A/cm^2 current

TABLE 3.1. Estimated MTF Values for Electromigration at 85°C, as a Function of the Exponent Used in Black's Equation

Current Density	Exponent	Ti: W/Al (years)	Ti: W/Al + Cu (years)
	$n = 1.0$	4	12
5×10^5 A/cm^2	$n = 1.5$	5	17
	$n = 2.0$	8	24
	$n = 1.0$	10	30
2×10^5 A/cm^2	$n = 1.5$	23	68
	$n = 2.0$	50	152

densities at 100% duty cycle. Even at the worst case $n = 1$, it has been anticipated [15, 16] that the actual time to failure will be greater than those predicted in Table 3.1. For current densities of 2×10^5 A/cm^2 temperatures of 125°C will not lead to failure in less than 10 years with a typical exponent of $n = 1.7$.

Few studies at lower temperatures have been performed on unpassivated stripes; they recorded the phenomenon of electromigration shifts from grain boundary migration to surface migration, and detachment of the stripe from the chip, at temperatures in the neighborhood of 223–347 K [34].

In order to improve the resistance of the conductor to surface electromigration, passivation consisting of glass overlays, metallic coatings, or natural oxides are used to cover the thin film conductor. Typically, the electromigration lifetime of most conductors increase by an order of magnitude or more with complete surface coverage [13, 29, 44]. The use of transition layers such as TiN [17], Cr [28], and Ti-W [14] have been reported to improve lifetime of thin film conductors by an order of magnitude. The transition layer provides a redundant structure that allows void healing and also acts as a metal diffusion layer.

3.3.4 Temperature Gradient Dependence of Electromigration Lifetimes

Temperature gradients exist both globally and locally in thin film conductors due to heat generation from Joule heating and power dissipation from active devices on the chip. The global temperature gradient is small except near electrodes or contact pads. Large local temperature gradients or hot spots can be caused by poor adhesion or contaminates at the interface between the metal film and the substrate, or by the thickness variation of the metal film.

Temperature gradients are important sources of atomic flux divergence, since the flux depends exponentially on temperature. For example, at 200°C in aluminum, a 5°C change in temperature results in a change of more than 10% in the electromigration flux [38]. Studies on electromigration damage due to temperature gradients have revealed that void formation occurs in regions where electron flow is in the direction of increasing temperature, and hillocks form in locations where electron flow is the direction of decreasing temperature [10]. Temperature gradients can also lead to thermomigration, although migration due to thermomigration is small for aluminum tracks in ICs compared to electromigration.

The temperature gradient dependence of electromigration failures has been confirmed by Lloyd and Shatzkes [30] and Schwarzenberger et al. [38]. Lloyd and Shatzkes [30] noticed that the electromigration failure location was typically nearer to the location of maximum temperature gradient, whereas the location of nontemperature-gradient induced failure was randomly distributed for Cr/Al-Cu conductors covered with polyimide passivation. Lloyd and Shatzkes [30] modeled the temperature of the stripe carrying current to produce significant Joule heating as the balance between the heat generated in the stripe, the heat conducted away from the stripe to the surrounding thermal sinks such as substrate, the passivation, and the portion of the metal stripe that is not conducting heat. The heat balance relation is given by [30]

$$\rho j^2 = K\left(\frac{d^2\,\Delta T}{dx^2}\right) - h\,\Delta T \tag{3.23}$$

per unit volume, where ρ is the metal resistivity, K is the metal's thermal conductivity, h is the parameter that characterizes the thermal efficiency of the heat sinks, and ΔT is the temperature rise due to Joule heating. The term on the left of Equation (3.23) is the heat generated due to current passing through the metal element, the first term on the right is the heat conduction along the metal stripe away from the heating element, and the last term is the heat conduction to the environment. If the heat conduction through the oxide (of thickness l_{oxide}) is only considered, the value of h is of the order [30]

$$h \approx \left(\frac{K_{\text{oxide}}K}{l_{\text{oxide}}\,l}\right) \tag{3.24}$$

where K_{oxide} is the thermal conductivity and l is the stripe thickness. The heat sink is assumed to be at the ambient temperature. The solution to the heat conduction equation for a stripe with origin in the center was given by Lloyd and Shatzkes [30] as

$$T = A\left[1 - \left(\frac{\cosh\,(Bx)}{\cosh\,(BL/2)}\right)\right] \tag{3.25}$$

where

$$B = \sqrt{\left(h - \frac{\rho_0 \alpha j^2}{K}\right)} \tag{3.26}$$

$$\rho = \rho_0(1 + \alpha \Delta T) \tag{3.27}$$

$$A = \frac{\rho_0 j^2}{B^2 K} \tag{3.28}$$

The location of failure was argued to be near the position of maximum flux divergence. The atomic flux is

$$J = \frac{DF}{kT} \tag{3.29}$$

where D is the diffusivity, F is the driving force for diffusion, the point of maximum flux divergence is

$$\frac{dJ}{dx} = \frac{dC}{dt} = \left(\frac{dJ}{dT}\right)\left(\frac{dT}{dx}\right) \tag{3.30}$$

and the condition of failure is when

$$\frac{d^2J}{dx^2} = \frac{d}{dx}\left(\frac{dJ}{dT}\right)\left(\frac{dT}{dx}\right) = 0 \tag{3.31}$$

The location of electromigration failure calculated from Equation (3.31) is near the location of maximum temperature gradient [30]. The points of maximum flux divergence move from the edges for higher current densities to the middle for lower current densities. The change in the failure location was explained to be due to an exponential dependence of flux divergence on temperature, and only linear dependence on temperature gradient.

Figures 3.7 and 3.8 show the variation of location of sites of failure versus distance along the stripe, for low and high current densities. At high current densities ($> 10^6$ A/cm^2) accompanied by high Joule heating, the highest flux divergence is very close to the location of highest temperature gradient, which is near the edge. The failure site is thus closer to the site of maximum temperature gradient. At lower current densities ($< 10^6$ A/cm^2), the failure location is more randomly distributed, since the effect of temperature gradient becomes small compared to other flux divergences, induced by structural discontinuities.

Variation of flux between the different regions of the conductor may lead to failure due to depletion in some regions. The time to failure is inversely proportional to the flux gradient within the region. The flux

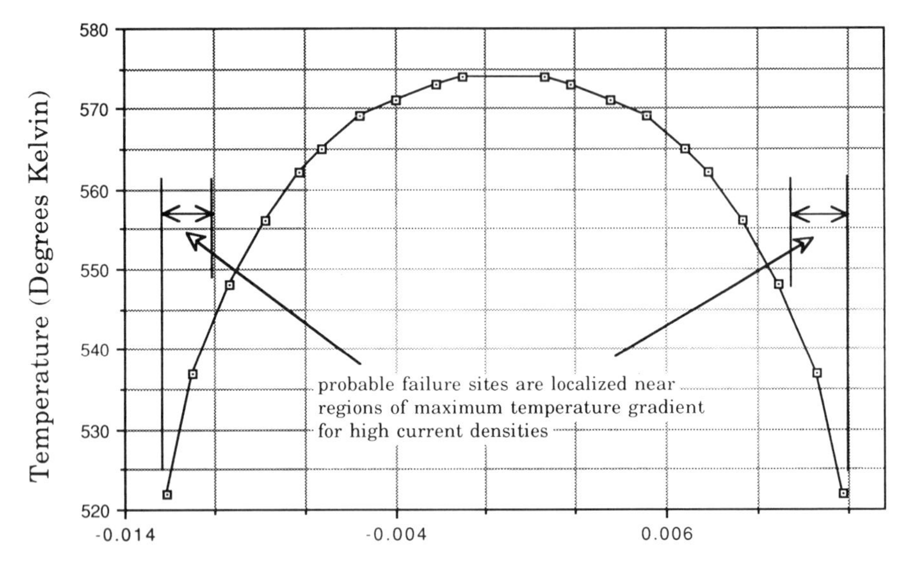

FIGURE 3.7. Location of failure sites versus temperature distribution along stripe for high current condition [30].

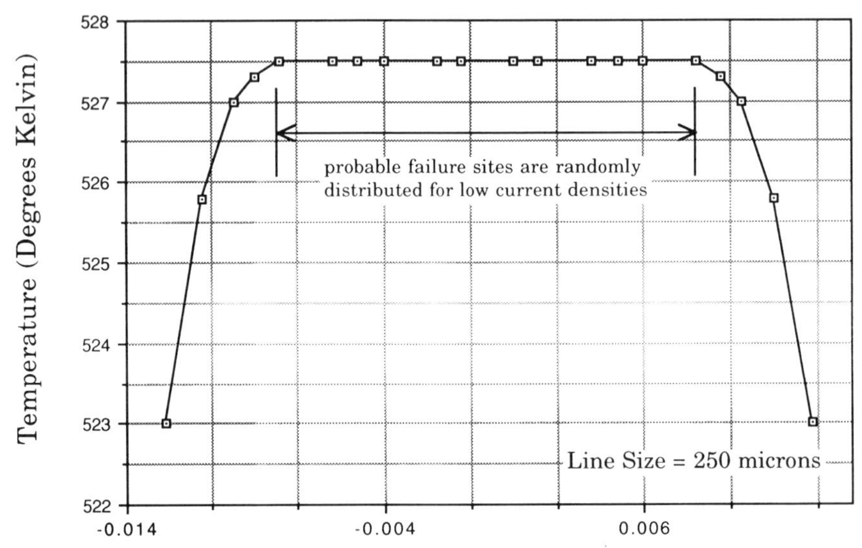

FIGURE 3.8. Location of failure sites versus temperature distribution along stripe length for low current condition [30].

gradient in terms of the temperature gradient is given as

$$\frac{dU}{dX} = \frac{N\epsilon ID_*}{k} e^{-E/kT} f(\rho, Z, T) \frac{dT}{dX} \tag{3.32}$$

where N is the density of the ions, ϵ is the electronic charge, ρ is the resistivity, Z is the effective ionic valency and represents the net effect of momentum exchange and electric field forces, $D_* \exp(-E/kT) = D$ is the diffusion coefficient for ions in the conductor, k is the Boltzmann constant, T is the average steady-state conductor temperature, and E is the activation energy. Schwarzenberger et al. [38] and Oliver and Bower [31] demonstrated the importance of temperature gradient as a source of electromigration flux divergence in metallization tracks and therefore the necessity of controlling the temperature profile in the integrated circuit to optimize its lifetime.

3.4 TEMPERATURE DEPENDENCE OF ELECTROMIGRATION IN MULTILAYERED METALLIZATIONS

The definition of electromigration failure has changed [32] with the introduction of multilayered interconnection metallizations with sensitive electrical circuits. Many studies on single-layer metallizations have used the opening of the conductor as a criterion for failure, ignoring the functional failures including the resistance change of the metallization. The open criterion for layered metal systems may not be achievable in a test environment if one of the layers is not susceptible to electromigration.

Onduresk et al. [32] derived a void formation model for multilayered metallizations that allows calculation of void length from measured resistance and temperature coefficients. The temperature coefficient of resistance was investigated to verify that the refractory layer remains undamaged throughout the voiding process. The temperature coefficient C for metal is defined by the following equation:

$$R = R_i[1 + C(T - T_i)] \tag{3.33}$$

where R is the resistance at temperature T and R_i is the resistance at some fixed initial temperature T_i. In a simplified case of void extending through the aluminum layer, all the current must pass through the refractory metal layer. The total resistance of the metal stripe is the sum of the initial component due to the refractory material metal/aluminum sandwich plus the additional series refractory metal resistance that results from the void. This is represented by the following expressions:

$$R(v) = R_1 + R_2 \tag{3.34}$$

where

$$R_1 = R_{im} \frac{L - v(t)}{L} [1 + C_m(T - T_i)] \tag{3.35}$$

and

$$R_2 = R_{ir} \frac{v(t)}{L} [1 + C_r(T - T_i)] \tag{3.36}$$

Taking the derivative of Equation (3.25) with respect to temperature gives

$$\frac{\partial R(v)}{\partial T} = R_{im} C_m \frac{L - v(t)}{L} + R_{ir} C_r \frac{v(t)}{L} \tag{3.37}$$

where R_{im} is the initial resistance of the composite, R_{ir} is the initial resistance of the refractory metal layer, C_m is the initial composite temperature coefficient, C_r is the refractory metal temperature coefficient, L is total metal line length, and v is the void length. Deviation from this model can occur due to the intermetallic compound formation.

3.5 REDUCING ELECTROMIGRATION DAMAGE IN METALLIZATION STRIPES

Solutions to reduction of electromigration do not lie in reducing steady-state temperature. The basic requirement for reducing the electromigration damage is to reduce the local divergence of atomic flux. This can be accomplished, in principle, by reducing the magnitude of the atomic flux and/or the inhomogeneity of the parameters controlling the mass transport. The magnitude of atomic flux is determined by the electromigration driving force and the grain boundary diffusivity. Thus to reduce the atomic flux, the choice is to reduce the driving force and/or the diffusivity. Reduction of the driving force has some basic difficulties, since it requires either a change in the scattering process responsible for the effective charge or a reduction in the current density. Because the current density is dictated by the device functional requirements, the only choice is to reduce the grain boundary diffusivity. The most common approach is by solute addition, which also results in improvements in conductor properties in the form of grain structure modification. The common examples are the addition of Cu to Al stripes, or other solute elements such as Mn, Mg, Si, and Ti [18, 37].

There is a strong correlation between microstructure and electromigration lifetime in thin metal lines, which is particularly noticed in VLSI technology when the linewidth and thickness of the metal lines are reduced to the submicrometer range comparable to the grain size. Larger lifetimes in large grained Al films [22, 23, 24] and large grained Al-Cu films with bamboo structure [33, 42] have been reported (Figure 3.9). Annealing Al-Cu lines at elevated temperatures has been found to induce grain growth, increasing the electromigration lifetime. Structural modification of the Al-Cu films by addition of Ti and Cr (*at about 400°C results in the formation of intermetallic compounds such as* Al_3Ti *and* Al_7Cr) improves electromigration lifetimes due to changes in microstructures, which reduce damage formation and block void growth by the presence of a redundant barrier that maintains current continuity [21].

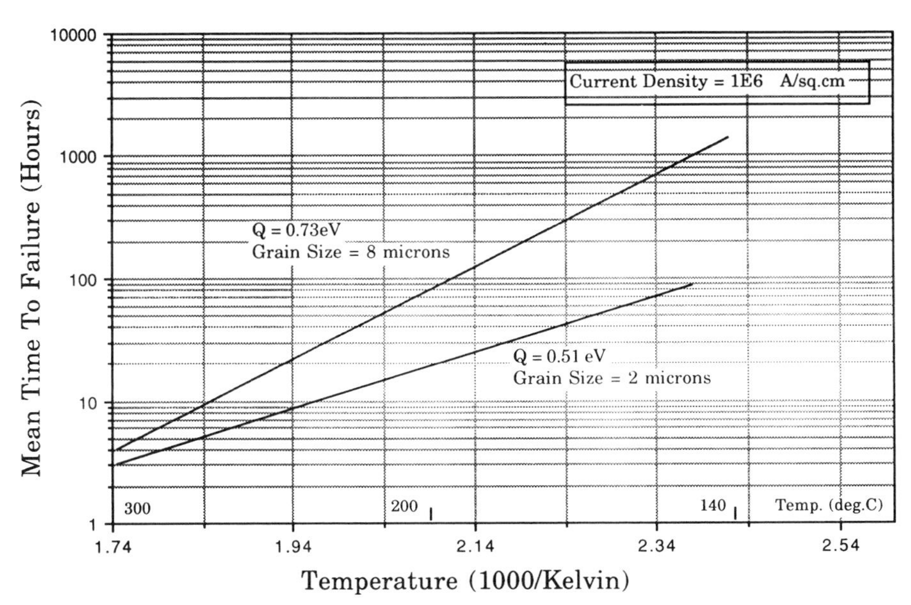

FIGURE 3.9 Effect of temperature on MTF versus grain size [2].

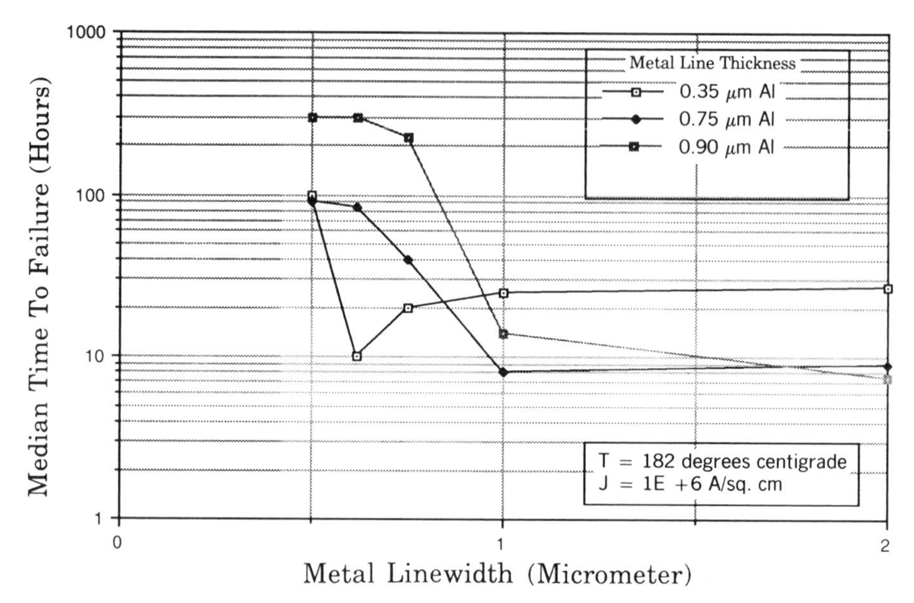

FIGURE 3.10. MTF due to electromigration versus aluminum linewidth at 182°C [21].

Linewidth has a strong influence on electromigration lifetimes [21]. Agarwala et al. [1] reported a linear relationship between linewidth and lifetime of aluminum wide lines with linewidths down to 5 μm (Figure 3.10). Subsequent studies have indicated that as linewidth reduces to below a critical value, lifetime is found to level off or go through a minimum and then increase, reversing the trend in wider lines [21, 22–26]. The critical width decreases with decreasing film thickness. Kwok [21, 22–26] found the critical linewidth to be about 0.75 μm, for the 0.5-μm line thickness, for Al-Cu lines (Figure 3.11). The critical linewidth is sensitive to thickness of the metal lines. Lifetime increases by a factor of 5 with increasing linewidth from 1 to 2 μm. The lifetime of Al-Cu-Si, Cr-Ag-Cr, and Al/Ti lines is found to level off beneath the critical linewidth of around 2.0, 1.5, and 1.2 μm, respectively. The linewidth dependence is much stronger for Al-Cu and Al-Cu-Si lines than for Cr-Ag-Cr lines. The probability of aligning failure causing defects across a wide line is lower than a narrow line, thus making it more difficult for a crack to propagate across a wide line, increasing the expected lifetime. The dependence of lifetime on linewidth is also a function of pattern technique, metallurgy, and metal deposition conditions. Scoggan et al. [39] found that Al-Cu lines patterned by chemical etch reached a minimum lifetime for linewidths in the neighborhood of 5.5

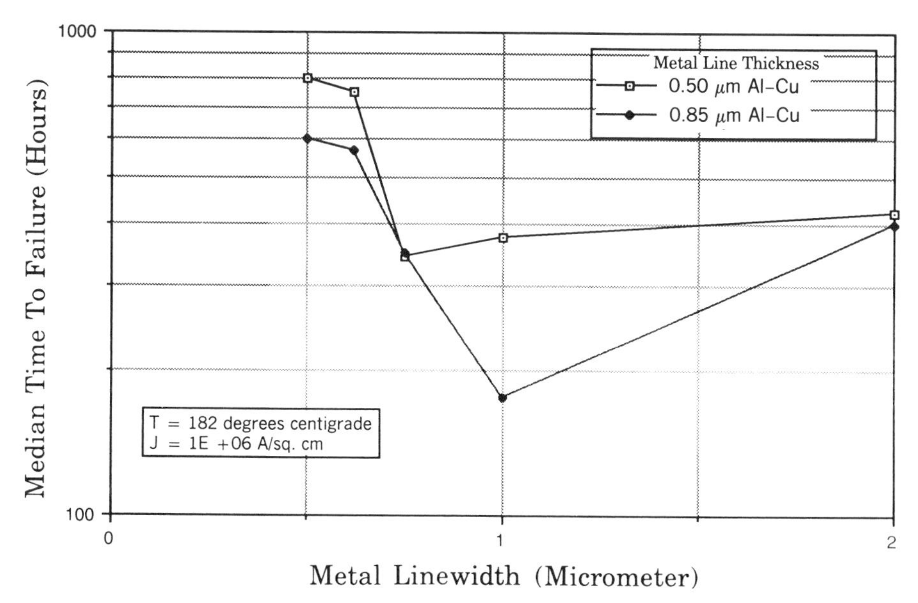

FIGURE 3.11. MTF due to electromigration versus Al-Cu linewidth at 182°C [21].

μm. The lifetime of Al-Cu lines of the same dimension but patterned by metal lift-off show a minimum in lifetime for widths in the neighborhood of 3.5 μm. Electromigration lifetime is found to increase with decreasing thickness. The critical width decreases from 2.5 to 1.5 μm when the film thickness decreases from 1.1 to 0.8 μm in Al-Cu-Si lines [39].

3.6 SUMMARY

Predicted interconnection failure rates at use conditions of current density and temperature typically vary several orders of magnitude, because they strongly depend on accelerated test data and model parameter selection. Generally, the failure time depends inversely on the temperature gradient and current density. Temperature acts as a strong accelerator of the electromigration above temperatures of 150°C. Electromigration failures in structurally uniform conductors cannot be accelerated in reasonable time frames at temperatures lower than 150°C. The simple Black type equation used for extrapolation of failure to use conditions can be used over a small range of current densities. The current exponent changes over various densities. The actual dependence of lifetime has been shown to be a complex function of current density and temperature, which cannot be represented by an activation energy. The lifetime can, however, be represented by an apparent activation energy that changes with operating conditions.

Furthermore, solutions to the reduction of electromigration do not lie in reducing steady-state temperature. Electromigration damage forms at sites of maximum atomic flux divergence, of which the three main sources are structural defects, microstructural inhomogeneities, and local temperature gradients. Because typical field failures are characterized by structurally induced flux divergences, geometric configurations and metallization grain structures can reduce electromigration damage.

REFERENCES

1. B. Agarwala, M.J. Attardo, and A.P. Ingraham, ''The Dependence of Electromigration Induced Failure Time on the Length of Thin Film Conductors,'' *J. Appl. Phys.*, Vol. 41, 3954 (1970).
2. M.J. Attardo and R. Rosenberg, ''Electromigration Damage in Aluminum Film Conductors,'' *J. Appl. Phys.*, Vol. 41, 2381 (1970).
3. J.R. Black, ''Current Limitation of Thin Film Conductor,'' in *International Reliability Physics Symposium*, pp. 300–306, (1982).
4. J.R. Black, ''Electromigration of Al-Si Alloy Films,'' in *Proceeding of the IEEE International Reliability Physics Symposium*, pp. 233–240, (1978).
5. J.R. Black, *Proc. IEEE*, Vol. 57, 1587 (1968).
6. J.R. Black, ''Electromigration Failure Modes in Aluminum Metallization for Semiconductor Devices,'' *Proc. IEEE*, Vol. 57, 1587 (1969).
7. J.R. Black, ''Electromigration—A Brief Survey and Some Recent Results,'' *IEEE Trans. Electron Devices*, Vol. ED-16, 338 (1969).
8. J.R. Black, ''Physics of Electromigration,'' in *Proceedings of the IEEE International Reliability Physics Symposium*, pp. 142–149 (1983).
9. J.C. Blair, P.B. Ghate, and C.T. Haywood, ''Electromigration Induced Failures in Aluminum Film Conductors,'' *Appl. Phys. Lett.*, Vol. 17, 281 (1970).
10. I.A. Blech and E.S. Meieran, ''Direct Transmission Electron Microscope Observations of Electrotransport in Aluminum Films,'' *Appl. Phys. Lett.*, Vol. 11, 263 (1967).
11. D.S. Chabra and N.G. Ainslie ''Open Circuit Failures in Thin Film Conductors,'' IBM Component Div., East Fishkill Facility, New York, Tech. Report 22.419, July 1967.
12. K.A. Danso and T. Tullos, ''Thin Film Metallization Studies and Device Lifetime Prediction Using Al-Si, Al-Cu-Si Conductor Test Bars,'' *Microelectronics Reliab.*, Vol. 21, 513 (1981).

13. L.E. Felton, J.A. Schwarz, R.W. Pasro, and D.H. Norburg, *J. Appl. Phys.*, Vol. 58, 723 (1985).
14. L.J. Fried, J. Hares, J.S. Lecdeton, J.S. Logen, G. Paal, and P.A. Totta, *IBM J. Res. Dev.*, Vol. 26, 263 (1983).
15. P.B. Ghate, *19th International Reliability Physics Symposium*, pp. 243 (1981).
16. P.B. Ghate, ''Aluminum Alloy Metallization for Integrated Circuits,'' *Thin Solid Films*, Vol. 83, 195–205 (1981).
17. E. Grabe and H.U. Schreiber, *Solid State Electronics*, Vol. 26, 1023 (1983).
18. P.S. Ho and T. Kwok, ''Electromigration on Metals,'' *Rep. Prog. Phys.*, Vol. 52, 301–348 (1989).
19. P.S. Ho and J.K. Howard, *J. Appl. Phys.*, Vol. 45, 3229 (1974); quoted from P.S. Ho and T. Kwok, ''Electromigration on Metals,'' *Rep. Prog. Phys.*, Vol. 52, 301–348 (1989).
20. H.B. Huntington and A.R. Grone, ''Current-Induced Marker Motion in Gold Wires,'' *J. Phys. Chem. Solids*, Vol. 20, No. 1/2, 76–87 (1961).
21. T. Kwok, ''Effects of Grain Growth and Grain Structure on Electromigration Lifetime in Al-Cu Submicron Interconnects,'' in *Proceedings of the 4th International Multilevel Interconnection Conference*, Santa Clara, CA, p. 456 (1987).
22. T. Kwok, C.Y. Ting, and J.U. Han, ''Microstructure Studies of Al-Cu Submicron Interconnection,'' in *Proceedings of the 2nd International Multilevel Interconnection Conference*, Santa Clara, CA, p. 83 (1985).
23. T. Kwok, C. Tan, D. Moy, J.J. Estabil, H.S. Rathore, and S. Basavich, ''Electromigration in a Two-Level Al-Cu Interconnection with W Studs,'' *Proceedings of the 7th International VLSI Multilevel Interconnection Conference*, p. 106 (1990).
24. T. Kwok, ''Effects of Metal Line Geometry on Electromigration Lifetime in Al-Cu Submicron Interconnects,'' in *Proceedings of the 26th International Reliability Physics Symposium*, Monterey, CA, p. 185 (1988).
25. T. Kwok, P.S. Ho, and S. Yip, *Surf. Sci.*, Vol. 144, 44 (1984).
26. T. Kwok, ''Electromigration in Submicron Interconnects and Multilevel Interconnection,'' in *Proceedings of the Symposium of Semiconductor Devices and Interconnection and Multilevel Metallization, Interconnection, and Contact Technologies*, ECS ed., pp. 1–11 (1989).
27. T. Kwok, P.S. Ho, S. Yip, R.W. Ballufi, P.D. Bristow, and A. Brokman, *Phys. Rev. Lett.*, Vol. 47, 1148 (1981).
28. E. Levine and J. Kitcher, in *Proceedings of the 22nd International Reliability Physics Symposium*, p. 242 (1984).
29. J.R. Lloyd and P.M. Smith, *J. Vac. Sci. Technol.*, Vol. A1, 455 (1983).
30. J.R. Lloyd, M. Shatzkes, and D.C. Challaner, ''Kinetic Study of Electromigration Failure in Cr/Al-Cu Thin Film Conductors Covered with

Polyimide and the Problem of the Stress Dependent Activation Energy,'' in *Proceedings of the IEEE International Reliability Physics Symposium*, pp. 216–225 (1988).

31. C.B. Oliver and D.E. Bower, ''Theory of the Failure of Semiconductor Contacts by Electromigration,'' in *Proceedings of the 8th Annual Proc. Reliability Physics Symposium*, pp. 116–120 (1970).
32. J.C. Onduresk, A. Nishimura, H.H. Hoang, T. Sugiura, R. Blumenthal, H. Kitagawa, and J.W. McPherson, ''Effective Kinetic Variations with Duration for Multilayered Metallizations,'' in *Proceedings of the 26th Annual International Reliability Physics Symposium*, pp. 179–184 (1988).
33. J.M. Pierce and M.E. Thomas, *Appl. Phys. Lett.*, Vol. 39, No. 2, 165 (1981).
34. W.E. Rhoden, D.W. Banton, D.R. Kitchen, and J.V. Maskowitz, ''Observation of Electromigration at Low Temperatures,'' *IEEE Trans. on Reliability*, Vol. 40, No. 5, 524–530 (1991).
35. R. Rosenberg and M. Ohring, ''Void Formation and Growth During Electromigration in Thin Films,'' *J. Appl. Phys.*, Vol. 42, 5671 (1971).
36. R. Rosenberg and L. Berenbaum, ''Resistance Monitoring and Effects of Non-Adhesion During Electromigration in Aluminum Films,'' *Appl. Phys. Lett.*, Vol. 12, 210 (1968).
37. G.L. Schnabble and R.S. Keen, ''Aluminum Metallization—Advantages and Limitations for Integrated Circuit Applications,'' *Proc. IEEE*, Vol. 57, 1570 (1969).
38. A.P. Schwarzenberger, C.A. Ross, J.E. Evetts, and A.L. Greer, ''Electromigration in Presence of a Temperature Gradient: Experimental Study and Modeling,'' *J. Electron. Mater.*, Vol. 17, No. 5, 473–478 (1988).
39. G.A. Scoggan, B.N. Agarwala, P.P. Peressini, and A. Brouillard, in *Proceedings of 13th International Reliability Physics Symposium*, pp. 151 (1975).
40. M. Shatzkes and J.R. Lloyd, ''A Model for Conductor Failure Considering Diffusion Concurrently with Electromigration Resulting in a Current Exponent of 2,'' *J. Appl. Phys.*, Vol. 59, 3890–3893 (1986).
41. R.A. Sigsbee, ''Electromigration and Metallization Lifetimes,'' *J. Appl. Phys.*, Vol. 44, No. 6, June (1973).
42. S. Vaidya, T.T. Sheng, and A.K. Sinha, *Appl. Phys. Lett.*, Vol. 36, No. 6, 464 (1980).
43. J.R. Venables and R.G. Lye, ''A Statistical Model for Electromigration-Induced Failures in Thin Film Conductors,'' *Proceedings of the 10th Annual Reliability Physics Symposium*, pp. 159–164 (1972).
44. L. Yeu, C. Hong, and D. Crook, ''Passivation Material and Thickness Effects on the MTTF of Al-Si Metallization,'' in *Proceedings of the 23rd International Reliability Physics Symposium*, p. 115 (1985).
45. English, A.K., K.L. Tai, P.A. Turner, ''Electromigration of Ti-Au Thin-Film Conductors,'' *J. Appl. Phys.*, Vol. 45, pp. 3757–3767 (1974).

4

ELECTROMIGRATION AND RELATED FAILURE MECHANISMS IN VLSI METALLIZATIONS

ARIS CHRISTOU
CALCE Electronic Packaging Research Center, University of Maryland, College Park

M.C. PECKERAR
Naval Research Laboratory, Surface and Interface Science Branch, Electronics Science and Technology Division, Washington, DC

4.1 INTRODUCTION

In this chapter the basic metallization failure mechanisms are reviewed for both the aluminum thin film system and for the gold refractory systems. The development of reliable metallizations is due in large part to our understanding of failure mechanisms. This chapter initially reviews the common failure mechanisms related to corrosion metal migration and to bond interface degradation. Following the examination of these failure mechanisms, the theory of electromigration is presented along with specific experimental details. Based on the understanding of failure mechanisms, the design and manufacturing guidelines for electromigration-resistant metallizations are discussed.

Electromigration and Electronic Device Degradation, Edited by Aris Christou.
ISBN 0-471-58489-4

4.1.1 Corrosion of Al Interconnects

The development of metallizations for VLSI circuits and for GaAs devices must follow an initial understanding of failure mechanisms. We first examine corrosion as a metal migration phenomenon and then review electromigration.

In discussing corrosion we note the following.

1. Corrosion occurs when there are two electrodes and an electrolyte, forming a classic galvanic cell.
2. Corrosion occurs at the anode, and the corrosion rate is a complex function of several variables, such as electrode materials, electrolyte, cell potential, current density, temperature, polarization effects, and passivation. The corrosion rate is affected by circuit passivation such as SiO_2 or Si_3N_4. The most common metallization is aluminum.

Aluminum is susceptible to corrosion both in acidic and basic solutions. Corrosion of Al film interconnects has been observed to depend on the environment, relative humidity, phosphorus content of the underlying SiO_2, chlorine contamination, passivation layer, pinholes in protective coatings that provide easy passage to H_2O molecules, bias condition, packaging, and use conditions.

Corrosion of Al film interconnects is minimized by:

- Reducing and/or eliminating surface contamination (such as Cl_2 molecules). The presence of halogens enhances surface migration and therefore increases leakage currents.
- Employing high-integrity pinhole-free protective layers such as SiO_2 and Si_3N_4 further eliminates paths for moisture migration.
- Using the minimum phosphorus content in SiO_2, but adequate to assure device stability.
- Hermetically sealed packages present a further level of moisture reduction.

The above factors must all be considered when designing a corrosion-resistant system.

Corrosion resistance of Al-Cu alloys is best in the heat-treated and naturally aged condition when all the copper is in solution or in the form of GP1 zones. This factor indicates that Al microstructure to a first order can minimize corrosion.

Copper, when present as $CuAl_2$, does not increase corrosion severely, since the potential difference between $CuAl_2$ and the copper saturated matrix is only a few hundredths of volt. Thus there is no substantial change in corrosion resistance in increasing the copper concentration from 5 to 15%. The maximum amount of copper in the aluminum is therefore 5%. Aluminum, however, presents a number of corrosion paths as shown in Table 4.1.

4.1.2 Metallization Reliability Issues

The following is a discussion of all possible reliability problems in metallizations, for circuits both in the silicon and GaAs technologies.

Contacts. Leakage problems occur by surface layers, especially when the metal has been incorrectly removed.

Barrier Layers. Diffusion barriers can degrade and introduce diffusion paths.

Pinholes in Interlevel Dielectrics. Dielectrics between first and second level metallizations may fail via pinholes.

Electromigration. This failure mechanism is the key wearout mechanism for metallizations.

Corrosion of Leads. This mechanism is driven mainly by stress effects and is present mainly in plastic packages.

Internal Stresses. The presence of high stress points within a package can lead to debonding effects that result in lead separation or bond pad lifting.

TABLE 4.1. Possible Aluminum Corrosion Reactions

Reaction	Remarks
a. $Al \rightarrow Al^{3+} + 3e$	$E_0 = 1.66$ V
b. $Al \rightarrow Al^{+} + 1e^{-}$	(Mechanisms for hydrogen evolution at anode)
c. $Al^{+} + 2H_2O \rightarrow Al^{3+} + 2OH^{-} + H_2$ (g)	
d. $2Al + 6HCl \rightarrow 2AlCl_3 + 3H_2$ (g)	(Acidic solutions and reactions involving chloride ions)
e. $Al + 3Cl^{-} \leftrightarrow AlCl_3 + 3e^{-}$	
f. $AlCl_3 + 3H_2O \rightarrow Al(OH)_3 + 3HCl$	
g. $Al + NaOH + H_2O \rightarrow NaAlO_2\ \frac{3}{2}\ H_2$ (g)	(Basic solutions and reactions involving oxides and hydroxides)
h. $Al + 3OH \rightarrow Al(OH)_3 + 3e^{-}$	
i. $2Al(OH)_3 \rightarrow Al_2O_3 + 3H_2O$	
j. $2AlO_2^{-} \rightarrow Al_2O_3 + H_2O$	

Cracks at Chip Corners. This is also related to internal stresses.

Bonds–Interface Problems. This mechanisms leads to bond lifting and failures and will be discussed in the next section. The bonding problems are essentially related to the presence of intermetallics such as Al/Au at the interfaces.

4.1.3 Gold Bonding and IC Failures

Gold bonding failure mechanisms are usually related to the gold to aluminum bonds on the IC chip. In cases where gold–aluminum beams are present, carbon impurities may lead to cracked beams. Such has been the case with power transistors and with analog circuits due to the elevated temperature environment. In this section we examine IC bond failures as an extension of the discussion on metallization failure mechanisms.

The failure of aluminum–gold bonds has been observed and has been reported extensively in the literature [1–5]. A number of conflicting reports have resulted from these investigations stating that (a) impurities inducing Kirkendal voiding are the primary cause of failure or (b) intermetallics at the interface are responsible for bond failures. The work by Philofsky [6], Goldfarb [7], and Horsting [8] indicates that bond failure is connected with impurities in the gold.

In contrast with aluminum–gold bonds, beam lead failures occur as a result of encapsulant expansion and thermal cycling of the beam lead devices. The encapsulant expands with temperature and beams crack after a minimum of two cycles of ΔT = 180–250°C. However, even though contrasting failure modes are present, the current investigation shows that the failures are both due to carbon contamination in the plated gold.

The work described in this chapter confirms the viewpoint that impurities in the plated gold is the predominant cause of failure. The work also introduces a new factor: subsurface carbon impurities can make a difference between bonds failing or remaining integral. In order to unambiguously identify the impurities, electron and nuclear scattering techniques were used. The following techniques were concurrently used: scanning electron microscopy (SEM), electron microprobe analysis, Auger electron spectroscopy (AES), and nuclear deuteron probe analysis (DPA). The AES techniques when used with ion sputter etching [9] can result in a composition depth profile. However, this technique has a potential problem in that it relies on surface layer removal by sputtering. The nuclear deuteron probe analysis relies on a carbon 12-

deuteron to proton reaction and the depth profile is obtained by nuclear resonance reactions [10]. This technique is totally nondestructive, since sputter-related problems are not present, and is quantitative in both depth determination and total concentration.

4.2 ANALYSIS OF BOND-RELATED FAILURES

4.2.1 Aluminum–Gold Bonds and Beam Lead Bonds

Power transistors from two different processing lots were observed to have at least one defective bond. The bonding system for these devices consisted of three gold-plated nickel posts (emitter, base, collector) to which bond wire connected the aluminum pads on the transistor to the posts. The bonding wire was 40 μm in diameter and made of standard 1% silicon–aluminum alloy material.

Bonding was done by thermocompression, resulting in as-bonded pull strength of the order of 20 g. The life test failure resulted in a separation of the bond at the post upon the 1000-hour shelf life at 180°C. Approximately 50% of the bonds failed due to "bondlifting" without evidence of an applied force such as bending or twisting. In many cases the bond was physically intact but was so weakened that it separated at the initial application of force. In order to further accelerate the process, additional transistors (sample size 15) were exposed at 250°C for up to 48 hours. These transistors had good bonds prior to testing as determined by a destructive analysis test performed on bonds prior to the 250°C life test.

The objective of the investigation was therefore to (a) measure the physical and chemical differences between "good" and "bad" bonds to the transistor posts and (b) estimate the effect of these differences on bond strength as a function of time.

Beam lead bonds of large-scale integrating circuits were observed to have cracks or microcracks in the beams after 250°C high temperature reverse bias tests. Two types of openings were observed on the failed IC. The first type was a combination of silicon to beam interface separation and broken beam of the edge of the silicon chip. The second type was that of a broken beam at the heel or midspan. Electron probe analysis was applied to all cracked areas in order to identify the presence of contaminants. The gold plating of the beams was measured to have 4% elongation. Typical elongation should be in the 20–25% range. The higher elongation value would result in a larger elastic region and hence greater plastic deformation before cracking.

4.3 FAILURE MECHANISM RELATED TO CONTAMINATION

4.3.1 Scanning Electron Microscopy (SEM) and Auger Electron Spectroscopy (AES)

Typical bad bonds were characterized by an incomplete Au-Al intermetallic formation with gold still being present at the interface or at localized spots around the periphery of the bond. It is evident that bond separation has occurred during the life test since initial visual inspection and pull tests showed them to be acceptable. These bonds when pulled also showed the presence of pure gold and the Al-Au intermetallic.

Analysis of the exposed bond area at higher magnification indicated the presence of four types of defects. Type I defects indicates the presence of darkened inclusions in the bond area. These inclusions can act as stress concentration centers, resulting in a degraded bond strength. Type II defects consisted of a carbon buildup region on the posts, probably in the form of graphite particles that adhered to the gold plating either before or during processing. Type III defects were identified to be absorbed carbon surface films, while Type IV defects were gross discontinuities in the gold plating. Any one of these defects will result in weakened bonds, and in particular, defects I and II will result in a time-dependent stress-related failure mode.

A series of microsections of good bonds also showed that the gold underneath the area where the aluminum had been bonded had been completely consumed by the formation of an intermetallic compound. In failed bonds, defects I–IV were always present in addition to unconsumed gold.

SEM examination of the beam lead failures indicated cracks at grain boundaries of the plated gold. Intergranular failures were observed at the midspan of the beam while shear type failures were observed at the upper portion of the beams. The shear failures have been related to side movement of the chip during bonding and will not be further discussed in this investigation.

AES analysis was performed on Type I, II, and III defects. The inclusions of Type I were identified as carbon with a depth of as much as 1 μm. Typical AES spectra are shown in Figure 4.1. The surface carbon peak at 272 eV is shown. The carbon is so prevalent that it obscures the lower energy gold peaks. At a depth of 0.2 μm the carbon peak at 272 eV has remained at the same intensity, indicating very little variation in concentration. After sputter etching for 3 hours, the carbon peak started to decrease in intensity. These results indicate that the carbon is

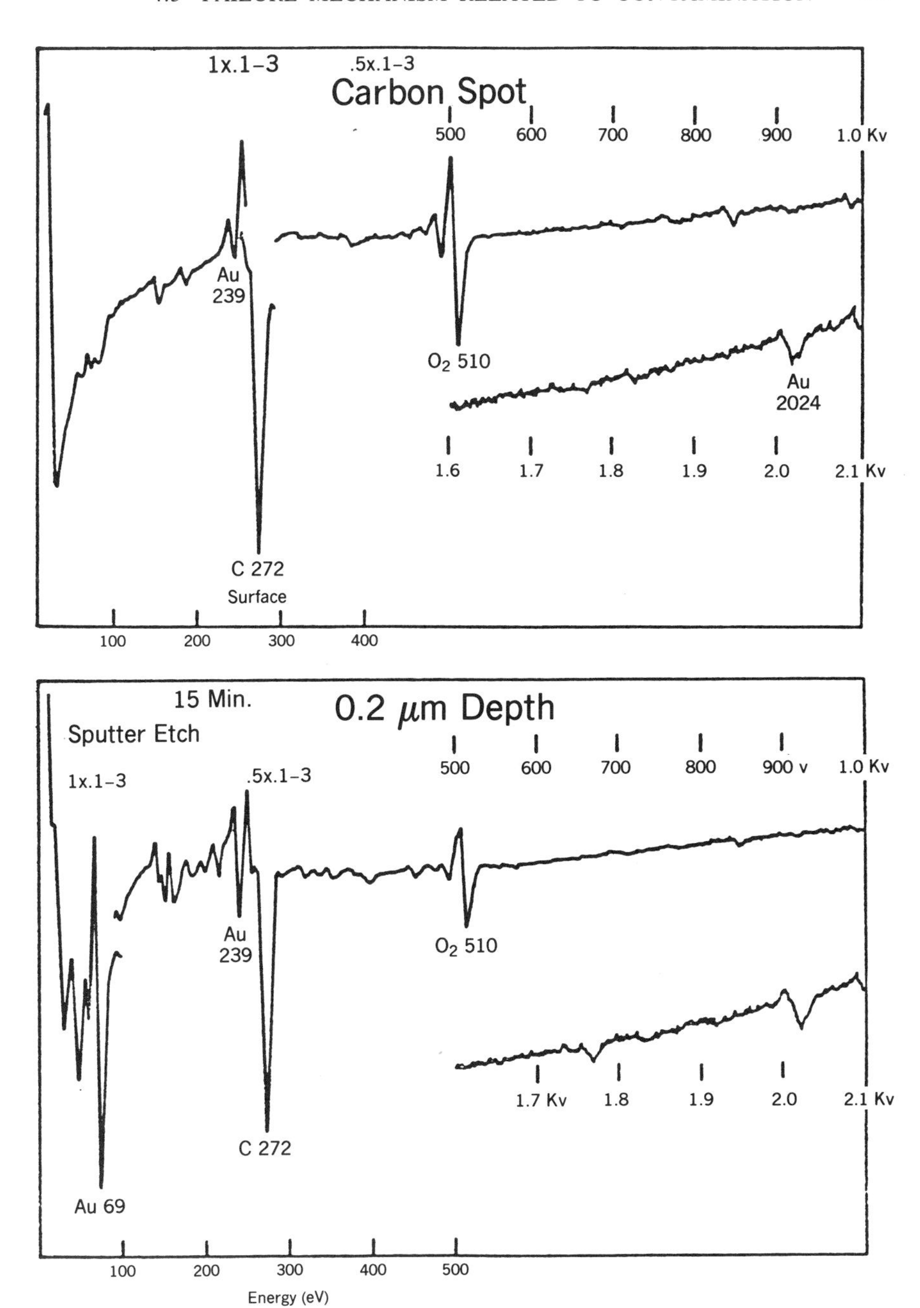

FIGURE 4.1. AES spectra of carbon spot on the surface and at 0.2 μm depth.

subsurface and further exposure to time–temperature stress may result in cracks at the interface. AES spectra for Type II defects indicated that carbon is in the form of agglomerated surface and near-surface regions, since a decrease in carbon intensity was observed after 45 minutes of sputter etching. Type III defects, however, were shown to be essentially surface defects.

4.3.2 Deuteron Probe and Electron Microprobe Analysis

The same transistor posts that exhibited the Type I–IV defects and were analyzed by AES were also examined by means of nuclear reactions induced by a 1.6-MeV deuteron beam. Measurements were performed on several different spots on each post by rotating the sample. The beam of heavy hydrogen ions from a Van de Graaff accelerator was directed onto the target in order to detect carbon. Nuclear reactions are produced in the carbon, resulting in protons being emitted. These emerging protons were detected by a solid state detector, which gave electrical pulses proportional in height to the energy of the detected protons. For a thin layer of carbon on the surface, the protons have the same energy except for some lower energy protons, which can be neglected. If the carbon layer is underneath the surface, the protons leaving the metal will have less energy. The change in measured energy between protons coming from a surface layer to protons coming from a buried layer can be used to determine the depth of the buried layer.

The results of the analysis of the data are shown in Figures 4.2 and 4.3. The concentration of carbon is given in atomic percent as a function of distance below the surface of the gold layer. All samples showed evidence of surface and subsurface carbon. Since the depth resolution obtained in these measurements was about 0.35 μm near the surface, it is not possible to determine whether the near-surface carbon is exactly on the surface or within 0.35 μm of the surface. Figure 4.2 shows a large near-surface carbon peak greater than 15% and a subsurface peak at 5.4 μm. Separate measurements were made in the angular regions from 0° to 90° and from 180° to 270° and a variation in near-surface carbon of a factor of 2 was observed.

Subsurface carbon was observed in all the posts analyzed with concentrations varying from 0.05 to 0.5 at.% as shown in Figure 4.3. At a depth of greater than 5 μm, subsurface carbon concentration was observed to increase probably due to the gold–nickel interface. In conclusion, the deuteron probe (DP) analysis indicates that both buried and near-surface carbon was present in all the transistor posts examined. The

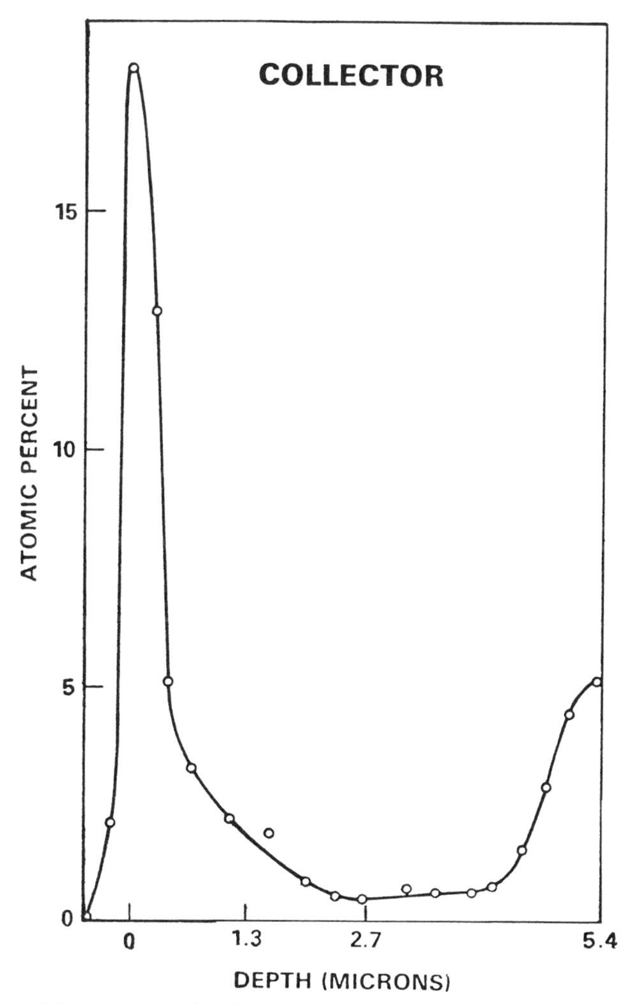

FIGURE 4.2. DP analysis of a collector post showing subsurface carbon.

interior peaks may be due to very thin layers containing appreciable amounts of carbon or might be due to large quantities of carbon particles concentrated at these levels. DP analysis of goods posts did not indicate the presence of subsurface carbon, and the surface or near-surface carbon was always below 0.05 at.%.

Deuteron probe analysis of failed beam leads also indicated the presence of subsurface carbon with concentration varying from 0.05 to 5.0 at.%. A near-surface carbon peak of concentration of 0.05–1.0 at.% was also present. The interior carbon may be due to carbon segregation of grain boundaries resulting in intergranular failures. Control inte-

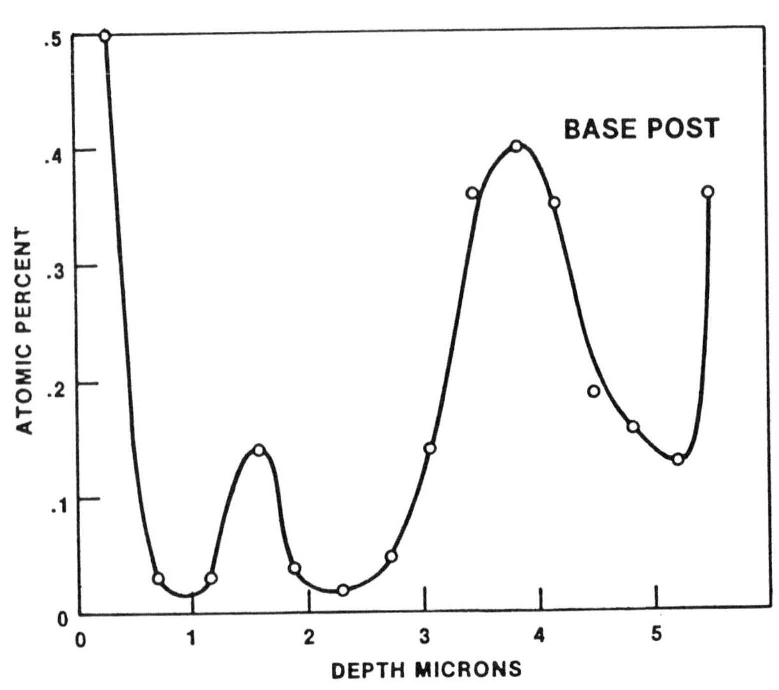

FIGURE 4.3. DP analysis of base post showing subsurface carbon.

grated circuits that did not fail, however, did not show the presence of subsurface carbon.

The electron microprobe analysis of failed transistors detected three different defective type areas. Types I–III were clearly identified by electron probe analysis as well as the Type IV, with large discontinuities in the gold. The subsurface carbon inclusions were detected by performing a line scan analysis on cross sections of the emitter and base posts, as shown in Figure 4.4. The subsurface carbon of Figure 4.4 was at the 6–10 at.% range. The surface carbon was detected at the 18–20 at.% level. The Type IV defects—discontinuities in the gold plating—are shown in Figure 4.5, indicating nickel interdispersed within the gold plating. These areas did not indicate the presence of carbon contamination. The results of Figures 4.4 and 4.5 are in good agreement with both the AES and the DPA results.

Electron probe analysis of cracked beams was conducted using the line scan and spot analysis modes. A typical line scan indicated carbon buildup in the area of a cracked beam. In various regions as shown in

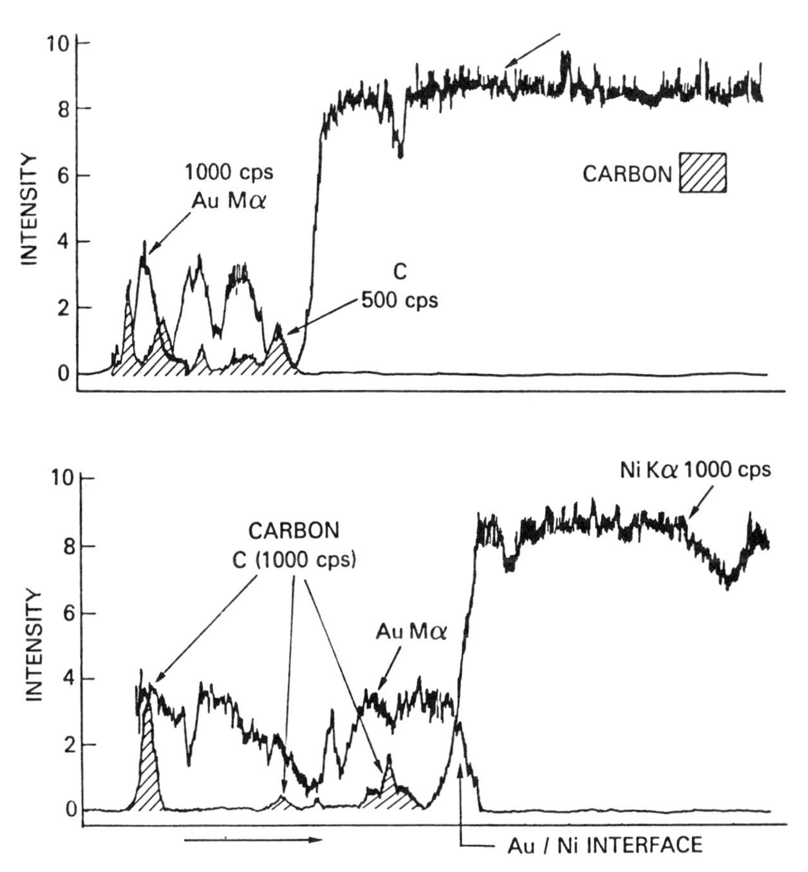

FIGURE 4.4. Electron microprobe line scan analysis of Au, C, and Ni.

Figure 4.6, the buildup of carbon was discontinuous. Spot analysis of regions showing carbon indicated that the electron beam was located on a grain boundary region. This result supports the hypothesis that carbon impurities in the plated gold have resulted in weakened grain boundaries.

Energy dispersive x-ray analysis has indicated the following contaminants present at levels of 0.5–1.0 at.%: copper, iron, manganese, nickel, iron, chlorine titanium, and sulfur. All these contaminants present at the bond interface will result in a time-dependent failure mode. The contamination level present is in excess of expected contamination levels in "good" gold-plated layers. Table 4.2 compares typical stereographic analysis of the present gold plating from a new semiconductor grade gold bath.

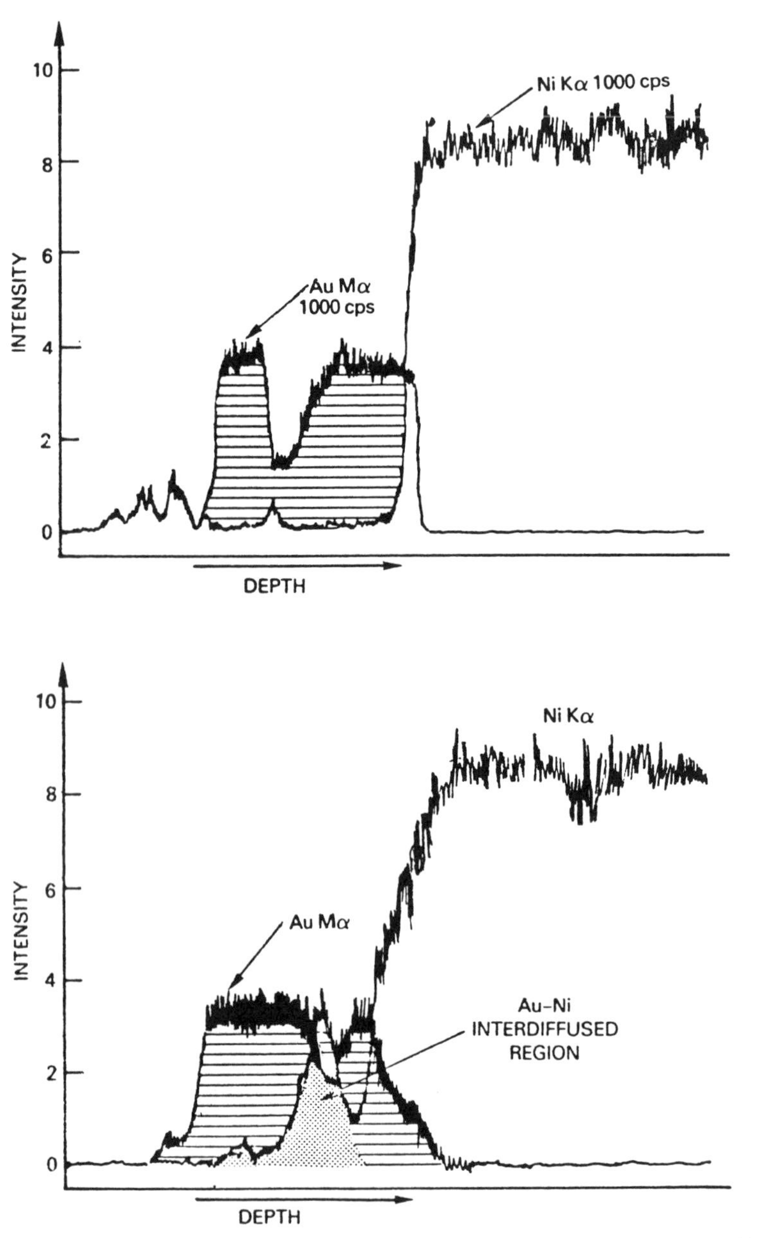

FIGURE 4.5. Electron microprobe data showing lack of uniformity in the gold plating. Also shown is interdiffusion between gold and nickel.

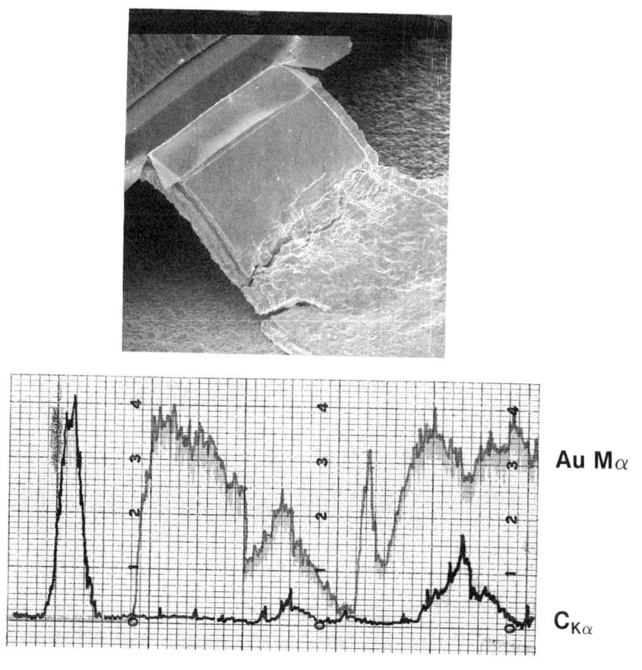

FIGURE 4.6. Electron probe analysis of cracked beams showing a line scan analysis of an area near a crack.

TABLE 4.2. Typical Stereographic Analysis of Vendors' Gold Bath

	Mn	Ni	Co	Fe (ppm)	B	Cu	Ag	Na	Cl	S
Present gold plating	10^4	10^4	10^3	10^2	100	50–100	10–50	10–50	100	100
New bath	N.D.	1–10	N.D.	N.D.	N.D.	1–10	10–50	1–10	1–10	1–10

In addition, energy dispersive x-ray analysis of plating from the new bath did not detect the above impurities, showing that an uncontaminated gold bath has impurity levels below 100 ppm.

4.4 THE FAILURE MODES

4.4.1 Gold to Aluminum Bonds

Bond pull data were taken on a set of control transistors subjected to the 250°C, 48-hour life test and compared to the bond pull data from the 180°C, 1000-hour life test. Both sets of data indicated the existence of

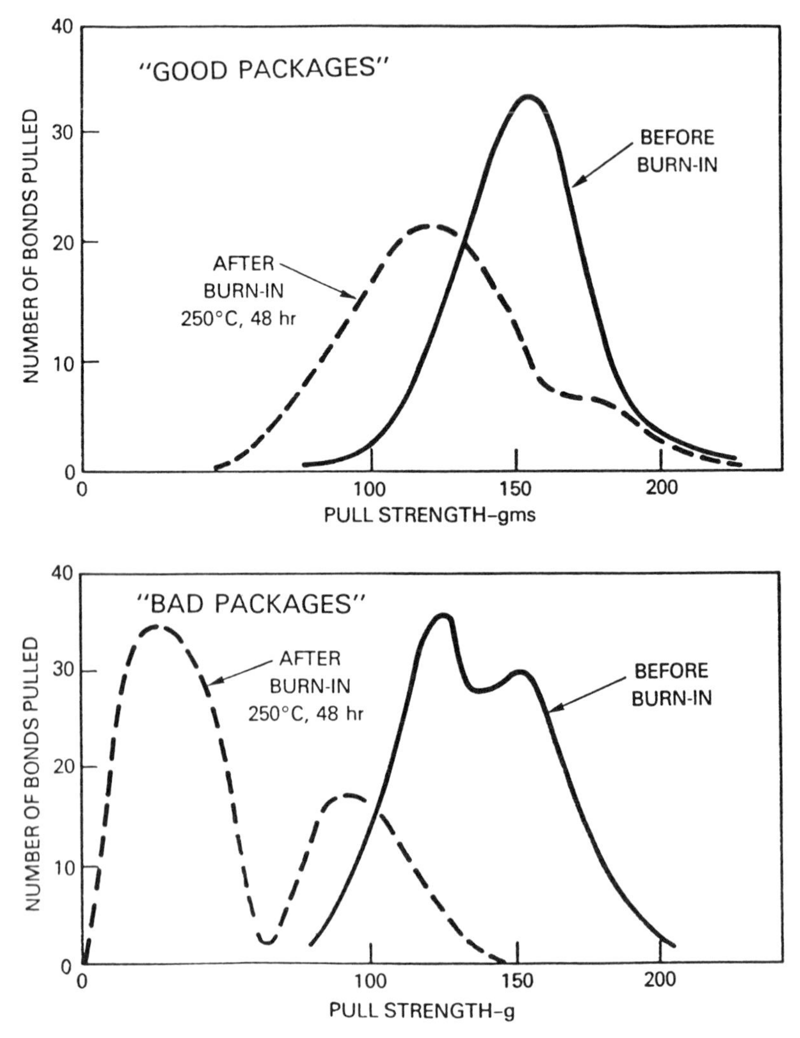

FIGURE 4.7. Pull strength distributions for good and bad packages showing a bimodal distribution after the burn-in.

a bimodal distribution of pull strengths. Figure 4.7 summarizes the data for the 250°C, 48-hour life test. The presence of the bimodal distribution along with the (a) carbon subsurface inclusions and (b) high levels of contamination found indicates a contamination-related failure mode exists. In good packages, which were characterized by contamination levels of less than 100 ppm, the pull strength distribution remains a Gaussian function even after extensive burn-in. In packages with extensive carbon contamination and gold bath contamination, after burn-in, a bimodal distribution results. The carbon inclusions act as stress concentration regions, resulting in microcracks that grow by a microvoid diffusion process [11] when exposed to temperature–time stress. The remainder of the impurities also induce the classical Kirkendal voiding [6, 8].

During the bonding operating, for pure gold, a diffusion front moves unimpeded through the gold layer and stops at the nickel interface. In the case of contaminated gold, the diffusion front starts moving into the gold but stops at carbon inclusions or at small microvoids. Failure usually occurs where the diffusion front stops. Impurities in the gold are expelled by the moving diffusion zone and are concentrated directly ahead of the moving diffusion front. When the concentration becomes critical, the precipitation occurs and small voids develop. The voids that form by a time–temperature process are further augmented by microcracks around carbon inclusions resulting in bond failure. The combination of the two processes has resulted in the perfect bimodal failure distribution shown in Figure 4.7.

4.4.2 Gold Beam Lead Devices

The present study indicates that when packaged silicon integrated circuits undergo temperature changes for a variety of reasons, beam fatigue and intergranular cracking occur as a failure mechanism. Beam fatigue has been accelerated due to the presence of carbon contamination in the gold plating. This observation agrees with previous studies [12] of beam fatigue. The usual experience is that devices with subnormal quality of gold fail early in temperature cycling. Devices with good connections and greater ductility in the plated gold (40–60% instead of 4–5%) will eventually fail by beam fatigue if the magnitude of the temperature cycle is large enough. In order to further test this conclusion, strength analysis of gold plated from a contaminated beam lead electroplating solution was compared with a pure (less than 100-ppm contamination) solution. Table 4.3 shows that for a carbon-contaminated gold plating, an elon-

TABLE 4.3. Tensile Test of Gold Plating

Sample	Thickness	Width	Length	Tensile (psi)	Elongation (%)
C contaminated	4 mm	0.5 mm	0.5 cm	18,000	4%
C contaminated	5 mm	0.5 mm	0.5 cm	14,000	5%
Pure Au	7 mm	1 mm	1 cm	19,000	20–25%

gation of 4–5% as obtained, while 20–25% elongation was obtained from the control solution.

The combination of diagnostic techniques consisting of electron scattering (AES, SEM, electron microprobe) and nuclear scattering (deuteron probe) has unambiguously detected the presence of bond surface and subsurface contamination in both Au-Al bonds and beam lead bonds. The predominant contaminant was carbon, although a significant number of other contaminants were present at levels above 100 ppm. These contaminants in the gold–aluminum bonds resulted in microcracks and voids, which formed by a temperature–time process. In beam lead bonds, carbon contamination resulted in accelerated fatigue failure during temperature cycling. The presence of carbon segregating at gold grain boundaries resulted in intergranular cracks.

4.5 ELECTROMIGRATION IN MICROCIRCUIT METALLIZATIONS

In addition to corrosion metal migration and interface degradation, microcircuit metallizations are susceptible to many other failure mechanisms, of which the major one is electromigration. Such a mechanism can result in open and short circuit modes of failure. This chapter explores the electromigration failure mechanism that results from electron transport. The mechanism also impacts the design of metallizations and the manufacturing of metallizations. Electromigration is characterized by an atomic flux (J_A) related to the ion flux (j_K) through the equation

$$J_A = \frac{ND}{kT} Z^* e \rho j_K$$

where e is the electron charge, ρ is the resistivity, D is the atomic diffusivity, and Z^* is effective charge density. This equation will be discussed in detail later.

A nonvanishing divergence of atomic flux is a requirement for electromigration to occur. The divergence may be due to a temperature gradient or to microstructural inhomogeneities. Since electromigration is cumulative, it affects the failure rate. Therefore the median time to failure (MTF) in the presence of electromigration is given by the equation

$$\text{MTF} = AJ^{-n} \exp (Q/kT)$$

where A = constant
J = current density
Q = activation energy $\approx$ 0.6 eV Al
$\approx$ 0.9 eV Au
T = temperature
k = Boltzmann's constant

The electromigration exponent ranges from $n = 1$ to $n = 6$. The failure mechanism results in voids, cracks, and hillocks. Two stages of electromigration are shown in Figure 4.8, showing hillocks near the bond pad with undetected cracks at the opposite interconnect. At the final stage, voids combine, forming a large cracked region. The median life is accompanied by a characteristic σ, a standard deviation. As σ gets larger, t_{50} also gets larger; the failure rate curve is shown to spread and become diffused.

4.5.1 Electromigration Under Circuit Conditions

This section summarizes the predominant theoretical views of electromigration as related to actual dynamic circuit conditions. Two major areas have been covered in the theory: steady-state current applications and pulsed current applications. However, before explaining the theory of electromigration, this section summarizes the pros and cons of alu-

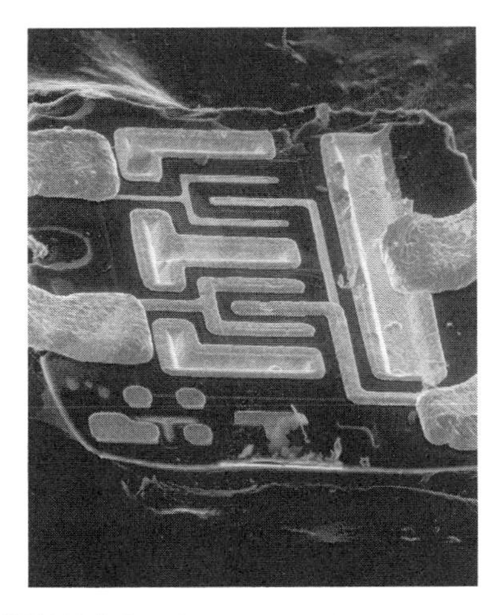

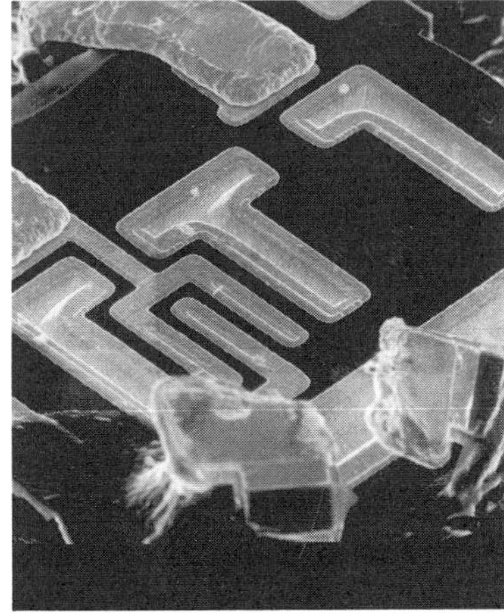

FIGURE 4.8. Electromigration showing hillocks and then metallization separation.

minum, which, in both pure and alloy form, has been the subject of most of the electromigration work.

Aluminum is the preferred metallization in the planar process for silicon semiconductors for several reasons [13]:

1. Ease of deposition: it can be sputtered or evaporated at fairly low temperatures.
2. Ease of patterning: phosphoric acid is usually used as the removal agent.
3. Aluminum forms ohmic/nonrectifying contact with silicon.
4. Schottky barrier contacts can be formed through careful process control.
5. Aluminum adheres to the die because it reacts slightly with SiO_2.
6. Aluminum will slightly reduce SiO_2 in thin layers and can use up the very thin skin of native oxide that forms contact windows.
7. Glass deposited onto an aluminum surface sticks easily.
8. Aluminum is a good electrical conductor.

However, there are several drawbacks to the use of aluminum:

1. Aluminum migrates easily at low current densities and temperatures. Gold is at least 10 times better in resisting electromigration.
2. Aluminum can absorb up to approximately 3% silicon. This creates alloying and aluminum spikes.
3. Aluminum corrodes easily in ionic solutions.
4. Addition of silicon is risky because silicon can precipitate out as nodules in an aluminum film.

Electromigration is the transport of microcircuit current conductor metal atoms due to electron wind effects. If, in an aluminum conductor, the electron current density is sufficiently high, an electron wind effect is created. Since the size and mass of an electron are small compared to an aluminum atom, the momentum imparted to an aluminum atom by an electron collision is small. If enough electrons collide with an aluminum atom, the aluminum atom will move gradually. The movement of aluminum atoms will cause a depletion at the negative end of the conductor and will form voids or hillocks along the conductor, depending on the local microstructure of the conductor. Hillocks will form as aluminum atoms pile up on conductor defects. Figures 4.9 and 4.10

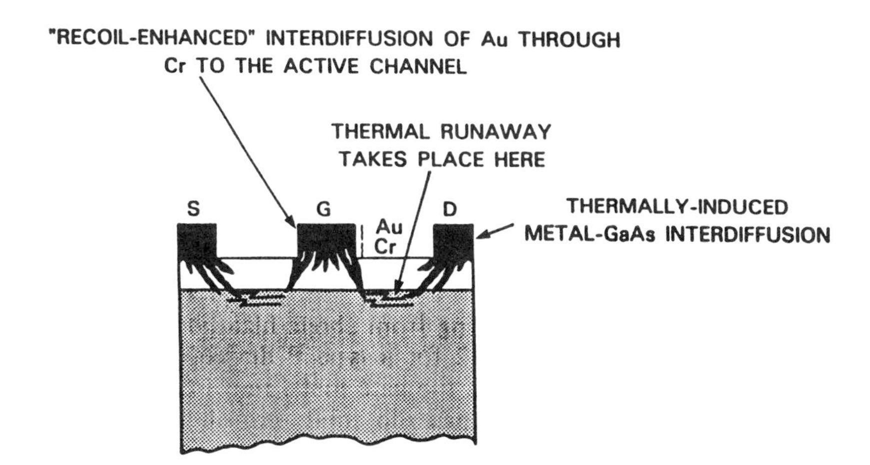

FIGURE 4.9. Schematic of electromigration in the presence of temperature gradients and interdiffusion.

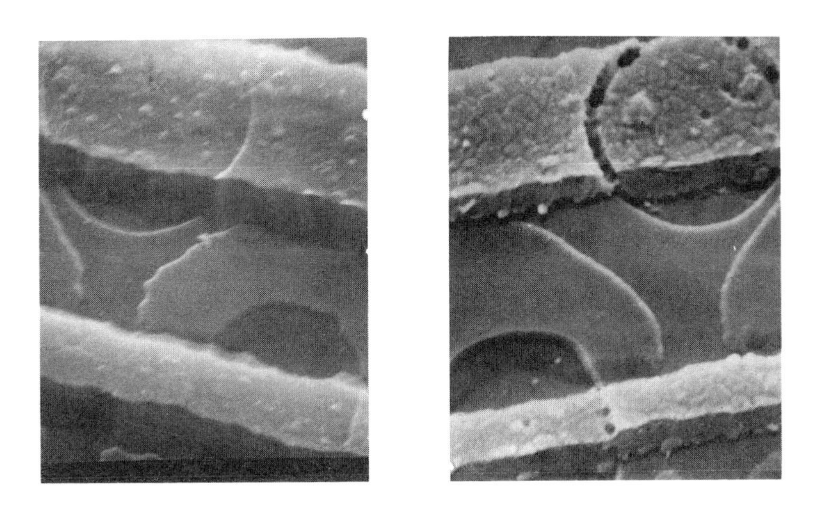

FIGURE 4.10. SEM micrograph showing void formation after reliability test as a result of electromigration in metallization lines.

illustrate electromigration and its characteristics. The creation of hillocks and voids is a time-dependent process and is shown in Figure 4.10.

There are several mechanisms for electromigration. The predominant one is the movement of atoms along the crystal grain boundaries. An activation energy of 0.45 eV has been calculated and used to characterize the temperature dependence of electromigration. Electromigration is heavily influenced by aluminum grain size. Smaller grain sizes are more easily transported.

Black [14] provides an excellent summary of the physics of electromigration. He notes that a thermally activated metal ion is essentially free of the metal lattice and is acted on by two forces in an electrically conducting single band metal: (a) the force created by an electric field applied the conductor and (b) the rate of momentum exchange between conducting electrons colliding with the activated metal ion heavily influenced by aluminum grain size.

Shielding effects reduce the electric field effect so that the predominant force is the momentum exchange. The metal ions are driven to the positive end of the conductor while vacancies move toward the negative end. The vacancies condense to form voids and the ions pile up to form hillocks and whiskers at discontinuities, such as a scratch or oxide that reduces conductor width and thinning or microcracking at oxide steps that reduces conductor thickness. The rate of mass transport R is described by

$$\begin{aligned} R = F &\times \text{(electron momentum)} \\ &\times \text{(number of electrons passing through a unit volume per second)} \\ &\times \text{(effective target cross section)} \times \text{(aluminum activated ion density)} \end{aligned}$$

The rate of mass transport is related to median time to failure and conductor cross-section area by

$$R = \frac{Fwt}{\text{MTF}}$$

where F = constant
w = conductor width
t = conductor thickness
MTF = median time-to-failure

Cross-sectional area is important because it determines minimum void size to create an open circuit. Momentum considerations and an exponential aluminum ion density yield the following result:

$$\text{MTF} = \frac{wt}{AJ^2} \exp (E_a/kT)$$

where the constant A includes the following physical properties:

- Volume resistivity of the metal
- Electron mean free time between collisions or the electron free path and average velocity
- Effective ionic scattering cross section for electrons
- Frequency factor for self-diffusion of aluminum in aluminum
- A factor relating rate of mass transport with median time to failure

The stress dependence of activation energy E_a was studied by McPherson [15]. He developed a generalized Eyring model to analyze thermally activated failure mechanisms. The general form of the Eyring model is the following:

$$\text{TTF} = AT^a \exp \left(\frac{E_a}{kT}\right) \exp \left[B + \left(\frac{C}{T}\right)\right] S$$

The model relates the time to failure to two different stresses: thermal and mechanical. His model predicts a stress-activated activation energy, E_a, provided that two conditions are met: (a) the applied stress must be of the same order of magnitude as the strength of the material; (b) a stress acceleration parameter g (see below) must be a function of temperature. The model is quite general and can be applied to dielectric breakdown as well as electromigration. Application of this model has suggested that the current density exponent n varies with current density from $n = 1$ at $j < 1 \times 10^5$ A/cm^2 to $n > 2$ at $j > 1 \times 10^6$ A/cm^2. The mathematical analysis begins with consideration of free energy of activation and stress functions. The stress function is represented as the difference between the application stress and a "breakdown strength." A time-to-failure (TTF) equation is derived:

$$\text{TTF} = A_0[\sinh (gz)]^{-1} \exp [Q/kT]$$

where $Q = \Delta H + az$
z = generalized applied stress
ΔH = free enthalpy
a = temperature dependent constant
A_0 = a constant
g = stress acceleration parameter

Even if the breakdown strengths are normally distributed, the observed TTF under stress z will be lognormally distributed. As is mentioned elsewhere, TTFs do exhibit this behavior in many cases. The basic physics are the same as described by Black above with the following additional consideration provided by McPherson [15]:

> For a vacancy to move, certain localized distortions of the metal lattice (or along grain boundaries) must occur. The energy required to create this distortion is the activation energy of the associated diffusion process. This energy can be obtained from lattice phonons by raising the temperature of the metal. At room temperature, vacancies in a metallization can diffuse freely and in an isotropic fashion. However, once a current is introduced, the vacancy diffusion becomes directional.

This model recognizes that the free energy of activation controls the process rate and that this energy depends on both the work contribution and the entropy change associated with the application of an external stress. The role of stress in electromigration is discussed more in the next section.

Partridge and Littlefield [16] note that electromigration models can be consolidated by including a variable power of the current density and that the variable power of J can be expressed as a function of the activation energy such that a factor directly proportional to the current density is subtracted from bulk values. As current density approaches zero, activation energy approaches bulk values. They also note that their empirical data agree with a vacancy diffusion mechanism where grain boundaries are only one of the contributions to vacancy concentrations. It is noted that Schafft et al. [17] report an empirical result that the current density exponent should be 1.53.

Kwok [18] notes that the electromigration atomic flux is due to electromigration inside metal grains and at the grain boundaries. The flux can be expressed by

$$J \propto (1/Kt) D j \rho e Z^*$$

where D is the diffusivity, j is the current density, ρ is the resistivity, and eZ^* is the effective charge (sum of ionic charge from electrostatic force and charge from electron wind force). At device operating temperatures, grain diffusion is the predominant mode of mass transport.

The above information focused on electromigration in the constant-current environment. There has been considerable research conducted with respect to pulsed or time-varying currents, which would be present in a microwave circuit. For example, Liew et al. [19] have developed a model for predicting interconnect electromigration time-to-failure under arbitrary current waveforms. The model is based on vacancy recombination. Under an arbitrary current condition, electromigration time-to-failure (TTF) is given by the following expression, which is valid for frequencies above 1 kHz:

$$\mathrm{TTF}_{\mathrm{ACDC}} = \frac{A_{\mathrm{DC}}(T)}{|J|^{m-1}J(1 + R_1 R_2)}$$

where J is the average current density, m, A_{DC}, and A_{AC} are experimentally defined parameters, and $R_1 = A_{\mathrm{DC}}\, A_{\mathrm{AC}}$ and $R_2 = (|J| - J)/J$. If $m = 2$ and $A_{\mathrm{AC}} >> A_{\mathrm{DC}}$, for DC and unidirectional current (pulse DC),

$$\mathrm{TTF}_{\mathrm{DC/pulse\ DC}} = A_{\mathrm{DC}}(T)/|J|^2$$

An expression also is derived for TTF under AC conditions. Liew et al. have used this model in a circuit electromigration reliability simulator, which is described in the design section of this chapter.

Suehle and Schafft [20] have noted that the median TTF enhancement (due to pulsed current) depends on current density and Joule heating (a decrease in enhancement). They note that metallizations will survive longer under pulsed current than under continuous current. A relaxation phenomenon is considered to be the operative mechanism.

A somewhat simpler formulation of electromigration in pulsed-current situations was constructed by Brooke [21]. The proposed model is

$$\mathrm{MTF} = A_0(J_{\mathrm{DC}}) \exp\,(E_A^{-n}/kT)$$

where A_0 is a constant and J_{DC} is the time-average current density. The value of n was determined by experiment and found to be dependent on the sample metal and to range between 2.6 and 5.4, close to constant-current values for the same metals. This model applies for a rectangular current waveform for a duty cycle between 20 and 100% and 175°C.

The formulation by Black and its application to pulsed currents by Brooke are the most commonly invoked theoretical formations of electromigration. A general form, the above equation provides a simple basis for relating the median time-to-failure (MTF) of a conductor to current and temperature.

$$\text{MTF} = AJ^{-n} \exp (E_a/kT)$$

Theoretical treatments of grain size and grain boundary orientation have not been developed. Other factors not addressed in theoretical terms are defects, layering of conductors, and conductor steps. All have been explored experimentally.

4.5.2 Empirical Electromigration Results

The theory associated with electromigration has addressed only some of the factors that govern electromigration. Fortunately, experimentation in this field is not very complicated so that gaps in the theory can be filled with empirical evidence. For the most part, an electromigration experiment consists of setting up a sample, running a current through it, and measuring either TTF or void concentration. This, of course, is an oversimplification because a considerable amount of care goes into the preparation of the test samples. Geometry, topography, and chemical composition must be carefully controlled. Similarly, data collection must be performed with care. Collection of TTFs is relatively straightforward. However, the analysis of void concentration and distribution (as well as hillock distribution) requires electron microscopy, which requires great care in sample preparation. The analysis of TTF data will be discussed at the end of this chapter.

It is not the purpose of this section to discuss experimental techniques in detail. The reader can find the details of experiments in the references. In this section, the focus will be on the results of electromigration experimentation, especially in areas not covered well by the theory. The reader will find discussions of grain size and orientation, multilayer technology, topography, line geometry, and voids.

Verification of the theory of electromigration and conductor failure was performed by Black [14]. He explored the effects of gradients in temperature, current density, and the ion diffusion coefficient. As expected, voids formed in the areas where positive gradients existed and hillocks formed where there were negative gradients. It was noted that conductors with a positive gradient in metal ion diffusivity will fail earlier than gradient-free conductors. Black also confirmed the existence of etch pits, which occur under the influence of elevated temperature and high current density at aluminum–silicon contacts. It was noted that device failure can occur when an etch pit (filled by aluminum) grows across an underlying junction, resulting in a short. His experiments explored the effects of aluminum structure in three forms: small crystal films, large crystal films, and large crystal films with a glass overcoat (to reduce surface diffusion).

The experiments were based on the premise that small crystal aluminum had a lower activation energy than large crystal aluminum. The experiments supported this premise with the large crystal activation energy estimated at almost twice that of the small crystal activation energy. Activation energy of the glassed large crystal aluminum was even higher.

Baerg et al. [22] demonstrated the usefulness of RR, the ratio of electrical resistances at 298 K and at 77 K, for monitoring the quality of Al-Si metal films. They established a correlation between RR, electromigration MTF, and median grain radius (MGR). They noted the use of the lognormal distribution to plot MTF data. The RR technique is intended as a postdeposition in-line process monitor. As other authors have noted [23], grain boundaries, film surfaces, and impurity sites play important roles in electromigration. In this case, RR is reduced from the bulk single crystal value of 12.6. They found that the RRs of sputtered thin films ranged between 6.0 and 9.5, with the highest values obtained from sputtering under high vacuum and thorough control of deposition temperature. MGR also was measured. Both RR and MGR were found to be correlated with MTF, as shown below. Although RR and MGR can be used as a quality check on metallization, Baerg et al. prefer RR because it is simpler. RR is measured using Van der Pauw four-probe method while MGR requires use of scanning electron microscope.

Towner [24] found that, for aluminum alloy films, lifetimes could be represented by lognormal distributions when grain size is smaller than linewidth and by either lognormal or logarithmic extreme value distributions if grain size exceeds linewidth. These results are attributed to the relative orientation of grain boundaries within a line. Series orientation leads to extreme value distributions. Parallel as well as series orientation leads to the lognormal distribution.

Hinode and Homma [25] explored the impact of refractory metal (titanium nitride and tin) on electromigration in layered conductors and found that layering with refractory metals degraded the electromigration immunity of the aluminum layer. They also found that grain growth was suppressed and that reducing this suppression increased conductor lifetimes by one or more orders of magnitude. Their technique for arriving at their conclusions was the monitoring of conductor resistance in specially constructed samples. They noted that there were three influencing factors: diffusion of the refractory metals into aluminum, crystal orientation affinity, and interface reaction. Table 4.4 indicates that effect of aluminum grain size on conductor lifetime, as reported by Hinode and Homma [25].

Towner [26] studied electromigration-induced short circuit failures in multilayered conductors. The layering examined was aluminum on titanium–tungsten. His results are summarized as follows:

1. The probability of a short circuit failure in conductors increases with the thickness of the titanium–tungsten layer.
2. Interlayer short circuit failures can be characterized by a lognormal distribution of lifetimes.
3. Both open and short circuit failures exhibit similar activation energies (0.5–0.6 eV).
4. In tungsten–titanium–aluminum conductors, short circuit lifetime increases linearly with increasing dielectric thickness.
5. The current density exponent for short circuit failure in tungsten–titanium–aluminum conductors was between 2 and 4.

Oates [27] noted that electromigration in aluminum conductors is influenced by the underlying topography. He investigated step spacing in two-level metal technology and showed that regions with poor coverage (10–20%) are particularly vulnerable to electromigration failure. At the step, the current density may be increased significantly due to metal thinning. Consequently, it is important that the steps are covered ade-

TABLE 4.4. Summary of the Effect of Grain Size on Mean Time-to-Failure for Refractory Based Metallizations

Metallization System	Metallization Width (μm)	MTF (hr) (250°C, 2 mA)	Grain Size (μm)
AlSi/TiN	1.2	6×10^3	1
AlSi/W	1.0	1×10^4	2
AlSi/TiN (sputtered)	1.2	2×10^5	2.7
W/AlSi/W	1.5	6×10^5	2.5
AlSi (monolayer)	1.0	1×10^6	3.0
W/AlSi	1.0	2.5×10^6	5.0

quately and uniformly. Metal step spacing was found to be a critical factor. Oates notes that for critical step spacings, poor step coverage results, with a consequent vulnerability to electromigration. However, if spacing is adjusted to improve coverage, the benefit will not be as large as expected because for a certain range of spacings, a new failure mode is introduced. The exact mechanism for the new mode was not determined. The step spacings for the new failure mode were found to be from 3 to 4.375 μm.

The issue of step coverage also was addressed by Kim et al. [28]. They studied electromigration of DC bias-sputtered aluminum and found step coverage to be excellent. They found electromigration resistance to be inferior to aluminum sputtered without bias. Failure was caused by voids along grain boundaries and an extremely small grain size. A wide distribution of grains sizes also was reported. Strausser et al. [29] noted that the most significant material parameter that changes at steps is grain size. They reinforced the view that, in terms of topographical factors affecting electromigration, frequency and severity of steps are the primary variables affecting electromigration results. There is a high likelihood of the formation of a grain boundary at a step. As the grain grows on either side of the corner, different orientations occur on either side and a boundary is formed. The grain boundaries are points for the entrapment of voids.

Kwok [18] performed an excellent study of the effect that conductor line geometry has on lifetime due to electromigration in aluminum–copper submicron lines. He found the following:

1. Lifetime decreases slightly with increasing line length and then levels off after a critical value.
2. As linewidth decreases, lifetime decreases, goes through a minimum, and then increases for less than a critical width.
3. Lifetime increases with decreasing film thickness.
4. Critical length and width decrease with decreasing linewidth and film thickness, respectively.
5. Grain morphology is an important parameter to control the electromigration lifetime.

LaCombe and Parks [30] evaluated conductor failure distributions as functions of varying length and width. The distributions were found to follow the lognormal distribution down to the 0.3% point. All failures were open circuits resulting from voids, which formed at the edge of a line and grew inward. Hillocks did not occur anywhere except within

10 μm of a large void. They concluded that the length dependence of early failures was consistent with an independent segment model. Later failures were found not to be consistent with the segment model and may have been influenced by more global parameters, such as current, average temperature, or average void thickness. Earlier work by LaCombe and Parks [31] noted an overall increase in conductor resistance primarily due to the formation and growth of voids. They did notice that the resistance will increase and decrease several times before final failure. This was attributed to some self-healing. They cautioned against using early resistance data to assess conductor lifetime because failure mechanisms other than just void formation may become significant.

Several resistance change techniques for measuring electromigration effects have been developed [32, 33]. These are based on void formation models. The void formation models consider that voids form, enlarge, increase current density, and increase local temperature gradients. The change in conductor resistance is used as the indicator of change.

Sadana et al. [34] used transmission electron microscopy to study the location and shape of voids in aluminum and aluminum alloys. They found both to be dependent on the type of impurity that was added. They concluded that precipitation of added impurities at grain boundaries did not always ensure electromigration resistance. They explained void formation and shape by considering flux divergence at triple points.

Note that the void nucleation occurs upstream of a grain boundary. Once voids nucleate, erosion takes place from the aluminum surface surrounding the voids and results in irregularly shaped voids. If the grain boundaries are perfectly hardened and the grains contain fine impurity precipitates, then the voids are located on the downstream side of the grain boundaries.

A method for minimizing void formation was discovered by Koyama et al. [35]. They noted that void formation could be suppressed by mercury light irradiation of plasma-enhanced chemical vapor deposition of silicon nitride film coatings. The light beam induces stress relaxation of the nitride film. The voids can be created through annealing at relatively low temperatures, especially when the plasma-enhanced coatings are used. Koyama et al. tested the voiding in samples that had been annealed at 350, 400, and 450°C. One mechanism they proposed for the formation of voids was the presence of hydrogen in the aluminum lattice, which creates microvoids that aggregate into aluminum voids under the stress of the nitride film. The hydrogen lowers the surface tension of the aluminum and allows microvoids smaller than 1 μm to remain. Such small voids are not stable in pure aluminum. Hydrogen inside the microvoids prevents collapse of the voids. The microvoids aggregate and grow more stable after the applied stress of passivation. A particular finding of theirs was that heat treatments of 200–400°C alone do not reduce aluminum voiding after the final aluminum sinter. Light beam irradiation (200–2000 nm) onto the nitride films does suppress the voids after final annealing at 450°C for 30 minutes.

Rosenmeyer et al. [36] explored the effects of bending stresses on electromigration and found that void formation was greatest in the region of highest stress. They noted that stresses and stress gradients have only negligible effects on vacancy diffusion.

4.5.3 Reducing Electromigration Data

Experimentally, the parameters available for reduction are either the current density or temperature. The median life is given as

$$t_{50} = AJ^{-n} \exp (Q/kT)$$

For a fixed current, we can then extrapolate t_{50} for a desired temperature:

$$\frac{t_{50}(T_1)}{t_{50}(T_2)} = \exp \left[\frac{Q}{k} \left(\frac{1}{T_1} - \frac{1}{T_2} \right) \right]$$

As a second experiment, we can fix the temperature and vary the current density. The ratio of t_{50} is then given as

$$\frac{t_{50}(J_1)}{t_{50}(J_2)} = \left(\frac{J_1}{J_2} \right)^{-n}, \qquad n = 1.5$$

Table 4.5 shows the estimated MTFs for refractory based metallizations.

TABLE 4.5. Estimated MTFs of Ti:W/Al and Ti:W/Al + Cu Film Conductors at 85°C

Current Density	Exponent	Ti:W/Al (hr)	Ti:W/Al + Cu (hr)
1×10^6 A/cm^2		17,882	53,230
5×10^5 A/cm^2	$m = 1.5$	50,592	150,557
	$n = 2.0$	71,548	212,920
2×10^5 A/cm^2	$n = 1.5$	199,982	595,130
	$n = 2.0$	437,175	1,330,250

One may also calculate failure rates due to electromigration since t_f obeys a lognormal distribution:

$$f(t) = \frac{1}{\sqrt{2\pi}}\left(\frac{1}{\sigma}\right)\left(\frac{1}{t}\right)\exp\left[-0.5\left(\frac{\ln t - \ln t_{50}}{\sigma}\right)^2\right]$$

where $f(t)\,dt$ is the probability of observing a failure in time interval dt. The failure rates are given by

$$\lambda(t) = \frac{f(t)}{[1 - F(t)]}$$

and $F(t)$ is the cumulative failure density function. Finally, the stability of interconnects must also be included especially in terms of elevated temperature reliability. The trend is toward silicide gates and the enthalpy change at elevated temperatures is shown in Figure 4.11. The key systems for reliability are shown to be the refractory silicides.

4.5.4 Design and Manufacturing Guidelines

Both the theoretical and experimental work in electromigration have led to changes in design and manufacturing to mitigate the effects of electromigration. For example, manufacturers have taken several steps to control electromigration.

Ti(×)	V(×)	Cr(×)	Fe(×)	Co(×)	Ni(×)
$TiSi_2$	VSi_2 ✓	$CrSi_2$	Fe_3Si ✓	Co_2Si ✓	Ni_2Si ✓
Ti_5Si_3 ✓			FeSi ✓	CoSi ✓	NiSi ✓
	Nb(×)	Mo(×)		$CoSi_2$	$NiSi_2$ ✓
	$NbSi_2$ ✓	$MoSi_2$ ✓			
	Ta(×)	W			Pd(×)
	$TaSi_2$ ✓	WSi_2 ✓			Pd_2Si(×)
	Ta_5Si_3 ✓	W_5Si_3 ✓			PdSi(×)
	$ErSi_2$(×)				Pt(×)
					Pt_2Si(×)
					PtSi(×)

FIGURE 4.11. Stable and unstable silicide metallization systems based on the enthalpy change at elevated temperatures. × = $\Delta H < 0$, unstable; ✓ = $\Delta H > 0$, stable at 850°C.

1. Control the shape of the oxide steps.
2. Control the metal deposition process carefully. The resistance ratio technique of Baerg et al. [22] should be considered as a quality control device.
3. Add 3–4% copper to the aluminum to retard metal migration. Copper lines the grain boundaries and prevents their movement through the metal. Alloys of aluminum–copper, aluminum–silicon, and aluminum–copper–silicon have shown no failures due to electromigration at temperatures between 150 and 215°C.
4. Deposit a glass layer onto the metal to partially cool it and pin down metal atoms.
5. Factor the current density/metal migration mechanism into their designs. One worst case design limit of current density is not to exceed 500,000 A/cm^2. A current density of 10^6 A/cm^2 should not be exceeded in the metallization, although some manufacturers claim this value can be exceeded by a factor of 5 in some devices with no adverse effects.

Common failure modes associated with electromigration in discrete and integrated devices are the following:

1. High resistance or open circuits due to the formation of voids.
2. Enhanced alloying and shorting due to penetration of aluminum into silicon surfaces.
3. Increased contact resistance and nonuniform current flow due to hillock formation, which results from high current densities and thermal cycling.

Bond lifts also have been observed. Failures may result from defective metallization layers caused by flaws in the pattern mask or errors in the manufacturing process, and the inclusion of contaminants at metallization interfaces. The basic result of the electromigration research is the development of new design and manufacturing guidelines for metallizations.

REFERENCES

1. B. Selikson, *Phys. Failure Electron.*, Vol. 3, 35–377 (1965).
2. J.A. Cunningham, *Solid State Electron.*, Vol. 8, 735 (1965).
3. B. Selikson and T.A. Longo, *Proc. IEEE*, Vol. 52, 1638 (1964).

4. I.A. Blech and H. Sello, *J. El. Chem. Soc.*, Vol. 113, 1052–1054 (1966).
5. E. Philofsky, *8th Annual Proceedings, Reliability Physics*, pp. 117–185 (1970).
6. E. Philofsky, *9th Annual Proceedings, Reliability Physics*, pp. 114–119 (1971).
7. S. Goldfarb, *Proceedings of 21st Ecc. IEEE Conference*, pp. 295–302 (1971).
8. C.W. Horsting, *10th Annual Proceedings, Reliability Physics*, pp. 155–158 (1972).
9. A. Christou, L. Jarvis, W. Weisenberger, and J.K. Hirvonen, *J. Elec. Mater.*, Vol. 4, 329–336 (1975).
10. K.L. Dunning, *Thin Solid Films*, Vol. 19, 145–154 (1973).
11. H. Berg and E.L. Hall, *11th Annual Proceedings, Reliability Physics*, pp. 10–20 (1973).
12. J.L. Daus and F.L. Howland, *IEEE Trans. Comp. Hybrids MT*, Vol. CHMT-1, No. 2, 158 (1978).
13. J.R. Devaney, G.R. Hill, and R.G. Seippel, *Failure Mechanisms, Techniques, & Photo Atlas*, Failure Recognition & Training Services, Inc., p. 10–24 (1983).
14. J.R. Black, *Proc. IEEE*, Vol. 57, No. 9, 1587–1593 (1969).
15. J. McPherson, *Proc. IEEE IRPS*, pp. 12–18 (April 1986).
16. J. Partridge and G. Littlefield, *Proc. IEEE IRPS*, pp. 93–99 (April 1985).
17. H. Schafft, T. Grant, A. Saxena, and C. Kao, *Proc. IEEE IPRS*, pp. 93–99 (April 1985).
18. T. Kwok, *Proc. IEEE IRPS*, pp. 185–188 (April 1988).
19. B.K. Liew, P. Fang, N.W. Cheung, and C. Hu, *Proc. IEEE IRPS*, pp. 111–118 (March 1990).
20. J. Suehle and H. Schafft, *Proc. IEEE IRPS*, pp. 106–110 (March 1990).
21. L. Brooke, *Proc. IEEE IRPS*, pp. 136–139 (April 1987).
22. W. Baerg, K. Wu, P. Davies, G. Dao, and D. Fraser, *Proc. IEEE IRPS*, pp. 119–123 (March 1990).
23. E.A. Amerasekera and D.S. Campbell, *Failure Mechanisms in Semiconductor Devices*, Wiley, New York, 1987, p. 38.
24. J. Towner, *Proc. IEEE IRPS*, pp. 100–105 (March 1990).
25. K. Hinode and Y. Homma, *Proc. IEEE IRPS*, pp. 25–30 (March 1990).
26. J. Towner, *Proc. IEEE IRPS*, pp. 81–86 (April 1985).
27. A. Oates, *Proc. IEEE IRPS*, pp. 20–24 (March 1990).
28. M. Kim, D. Skelly, and D. Brown, *Proc. IEEE IRPS*, pp. 126–129 (April 1987).
29. Y. Strausser, B. Euzent, R. Smith, B. Tracy, and K. Wu, *Proc. IEEE IRPS*, pp. 140–144 (April 1987).

30. D. LaCombe and E. Parks, *Proc. IEEE IRPS*, pp. 1–6 (April 1986).
31. D. LaCombe and E. Parks, *Proc. IEEE IRPS*, pp. 74–80 (April 1985).
32. H. Hoang, J. Ondrusek, A. Mishimura, T. Sigiura, R. Blumenthal, H. Kitagawa, and J.W. McPherson, *Proc. IEEE IRPS*, pp. 179–184 (April 1988).
33. J. Maiz and I. Segura, *Proc. IEEE IRPS*, pp. 209–215 (April 1988).
34. D. Sadana, J. Towner, M. Norcott, and R. Ellwanger, *Proc. IEEE IRPS*, pp. 38–43 (April 1986).
35. H. Koyama, Y. Mashiko, and T. Nishioka, *Proc. IEEE IRPS*, pp. 24–29 (April 1986).
36. C. Rosenmeyer, F. Brotzen, J. McPherson, and C. Dunn, *Proc. IEEE IRPS*, pp. 52–56 (April 1991).

5

METALLIC ELECTROMIGRATION PHENOMENA

SIMEON J. KRUMBEIN
AMP, Incorporated, Harrisburg, Pennsylvania

5.1 INTRODUCTION

Metallic electromigration can be defined as the movement of metallic material, usually through or across a nonmetallic medium, under the influence of an electrical field. As such, it has assumed increasing importance in the performance and reliability of packaging systems that incorporate electrical contacts. The characteristics of the different electromigration processes are discussed, including descriptions of related phenomena. The primary emphasis is on electrolytically controlled processes that take place under low power and typical ambient conditions.

Metallic electromigration has long been recognized as a significant potential failure mode in many electrical and electronic systems [1, 2]. Unfortunately, the study and understanding of this field have been complicated by the large array of interrelated phenomena, as well as by a number of effects that have only a superficial similarity to electromigration. One of the two main purposes of this chapter then is to help clear up some of the prevailing confusion and misunderstanding by categorizing these phenomena and lending some degree of order to this important, but complex, field. The second purpose is to outline the role of the various factors that determine the occurrence and extent of electromigration.

Electromigration and Electronic Device Degradation, Edited by Aris Christou.
ISBN 0-471-58489-4

5.1.1 Definitions

As a first step, we include in the term ''metallic electromigration'' all migration phenomena that involve the transport of a *metal*—usually through or across a nonmetallic medium—under the influence of an *applied electrical field*.

Under this proposed definition, the material that migrates is considered to be in the metallic state, in that both its *source* (e.g., a plating, a clad stripe, a metal-loaded ink, or a base-metal substrate) and its *final form*, after migration, would function as metallic conductors. The migration of metallic elements in the form of corrosion-product compounds would not come under this definition, since such compounds do not exhibit full metallic conduction. Of greater importance, however, is the fact that this latter type of migration, as exemplified by copper tarnish creepage [39, 40], does not require an applied electrical field. Nevertheless, tarnish creepage effects are often confused with electromigration because both phenomena can exhibit dendritic morphologies—as shown by comparing the copper sulfide creepage dendrites in Figure 5.1 and the dendritic copper sulfide crystals of Figure 5.2 with the (metallic) silver migration dendrites of Figures 5.3 and 5.4.

In the same way, the requirement of an applied electrical field differentiates electromigration from most tin whiskering effects, since the latter have been shown to be caused by mechanical stresses in the plating [53]. This is so even though both phenomena can produce failures by shorting across closely spaced metallic conductors.

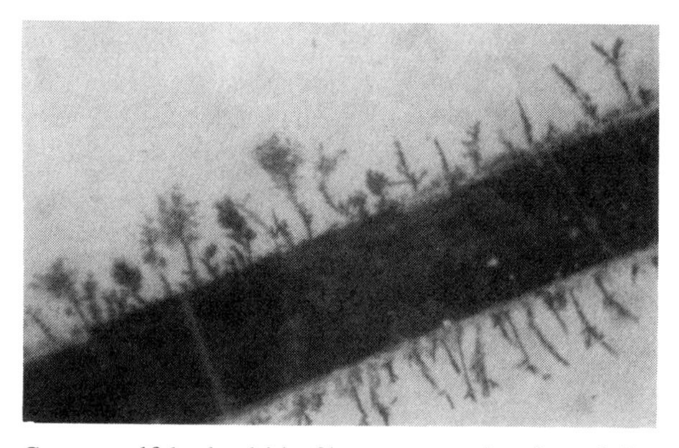

FIGURE 5.1. Copper sulfide dendritic filaments growing from 0.5-mm (0.020-in.) wide copper conductor between polyester sheets, without an applied voltage, after humid H_2S exposure (24×).

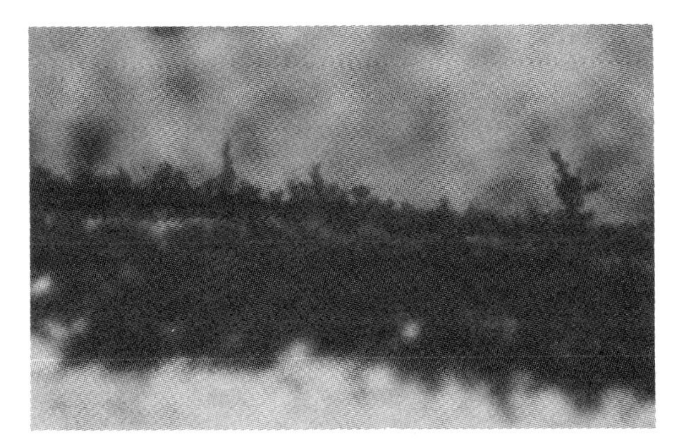

FIGURE 5.2. Silhouette of copper sulfide dendritic crystals after copper exposure to 23 days in 50 ppb sulfur vapor at 87% RH with no applied voltage (40×).

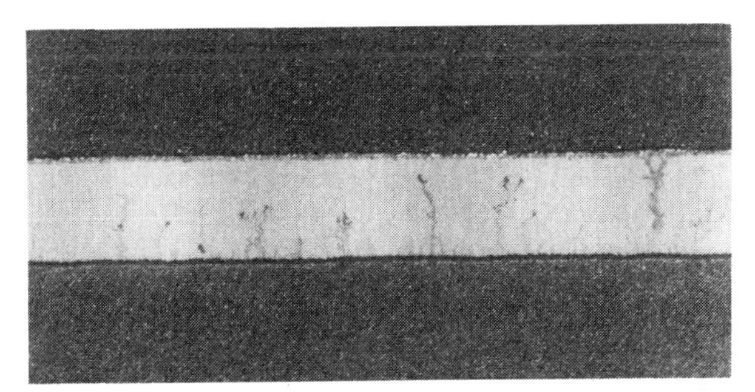

FIGURE 5.3. Various stages, including one dendritic bridge of silver electromigration across 0.38-mm (0.015-in.) spacing on epoxy–glass PC board.

FIGURE 5.4. Severe silver electromigration effects across 0.38-mm (0.015-in.) epoxy–glass spacing.

5.2 ELECTROLYTIC ELECTROMIGRATION

As outlined in Table 5.1, this type of electromigration is an *electrochemical* phenomenon that takes place primarily under normal ambient conditions, that is, when the local temperatures and current densities are low enough to allow water to be present on the surface (contrast with the section on solid state electromigration).

The actual mechanism of electrolytic electromigration is water-dependent and tends to occur whenever the insulator separating the conductors (as on PC boards, flexible circuitry, chip carriers, or IC ceramics) has acquired sufficient moisture to allow electrolytic (ionic) conduction when an electrical potential is applied.

Electrolytic electromigration is primarily a direct-current (DC) phenomenon, although the direction of the migration can be reversed by reversing the polarity of the applied field [52]. However, when a true AC voltage (60 Hz) was applied [3], "far less" migration was produced, and no electromigration could be detected at higher frequencies.

Although electromigration phenomena can occur with many metals, only silver—and to a very limited extent copper [3, 12–14] and perhaps tin [28, 45, 46]—normally undergoes these effects under noncondensing, but humid, conditions [1–6, 27]. The other potentially susceptible metals generally require a visible layer of water (i.e., bulk condensation) for electromigration to occur at ambient temperature [15–25]. From an "operating conditions" viewpoint therefore, these effects can be divided into two rough subcategories, according to whether the surface is covered by a thin invisible film of moisture ("humid" electromigration) or by a visible layer of condensed water ("wet" electromigration) (see Table 5.1).

The essentially unique ability of silver to undergo slow electromigration at only moderate elevated temperatures in the absence of condensation has been commented upon in the literature [2, 24, 28, 47], and

TABLE 5.1. Metallic Electromigration

A. Electrolytic (ionic) Ambient temperatures ($<100°C$) Low current densities ($<1\ mA/cm^2$) 1. Normal conditions—no visible moisture ("humid" or silver electromigration) 2. Visible moisture across conductors ("wet" electromigration)
B. Solid state (electron momentum transfer) High temperatures ($>150°C$) High current densities ($>10^4\ A/cm^2$)

various explanations have been proposed. This is also the reason why many workers have used the term "silver electromigration" to describe the humid migration phenomenon.

A third category, that of "dry electromigration" of silver through glass, has been reported [48, 49]. Although no moisture is present and higher temperatures (150–200°C) are involved, both the dendritic appearance of the migrating silver and the magnitudes of the electrical potentials, interconductor spacings, and failure times are essentially the same as those of normal silver electromigration. This would lead one to conclude that the function of the glass as a solid electrolyte is very similar to that of the invisible moisture films on other insulators (see below).

5.2.1 "Humid" Electromigration

In the simplest case, metallic silver on the conductor with the more positive applied potential (anode), is oxidized to a more soluble form. The resulting positively charged ions then move under the influence of the electrical field through moisture paths on or in the insulator toward the more negative conductor (cathode), where they are reduced back to silver metal [2–4, 6].

If the migrating ions were deposited in a uniform manner (as in electroplating), there would be no discernible effect on the electrical properties of the circuitry, since transport of metal ions at the very low currents extant across the insulator would be too small to produce any significant changes in the contact or conductor spacings.

In practice, however, the electromigration mechanism produces two separate (though not always distinct) effects that lead to impairment of the circuit's electrical integrity. These are colloidal "staining" and filamental or dendritic bridging, both of which are shown in Figure 5.4 Deposits of colloidal silver (or copper) often appear as brownish stained regions, which originate at the positively polarized conductor but do not necessarily remain in contact with it (Figure 5.5). They are assumed to result from reduction of the migrating ions, either by light or by chemical reducing agents on the insulator surface [2–5].

Dendritic growth, on other hand, results from the fact that the ions tend to deposit at localized sites on the cathode in the form of needles or spikes [35]. Once these nuclei have formed, the higher current density at their tips will greatly increase the probability of further deposition. This shows up as an accelerated growth outward from the tips in the form of thin black filaments of silver, extending from the cathode back toward the positive conductor (Figure 5.6). In the course of this

FIGURE 5.5. Colloidal silver "stain-front" from positively polarized silver conductor on epoxy–glass PC board (6.4×).

growth, branching usually occurs at definite crystallographic angles, resulting in a characteristic "dendritic" structure.

When a filament finally bridges the gap between conductors, a sudden drop in resistance will occur. Although the magnitude of the initial resistance drop will be small (because of the small cross-sectional area of a single filament), thickening of the filaments and a rapid increase in additional bridges—as would be expected when dendrites are present—will soon lower the resistance sufficiently to produce circuit failures [1–4, 24, 25, 45, 47–51].

Bridging can also be produced by colloidal deposits from the anode,

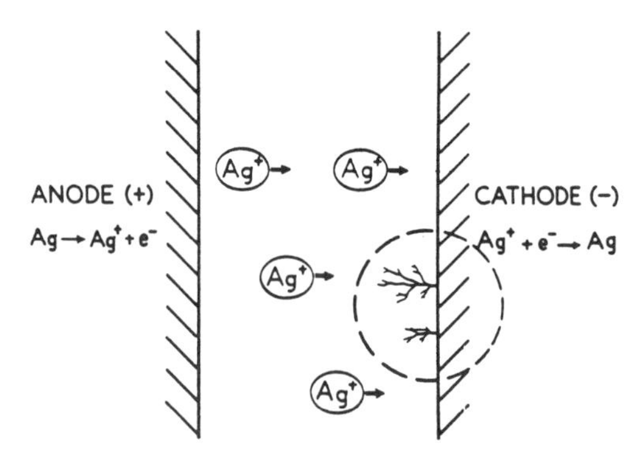

FIGURE 5.6. Schematic diagram of early stages of dendrite growth. (From Krumbein et al. [8].)

especially on contaminated surfaces [6, 7] and at higher temperatures and voltages [4]. A number of different phenomena have been reported in the literature, all of which appear to involve metallic electromigration from the anode [10–14, 21, 45, 46], often under conditions of comparatively high ambient temperature (75°C) or humidities (95%). In one series of papers [12–14], copper filaments were reported to be growing from anodic to cathodically polarized pads or plated-through holes after the glass–epoxy printed circuitry (with copper conductors) had been subjected to conditions that produced local physical separations of the epoxy resin from the glass fibers. The underlying mechanism for this effect—to which the authors [12–14] gave the name "conductive anodic filaments"(CAF)—was postulated as involving prior formation of moisture paths along those bare (hence water-wettable) glass fibers within the board that happened to bridge the oppositely polarized conductors [13]. In other studies, we have found anodic silver in filamental form bridging 0.38-mm (15-mil) glass–epoxy spacings (Figure 5.7), while higher voltage (100–300 V) work with aluminum-metallized integrated circuits produced practically instantaneous filamental bridging by the aluminum across 0.02–0.05-mm (1–2-mil) spacings on GaAs [10] substrates.

A related silver electromigration phenomenon appears to take place electrolytically in a nonaqueous "solvent" at high (presumably dry) temperatures. This is the reported migration of the metal from a silver-loaded polyimide adhesive that is used to bond diode chips to gold-plated headers. In this case [11], silver was found to migrate from the adhesive after 1000 hours at 300°C, when the diode was under a reverse bias of 10 V. Under these conditions, the gold/silver–adhesive system would be polarized positive (anodic).

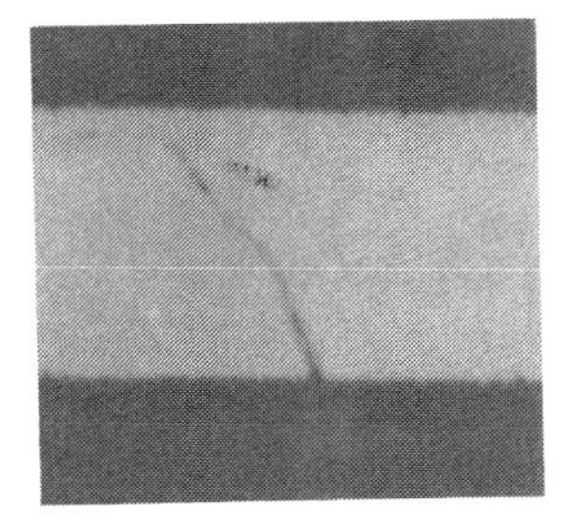

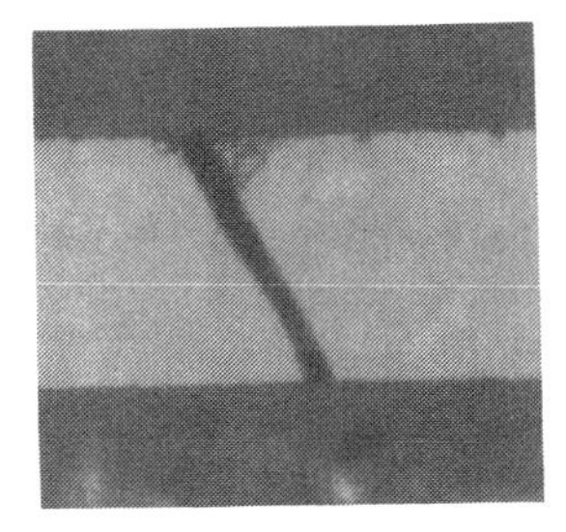

FIGURE 5.7. Colloidal-silver anodic filaments on 0.38-mm (0.015-in.) spacings of glass–epoxy PC boards. Anode is on top (36×).

5.2.2 Operating Conditions

The primary operating parameters that promote humid electromigration problems are as follows:

1. Moisture (i.e., high relative humidity).
2. Contamination on the insulator surface.
3. Voltage difference between conductors.
4. Narrow spacing widths.
5. Elevated temperatures (at high relative humidity).

Detailed examples of how these factors affect both the speed and severity of the electromigration reactions for different insulator surfaces can be found in a number of publications [2–7, 9, 24, 26, 51]. The discussion given below will be primarily for the purpose of showing how these factors interact with each other, as well as in clarifying some of their less understood aspects. In discussing these points, it will be helpful to bear in mind the three basic electromigration steps that were previously described: (a) oxidation or dissolution at the anode, (b) migration across or through the insulator, and (c) reduction and deposition.

Moisture For electromigration to occur, a comparatively low-resistance electrolytic path must exist between two conductors of opposite polarity. These paths need not necessarily be permanent. They may dry up and reappear with changing humidity, but their duration will be one of the factors determining the time to failure [4].

In the early experimental work with phenolic–paper laminates and other porous or fibrous substrates [1–3], the conductive solution was found to follow the pores or fibers, so that silver filaments would eventually be found within the insulator material itself. However, with nonporous substrates, such as glass–epoxy PC boards or ceramic microelectronic devices, the electrolytic path would be provided by, and generally restricted to, an adsorbed layer of moisture on the surface. Thus the presence of moisture and its effects will also be related to the nature of the insulator and to the type and quantity of surface contamination in its vicinity (factor 2 above).

Humidity and Insulator Material It has been shown experimentally [2–9, 11, 12, 18, 21, 27] that the higher the relative humidity, the faster the onset of electromigration and the greater the probability of low-resistance failures. However, for any one relative humidity, the tendency for moisture adsorption on the insulator surface will vary with the type of insulator material used and its surface condition. Insulator materials with strongly polar groups, like phenolics, nylons, alumina, or glass, have a much higher susceptibility to electromigration than "hydrophobic" materials with weak or no polar groups, like Mylar (polyester) or polyethylene [2, 3, 13, 26], since the latter materials have a much lower propensity for adsorbing moisture and retaining moisture films on their surfaces.

A particularly striking example of this effect is that of porcelain surfaces, as in porcelainized-steel circuit boards. When such boards with silver-ink circuitry were exposed to 75°C and 90% relative humidity (RH), with 30-V polarization, significant electromigration was observed at the 0.38-mm (15-mil) spacings as early as the first week of the test. Within two weeks, significant bridging had occurred (Figures 5.8 and 5.9), with a resulting drop in interconductor resistance of at least four orders of magnitude. The burnt-looking spots on the positive conductors in Figures 5.8 and 5.9 are due to the dissolution of the white silver ink (over the black porcelain) in the initial stages of electromigration.

The humidity effect can also be explained in terms of phase equilibria between the gaseous moisture and the moisture adsorbed on the insulator and conductor. The RH, by definition, is the fraction of the saturation vapor pressure of water that is actually present in the operating environment. This saturation pressure (i.e., 100% RH) is, in turn, de-

FIGURE 5.8. Silver electromigration on porcelainized PC board (4.8×). Note dendritic bridging of 0.38-mm (0.015-in.) spacing and orientation of dendrites on circular pad toward anodic conductor.

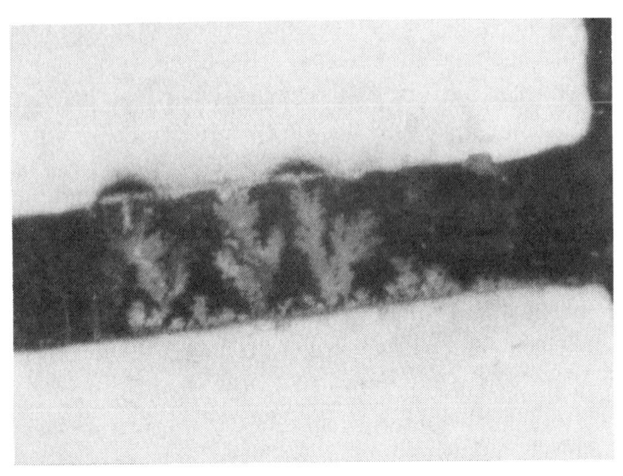

FIGURE 5.9. Higher magnification (16×) view of bridging across the 0.38-mm (0.015-in.) spacing in Figure 5.8.

fined as the pressure of the vapor that is in equilibrium with the bulk liquid phase at any one temperature. Thus clean and smooth hydrophobic surfaces (polyethylene or polystyrene) would not be expected to acquire more than a few molecular layers of water until 100% RH had been approached, whereupon condensation of bulk liquid (as droplets) would occur. On the other hand (as explained above), insulating materials with water-attracting polar groups could begin to adsorb water as invisible films at much lower RH values [38], the extent of the adsorption increasing with rising humidity. In the same way, capillary condensation could occur in scratches, pits, and other high-surface-energy areas, the net effect being the lowering of the minimum ambient RH required for adsorption, even on otherwise water-repellent materials.

This capillary effect was recognized during an extensive experimental program on silver migration on conventional epoxy-laminated glass–cloth, printed circuit (PC) boards [4, 7, 8], part of which involved examining the dendritic filaments under a scanning electron microscope (SEM). During these examinations, in which we found that the dendritic growth was always limited to the surfaces of these well-laminated boards [4], we also observed that the growths tended to follow low-lying channels or pathways on the epoxy-covered surface (Figure 5.10).

An analogous phenomenon, involving *anodic* filaments, was observed with silver-ink conductors on Mylar polyester. After seven weeks in a 65°C, 90% RH environment, with an applied voltage of 5 V, there was no significant resistance change and no sign of dendritic growth from the cathode. However, on some of the spacings, thin straight black

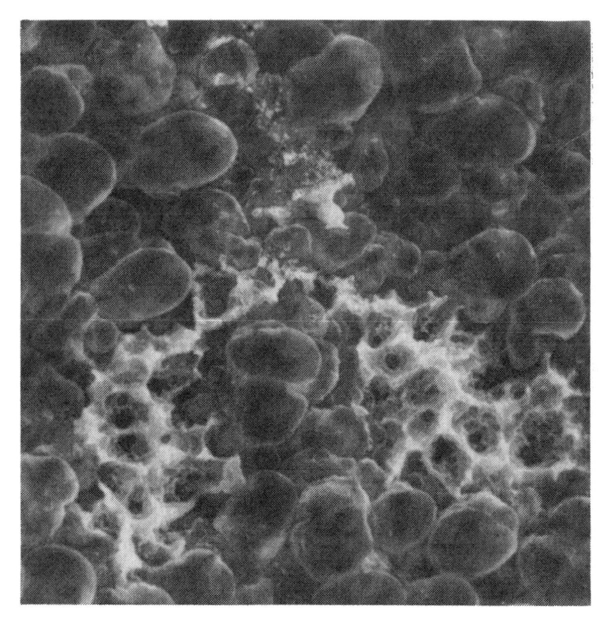

FIGURE 5.10. SEM micrograph (800×) of silver dendrite residue on epoxy–glass surface. Anode is at the bottom. (From Krumbein and Reed [4].)

needles were seen to be coming from the positively polarized conductors (Figure 5.11).

SEM microscopy and energy-dispersive spectroscopy (EDS) and x-ray mapping showed these needles to be composed of colloidal silver (Figure 5.12) in well-defined scratches or craze-marks on the polyester surface.

Contamination Many different types of contaminants and impurities may be present on the insulator and at the insulator–conductor interface. Although they can originate from a number of sources and may have varying effects on electromigration, most appear to function by encouraging moisture adsorption on the insulator surface, as well as by stabilizing the resulting surface solution and increasing its conductivity [2, 5–7, 12, 21].

In dealing with the diverse nature of the many potential contaminants, it is useful to divide them into two broad categories, according to their apparent origin: (a) materials-based impurities, originating from the contacts, conductors, and insulator materials, and (b) extraneous contaminants from the environment surrounding the contacts or conductor–insulator system.

In making this semi-arbitrary division, one accepts the fact that there

FIGURE 5.11. Needles of colloidal silver from anodically polarized silver-ink conductors on polyester film (15×).

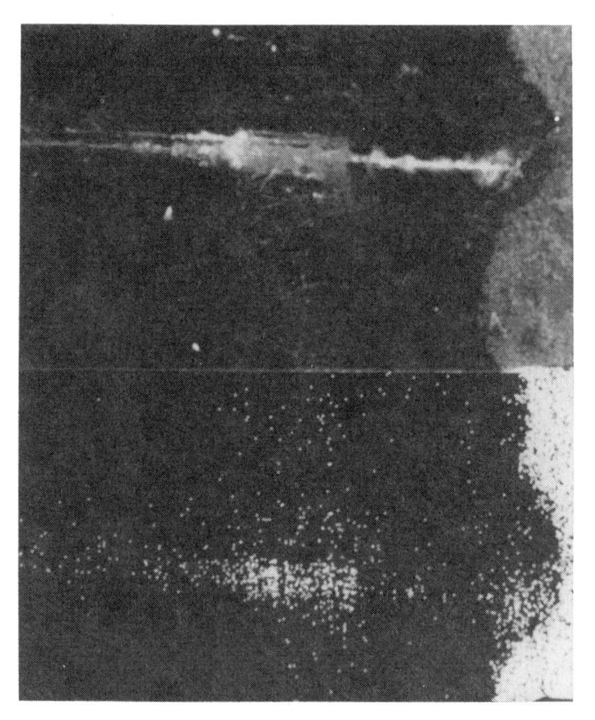

FIGURE 5.12. SEM micrograph (top, 150×) and silver x-ray map (bottom) of silver needle from Figure 5.11.

are some contaminants that do not fit into either category. One example is where the migrating silver itself originates as an impurity, as when silver-containing conductive adhesives are used to bond gold-plated surfaces in hybrid microcircuits [29, 32].

Materials-Based Impurities Among the most common—and harmful—of these contaminants are chemical residues from the plating and etching operations, which, fortunately, can often be minimized by proper and adequate rinsing. Ionic oxidants, such as ferric chloride etchant residues, have been found to encourage silver dissolution and a rapid buildup of colloidal silver on the insulator surface [7]. In other cases, wetting agents from the plating bath could decrease the hydrophobicity (or water repellency) of portions of the insulator surface, while ionic residues would increase the conductivity of the adsorbed moisture film.

We have also found that differences in free cyanide content between two different silver-plating bath formulations can have a profound effect on the relative susceptibilities (to electromigration) of the resulting plated conductors. In this particular case [4, 7], however, it was the "no free cyanide" neutral silver formulation—which is free-rinsing and leaves essentially no residual cyanides on the finished plating [8]—that had the highest rate of electromigration. This apparent contradiction can be explained by noting that in the presence of moisture any excess free cyanide would tend to convert practically all other silver species into the very stable, negatively charged Ag(CN) complex [36]. This would effectively inhibit dendrite growth, by preventing ionic migration to the negatively charged conductor, until the free cyanide concentration was reduced to a very small value.

The insulator material itself can also be a source of contamination. Many of these contain flameproofing agents or similar additives that can leach out to the surface. Thus, with conventional glass–epoxy PC boards, the purer G-10 type was found to be less prone to electromigration than the flameproof FR-4 variety [4].

Extraneous Contaminants These may include airborne dusts, flux residues, gaseous pollutants [6, 26, 28, 29, 41, 42], fingerprint residues from improper handling, metallic particles and their corrosion products (including wear particles from conductor tabs), as well as fibers, plastic scrapings, and other materials that are introduced during connection or assembly.

Contamination can also be produced within a package or container that is otherwise protected from the outside environment. In many such cases, volatile compounds are emitted by plastics or desorbed from other inside surfaces of the package, processes that are accelerated by temperature changes. Some of the more common examples are HCl, acetic acid vapor, and ammonia or volatile amines [29, 43, 44]. Water, however, is the most common emission and desorption product observed in such systems. In certain cases, local buildups of this closed-system water resulted in the onset of "wet" electromigration phenomena [16, 21, 30, 31]. This effect may have also been involved in copper and tin–lead migration under certain solder-mask coatings, where the *external* humidity was 90% or less [45, 46].

Although the functions of extraneous contamination in supporting or inhibiting humid electromigration are similar to those discussed in the previous section, a number of additional points should be added. Many types of dust particles and almost all common corrosion products are hygroscopic; they will increase water adsorption in their vicinity. In an extreme example of this effect, Der Marderosian [21] found that placement of salt crystals in the insulator spacing produced visible water droplet condensation (leading to "wet" electromigration) at relatively low relative humidities. He also observed that the "critical humidity" (for water condensation on a particular surface) for a particular contaminant salt was approximately related to the threshold RH for dendritic growth.

In addition, many acidic pollutants, such as HCl and the sulfur and nitrogen oxides, are very soluble in both types of surface moisture films ("humid" and "wet") and will encourage electromigration [6, 26, 28, 29, 41, 42]. One common gaseous contaminant that is often overlooked is ammonia [29]. Its importance to silver electromigration lies in its tendency to form stable positively charged complexes with silver ions [36], thereby aiding the initial oxidation–dissolution step at the silver anode and supporting the ionic migration of silver toward the negative electrode.

Voltages and Spacings A great many experimentalists have shown that the higher the polarizing DC voltage (or "bias"), the greater will be the rate and severity of the electromigration. In addition, an inverse relationship was almost always seen between the migration rate and the width of the interconductor spacing [24, 29, 48, 51]. In a related extensive series of experiments, with redundant sets of three or more different spacings on the same board (Figure 5.13), we found that the filaments or dendrites always began in the narrowest spacings [4] and, in most cases, were never found on any of the wider spacings until bridging had first "shorted out" a narrower spacing. These and similar observations would indicate that the critical voltage variable is not the absolute value

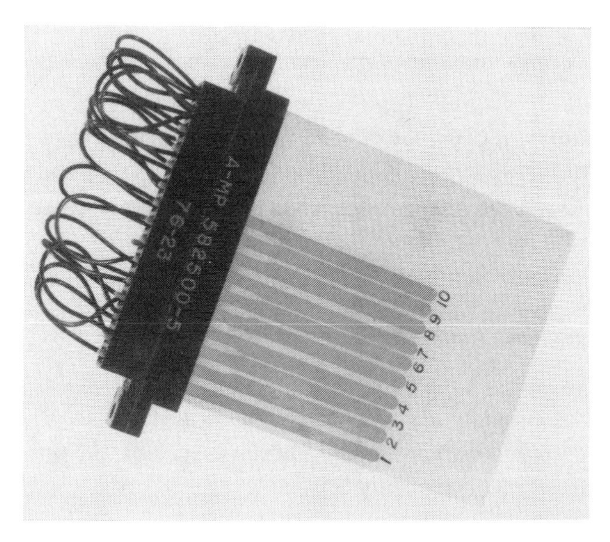

FIGURE 5.13. Test board with redundant sets of three different conductor spacings. (From Krumbein and Reed [4].)

of the applied voltage but rather the voltage *gradient* across the width of the spacing.

From the viewpoint of electrochemical theory, there are two fundamental ways in which the potential gradient can affect the electromigration. Both of these involve the rate at which the silver ions move from the anode to the cathode, and both assume that a film of moisture is continually present on the insulator spacing. In addition, both mechanisms are dependent on the temperature.

In the first mechanism [34], which assumes a bulk solution and is therefore more applicable to ''wet'' electromigration, the velocity of an ion in an electrical field is directly proportional to the voltage gradient; that is,

$$\boldsymbol{v}_+ = u_+\mathbf{F} \tag{5.1}$$

where u_+ is the ionic mobility of a specific ion (such as Ag^+) and $\mathbf{F}$ is the potential gradient, usually in units of volts per centimeter.

The second mechanism (which is superimposed on the first) involves the electrokinetic phenomenon of electro-osmosis and is only applicable when the electrolyte is confined to scratches, narrow channels, or thin surface layers—such as would be postulated for humid electromigration. Under these conditions [33] the moisture layer will tend to flow toward the negative electrode, carrying with it the soluble silver species. (This is because the flow direction depends upon the relative dielectric constants of the water and insulator phases. In this case, since water's dielectric constant is much higher than that of plastics or ceramics, the liquid will acquire a positive charge and move toward the cathode [33].) This additional migration mode is also a function of the potential gradient, since the flow rate of the liquid $\boldsymbol{v}$ is directly proportional to it; that is,

$$\boldsymbol{v} = \frac{D\mathbf{F}E_{\mathrm{ek}}}{4\pi\eta} \tag{5.2}$$

where D is the dielectric constant of the liquid, η is its viscosity, and E_{ek} is the electrokinetic potential (which depends on the properties of the solid surface and the solution).

From a practical point of view, however, it would be extremely difficult to evaluate the voltage gradient experienced by the migrating silver. This is because of the branching and tortuosity of the electrolytic paths as they actually exist on the insulator material. In addition, there are problems associated with the time-dependency of many of the properties associated with these surface electrolyte paths (such as cross-sectional area and solution concentration and conductivity). This would also include the ongoing decrease in the length of many of the paths once the silver filaments have begun to grow out from the cathode.

Temperature All other things being equal, the higher the temperature the more extensive the migration, particularly at very high relative humidity values. This correlation has been observed for a variety of insulating materials with different silver migration susceptibilities [3, 4, 9, 12]. Unfortunately, the number of reports dealing with the effects of different constant temperatures within the normal operating range of electrical contacts, boards, and hybrids is extremely limited. This is probably due to the long times required for each experimental run under these conditions, as well as the emphasis on standard testing (rather than operating) conditions. Thus most electromigration failure programs have been limited to comparatively high temperatures (as in 85°C/85% RH exposures) or to the thermal cycling—where the main effect would be to produce local humidity variations and condensations.

From a theoretical point of view [34, 35], the main effect of temperature would be to increase the conductivity of the electrolytic surface solution on the insulator and speed up the migration of silver ions toward the cathode (i.e., migration across or through the insulator). This increase in conductivity comes about through the increase in the value of ionic mobility [see Equation (5.1)], and the decrease in solution viscosity [see Equation (5.2)], with rising temperature. On the other hand, only a small part of any temperature effect would be attributable, theoretically, to the oxidation and dissolution processes at the anode. The anodic oxidation of silver is considered to be thermodynamically reversible, especially at the low current densities present across the insulator. One would therefore not expect any significant kinetically controlled processes (involving activation energies and exponential functions of the temperature) for this oxidation step. From a practical viewpoint, however, there seems to be doubt as to whether conclusions derived from electrodic oxidation equilibria in bulk solution can be extended to interfaces between silver conductors and a thin film of moisture on an insulator, although with "wet" electromigration (see below) they may be more applicable.

From our own observations, most of which have already been reported [4, 8], the decrease in time-to-failure at the higher temperatures seems to be connected to a comparatively rapid buildup of colloidal silver on the insulator and the movement of these "stains" toward the cathode. Examination of some of the failed spacings showed that the stains had bridged these spacings in some cases (Figure 5.5), while in others, dendrites had grown from the cathode to the stain (Figure 5.4). We had also previously reported [4] that at high temperatures (i.e., 75°C) a "stain front" could sometimes be seen parallel to the anodic conductors similar to the one shown in Figure 5.5. The advancing edge of this stain was often at the margin of farthest advance of the dendrites and may have served as a secondary source of silver ions for their growth [4].

5.3 "WET" ELECTROMIGRATION

When liquid water is present in the spacings between the conductors, the electromigration effects will increase in speed and severity by several orders of magnitude. In work carried out in the author's laboratory, for example, the time to silver dendrite inception at 45°C, with an applied voltage of 10 V across 0.38-mm (15-mil) spacings, was at least a week at 90% RH. In addition, several more days were required for the dendrites to bridge the spacing. On the other hand, when a drop of water was placed between the conductors at room temperature, the time of inception was reduced to minutes (Table 5.2), and subsequent bridging occurred in less than a minute. Other workers have reported similar accelerations [15, 19, 21, 24].

TABLE 5.2. Average Time Between Application of Voltage and First Definite Appearance of Metallic Filaments

Voltage (V)	Time (min)	
	0.64-mm Spacings	1.28-mm Spacings
6	0.92 ± 0.14 (3 runs)	2.4 ± 1.0 (3 runs)
5	1.00 ± 0.38 (2 runs)	3.3 ± 2.5 (4 runs)
4	0.93 ± 0.55 (4 runs)	
3	1.75 ± 0.35 (2 runs)	4.7 ± 1.0 (4 runs)
2	1.81 ± 0.59 (4 runs)	

A second pronounced effect of condensed moisture in the interconductor spacing is the extension of all the previously discussed electromigration effects to many other metallic systems [15–22, 24, 25, 28]. Filament bridging from cathode to anode, under these conditions, has been observed for copper, gold, tin, nickel, lead, palladium, and solder, but not for metals with protective oxide films, such as chromium, aluminum, and tungsten [15, 24].

Wet electromigration has also become a quality-control problem with integrated circuit (IC) devices and hybrid microcircuit systems, in general. Of particular importance was the finding that many "black-box failures" of gold-metallized circuitry were due to bridging of the narrow IC spacings by dendritic filaments of gold—which the reliability engineers who first discovered this effect [16–18] called migrated gold resistive shorts (MGRS). Although all the failures were caused by condensed moisture, the presence of chlorides or other halogen-containing compounds was also considered necessary in order to convert the gold (at the anode) into a soluble complex from [16–18, 21, 22]. In the absence of such compounds, dendritic filaments were usually not produced. However, bridging by gold can still occur across the narrow spacings through the formation and migration of colloidal gold from the anodes [23].

Figure 5.14 shows early bridging by silver dendrites of a 0.76-mm (30-mil) spacing of a conventional epoxy–glass PC board on which a drop of deionized water had been placed. This micrograph was taken only a few minutes after we had applied a 5-V potential across the wet spacing. The bubbles (presumably of hydrogen gas) that can just be seen at the cathode–insulator interface were found to be a common feature of wet electromigration. In Figure 5.15, for example, which shows a 0.76-mm (30-mil) spacing across which 10 V had been applied (and copper dendrites had bridged), the bubbling has increased considerably as a result of the increased voltage. In addition, the appearance of bub-

FIGURE 5.14. "Wet" electromigration: silver dendrites in deionized (DI) water on glass–epoxy PC board (24×). Note gas bubbles at cathodic conductor.

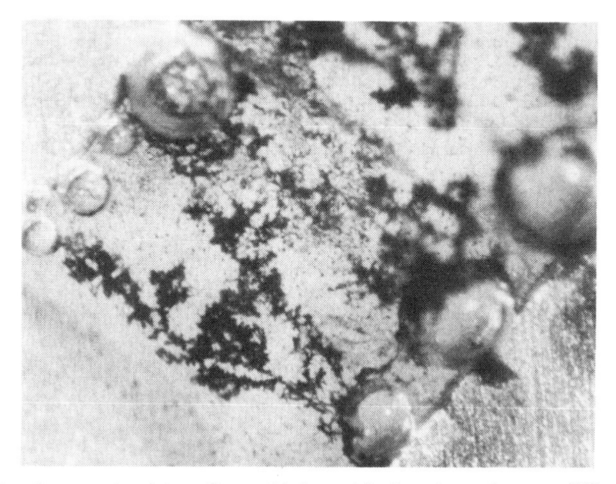

FIGURE 5.15. Copper dendrites from thinly-gold-plated conductors (32×). DI water on glass–epoxy board.

bles at both of the conductors (of thinly gold-plated copper) indicates that electrolysis of water is taking place at the higher potential. As explained below, the turbulence produced by bubble formation at the interface between the conductor and the water-covered insulator appears to play a major role in the apparent "saturation" of the failure rate when a certain level of applied voltage has been reached.

In a series of experimental programs with a relatively hydrophobic, Mylar-based flexible circuitry, the rapid onset of electromigration effects could be obtained at low applied potentials (down to 2 V) if a drop of deionized water was placed across either 25- or 50-mil spacings be-

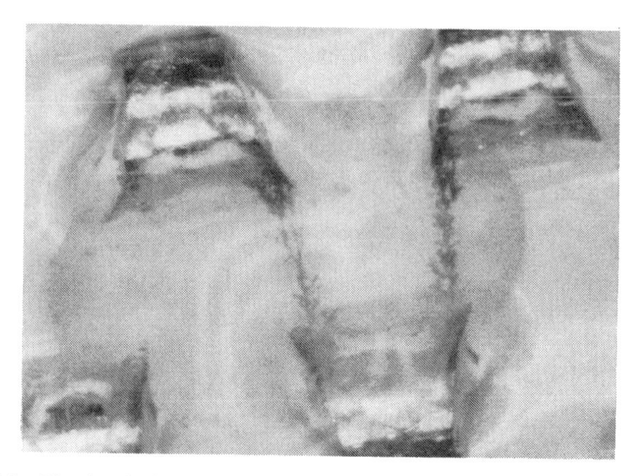

FIGURE 5.16. Tin–lead electromigration in drop of DI water on polyester film (23×).

tween tin–lead-plated conductors (Figures 5.16 and 5.17). Analysis of the dendritic fragments that remained after evaporation of the water (Figure 5.18) showed them to be composed of both tin and lead, with the same approximate composition as the tin–lead platings on the conductors. In addition, SEM micrographs of some of these fragments showed that the dendritic structure was present down to the micrometer range (Figure 5.19).

Table 5.2 lists the average time intervals between application of the voltage and the definite appearance of the first dendritic filaments of tin–lead at the negatively polarized conductors. It shows that these room-

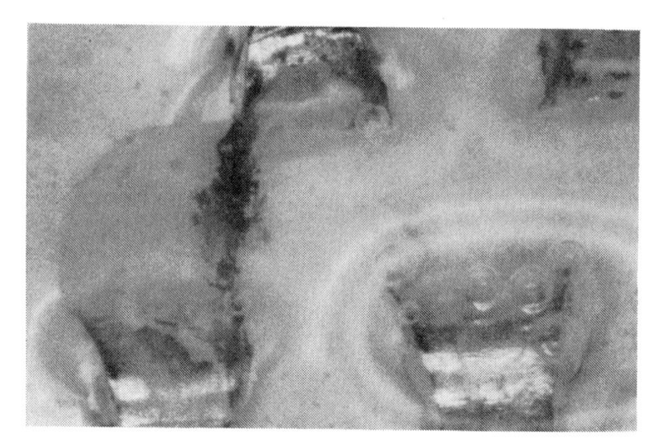

FIGURE 5.17. Tin–lead electromigration in DI water drop on polyester film (18.4×). Note gas bubbles on both anodic and cathodic contacts.

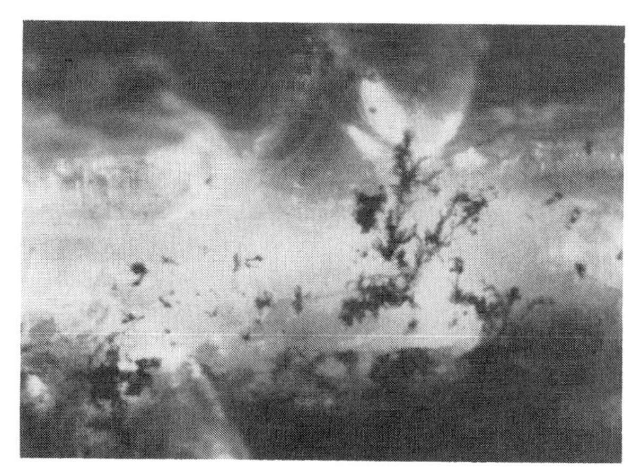

FIGURE 5.18. Dendritic fragments in 0.64-mm (0.025-in.) spacing after evaporation of water drop from polyester surface (32×).

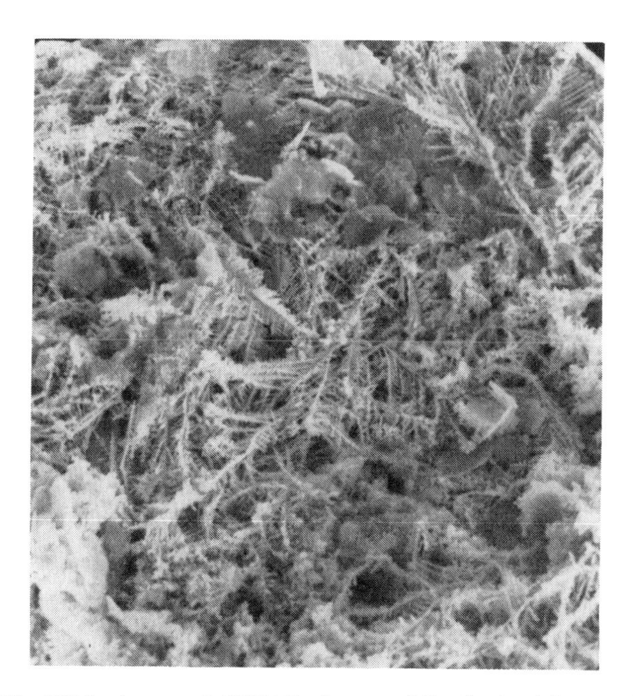

FIGURE 5.19. SEM micrograph (600×) of some of the tin–lead fragments in Figure 5.18.

temperature inception times depend, as expected, on both the magnitude of the applied potential and the interconductor spacing. The table also gives some indication of the spread in the values for the different experimental conditions.

Further examination of these results, particularly with the 0.64-mm (25-mil) spacings, shows what appears to be a maximum voltage above which no further acceleration in the electromigration effects would be obtained. In practice, this apparent leveling off in the rate of dendritic inception and growth is at least partly due to the increased turbulence in the liquid drop at the higher voltages, which, as seen through the microscope, was produced by rapid eruptions of the gas bubbles. These tended to interfere with the nucleation of the first thin filaments at the negatively polarized conductors and especially with the unidirectional growth of the dendrites toward the positively polarized conductors. These same phenomena also occurred with the 1.28-mm (50-mil) spacings. However, their effect on inception time was primarily in the large observed variations in the time intervals for the different runs.

It should also be pointed out that when a filament or dendrite moves through the three-dimensional water layer of "wet" electromigration, motion within the liquid prevents the filament's path from going straight toward the anode, even allowing for branching. Instead—as observed both by direct stereomicroscope viewing and by examination of real-time videotapes—the filaments tend to take more tortuous and wiggly three-dimensional paths in response to the random motions of the liquid medium.

5.4 A WORD ABOUT PREVENTION

Problems with both wet and high-humidity electromigration are becoming much more prevalent as interconductor spacings become narrower and board loadings more complex. In addition to increasing the probability of capillary condensation of water in the very small nooks and crannies that arise from such loading, these discontinuities can also serve as traps for deleterious contaminants. It is therefore essential that those involved in testing, qualifying, and trouble-shooting such systems should pay even more attention to assuring the cleanliness and freedom from moisture of the entire package, if electromigration problems are to be avoided.

Because of the critical effect of moisture on electrolytic electromigration, it was felt that one could inhibit, or slow down, its effects by applying water-repelling coatings to the conductors and the insulator surface. Extensive work by different workers [3, 8, 24, 45, 46, 50] has shown that such techniques can be effective for certain conductor–insulator systems but will also depend on the magnitude of the voltage, temperature, and humidity to which the system is exposed. Another common method for mitigating electromigration (particularly of silver) is to inhibit the initial oxidation/dissolution step by reducing the thermodynamic "activity" (concentration) of the migrating metal. Techniques that have been tried with varying degrees of success [3, 6, 22, 25, 28, 50, 51] are overplating the silver with less "active" metals (such as gold, palladium, or even tin), alloying the silver (particularly with palladium), and tying up the silver or reducing its surface concentration through tarnishing or other chemical conversions.

5.5 SOLID STATE ELECTROMIGRATION

The term "electromigration" is also used in solid state physics, where it describes mass transport in solids under the influence of high current densities. As such, it has received considerable attention in microelectronics because of its role in producing failures in integrated circuits [54–65], so that the *Annual Symposia on Reliability Physics*—sponsored by the IEEE Electron Devices and IEEE Reliability Societies—have included, over the past several years, special sessions on this type of electromigration.

In solid state electromigration, the passage of a direct current produces migration primarily through the force of electrons impinging on the metal atoms. However, because of structural inhomogeneities such as grain boundaries, the resultant atomic flux is not spatially uniform, and its divergence leads to material depletion and material accumulation. Material depletion results in holes and eventual open circuits (conductor failure). Material accumulation, on the other hand, leads to hillocks and whiskers, which can cause short circuits or even corrosion, such as by breaking through protective coatings. In a recent case, hillock formation was even found to produce cracking in a silicon dioxide (SiO_2) overcoating [61].

These two complementary phenomena may also be related to temperature differences on the metallization surface, with hillocks being associated with areas that are cooler than the average [65].

Although most solid state electromigration failures in the literature have involved aluminum-based conductors applied directly over silicon dioxide films, some have been reported for other substrate systems. One example is that of aluminum over titanium–tungsten (TiW) diffusion barrier films, where failures were ascribed to short-circuiting aluminum protrusions and whiskers [62]. In this regard, tin whiskers and filaments can also be produced by solid state electromigration, provided that the current density is sufficient to raise the local temperature close to the melting point of tin [66].

Finally, the mechanisms and operating conditions for *solid state* electromigration are entirely different from those of *electrolytic* electromigration. The former takes place under dry conditions and becomes important at local temperatures above 150°C, and, most importantly, at DC current densities of at least 10^4 A/cm^2 (Table 5.1). These conditions are commonplace in the microscopic world of integrated circuits and thin film conductors but would cause melting and other damage to PC board conductors and similar macroscopic components [63]. On the other hand, such conditions would completely prevent the presence of moisture, thereby eliminating the primary mode of *electrolytic* electromigration.

ACKNOWLEDGMENT

I acknowledge the participation of Dr. A. Reed and L. Lerner as coauthors in the early referenced work [4, 7, 8]. W. Miller and R. Stoner assisted in later experimental work, and R. Geckel did the SEM and EDS analyses.

REFERENCES

1. D.E. Yost, "Silver Migration in Printed Circuits," in *Proceedings of the Symposium on Printed Circuits*, Philadelphia, PA (1955).
2. G.T. Kohman, H.W. Hermance, and G.H. Downes, "Silver Migration in Electrical Insulation," *Bell Syst. Tech. J.*, Vol. 34, 1115 (1955).
3. S.W. Chaikin, J. Janney, F.M. Church, and C.W. McClelland, "Silver Migration and Printed Wiring," *Ind. Eng. Chem.*, Vol. 51, 299 (1959).
4. S.J. Krumbein and A.H. Reed, "New Studies of Silver Electromigration," in *Proceedings of the 9th International Conference on Electric Contact Phenomena*, p. 145 (1978).
5. A. Der Marderosian and C. Murphy, "Humidity Threshold Variations for Dendrite Growth on Hybrid Surfaces," in *Proceedings of the Reliability Physics Symposium*, p. 92 (1977).
6. G. DiGiacomo, "Metal Migration (Ag, Cu, Pb) in Encapsulated Modules and Time-to-Fail Model as a Function of the Environment and Package

Properties,'' in *Proceedings of the International Reliability Physic Symposium*, p. 27 (1982).

7. A.H. Reed and S.J. Krumbein, Abridged version of [4], but with some more recent experimental observations added. Presented at 3rd International Precious Metals Conference, Chicago, IL, May 1979.
8. S.J. Krumbein, L.B. Lerner, and A.H. Reed, ''A New Silver Plating Bath for Electronic Applications,'' presented at 67th Annual Technical Conference of American Electroplaters Society, Milwaukee, WI, June 24, 1980.
9. E. Tsunashima, ''The Sandwich Coating Between Conductive Layers for the Prevention of Silver-Migrations on a Phenolic Board,'' *IEEE Trans. Comp. Hybrids Manuf. Technol.*, Vol. CHMT-1, 182 (1978).
10. K.-H. Kretschmer and H.L. Harnagel, ''XPS Analysis of GaAs-Surface Quality Affecting Interelectrode Material Migration,'' in *Proceedings of the Reliability Physic Symposium*, p. 45 (1985).
11. R.J. Chaffin, ''Migration of Silver from Silver-Loaded Polyimide Adhesive Chip Bonds at High Temperatures,'' *IEEE Trans. Comp. Hybrids Manuf. Technol.*, Vol. CHMT-4, 214 (1981).
12. J.N. Lahti, R.H. Delaney, and J.N. Hines, ''The Characteristic Wearout Process in Epoxy–Glass Printed Circuits for High Density Electronic Packaging,'' in *Proceedings of the 17th Annual Reliability Physics Symposium*, p. 39 (1979).
13. D.J. Lando, J.P. Mitchell, and T.L. Welsher, ''Conductive Anodic Filaments in Reinforced Polymeric Dielectrics: Formation and Prevention,'' in *Proceedings of the 17th Annual Reliability Physics Symposium*, p. 51 (1979).
14. T.L. Welsher, J.P. Mitchell, and D.J. Lando, ''CAF in Composite Printed-Circuit Substrates: Characteristics, Modeling, and a Resistant Material,'' in *Proceedings of the 18th Annual Reliability Physics Symposium*, p. 235 (1980).
15. C.W. Jennings, ''Filament Formation on Printed Wiring Boards,'' *IPC Tech. Rev.*, pp. 9–16 (Feb. 1976) (esp. pp. 11–14).
16. A. Shumka and R.R. Piety, ''Migrated Gold Resistive Shorts in Microcircuits,'' in *Proceedings of the 13th Annual Reliability Physics Symposium*, p. 93 (1975).
17. F.G. Grunthaner, T.W. Griswold, and P.J. Clendening, ''Migratory Gold Resistive Shorts: Chemical Aspects of a Failure Mechanism,'' in *Proceedings of the 13th Annual Reliability Physics Symposium*, p. 99 (1975).
18. A. Shumka, ''Analysis of Migrated-Gold Resistive Short Failures in Integrated Circuits,'' in *Proceedings of the Technical Program of the International Microelectronic Conference*, p. 156 (1976).
19. P.E. Rogren, ''Electro Migration in Thick Film Conductor Materials,'' in *Proceedings of the Technical Program of the International Microelectronic Conference*, p. 267 (1976).

20. E.B. Flower, ''Electromigration on Printed Wiring Boards,'' in *Transactions of the Symposium of California Circuits Association* (PC Boards for the '80's), p. 22 (1978).
21. A. Der Marderosian, ''Humidity Threshold Variations for Dendrite Growth on Hybrid Substrates,'' in *Proceedings of the 20th Annual Meeting IPC*, April 1977 (IPC-TP-156) (Tech. Paper 13).
22. N.S. Sbar, ''Bias Humidity Performance of Encapsulated and Unencapsulated Ti-Pd-Au Thin-Film Conductors in an Environment Contaminated with Cl_2,'' *IEEE Trans. Parts. Hybrids, Packag.*, Vol. PHP-12, 176 (1976).
23. R.P. Frankenthal, ''Corrosion Failure Mechanisms for Gold Metallizations in Electronic Circuits,'' *J. Electrochem. Soc.*, Vol. 126, 1718 (1979).
24. T. Kawanobe and K. Otsuka, ''Metal Migration in Electronic Components,'' in *Proceedings of the Electronic Components Conference*, p. 220 (1982).
25. R.W. Gehman, ''Dendritic Growth Evaluation of Soldered Thick Films,'' *Int. J. Hybrid Microelectron.*, Vol. 6, 239 (1983).
26. J.J. Steppan, J.A. Roth, L.C. Hall, D.A. Jeannotte, and S.P. Carbone, ''A Review of Corrosion Failure Mechanisms During Accelerated Tests: Electrolytic Metal Migration,'' *J. Electrochem. Soc.*, Vol. 134, 175 (1987).
27. R. Gjone, ''The Migration Failure Mechanism on Pin Grid Array VLSI Packages,'' presented at the 1983 International Forum of the National Association of Corrosion Engineers, Anaheim, CA, Apr. 18–22, 1983, paper 231.
28. P. Dumoulin, J.-P. Seurin, and P. Marce, ''Metal Migration Outside the Package During Accelerated Life Tests,'' *IEEE Trans. Comp. Hybrids Manuf. Technol.*, Vol. CHMT-5, 479 (1982).
29. R.C. Benson, B.M. Romenesko, B.H. Nall, N. deHaas, and H.K. Charles, Jr., ''Materials-Related Current-Leakage Failure in Hybrid Microcircuits,'' in *Proceedings of the Electronic Components Conferences*, p. 111 (1986).
30. J.E. Ireland, ''A Performance Evaluation of a High Density Hermetic Assembly Using Epoxy Die Attachment,'' *Int. J. Hybrid Microelectron.*, Vol. 6, 352 (1983).
31. R.W. Thomas, ''Moisture Myths and Microcircuits,'' in *Proceeding of the Electronic Components Conference*, p. 272 (1976).
32. F.N. Lieberman and M.A. Brodsky, ''Dendritic Growth of Silver in Plastic I.C. Packages,'' presented at International Symposium on Microelectronics, Sept. 17–19, 1984.
33. G. Kortum, *Treatise on Electrochemistry*, 2nd ed., Wiley, New York, 1965, Chap. XI.

34. G. Milazzo, *Electrochemistry, Theoretical Principles and Practical Applications*, Elsevier, New York, 1963, esp. Chap. II.
35. J.O. Bockris and A.K. Reddy, *Modern Electrochemistry*, Plenum, New York, 1973, Chaps. 4 and 10.
36. E.J. King, *Qualitative Analysis and Electrolytic Solutions*, Harcourt, Brace, New York, 1959.
37. S.M. Maron and J.B. Lando, *Fundamentals of Physical Chemistry*, Macmillan, New York, 1974, Chap. 9.
38. N.O. Tomashov, *Theory of Corrosion and Protection of Metals*, Macmillan, New York, 1966, Chap. XIV.
39. V. Tierney, ''The Nature and Rate of Creepage of Copper Sulfide Tarnish Films Over Gold Surfaces,'' *J. Electrochem. Soc.*, Vol. 128, 1321 (1981).
40. W.H. Abbot and W. Campbell, ''Recent Studies of Tarnish Film Creep,'' in *Proceedings of the 9th International Conference on Electric Contact Phenomena*, p. 117 (1978).
41. S.P. Carbone and E.A. Corl, ''Atmospheric Active Pollutant Indicator,'' in *Atmospheric Corrosion*, W.H. Arbor, ed., Wiley, New York, 1982, Chap. 12.
42. L.G. Feinstein and N.L. Sbar, ''Performance of New Copper-Based Metallization Systems in an 85°C, 78% RH, SO_2 Contaminated Environment,'' *IEEE Trans. Comp. Hybrids Manuf. Technol.*, Vol. CHMT-2, 159 (1979).
43. L.G. Feinstein, ''Failure Mechanisms in Molded Microelectronic Packages,'' in *Proceedings of the International Microelectronics Conferences*, p. 49 (1979).
44. S.J. Krumbein and A.J. Raffalovich, ''Corrosion of Electronic Components by Fumes From Plastics,'' Res. Dev. Tech. Rep. AD733903, U.S. Army Electronics Command, Ft. Monmouth, NJ, Sept. 1971.
45. P.J. Dudley, ''Electrical and Environmental Testing of UV Curable and Dry Film Solder Masks,'' presented at PC Fabrication Technical Seminar, Atlanta, GA, Dec. 5–7, 1983.
46. R.R. Sutherland and I.D.E. Videlo, ''Accelerated Life Testing of Small Geometry PCB's,'' presented at the 7th Northern Symposium of the Institute of Circuit Technology, Edinburgh, Scotland, Nov. 22, 1984; reprinted in *PC Fabrication*, p. 24, Oct. 1985.
47. H. Stastna and V. Gerlich, ''Silver Migration in Plastic IC's,'' presented at 4th Symposium on Reliability in Electronics, Oct. 4–7, 1977.
48. G.J. Kahan, ''Silver Migration in Glass Dams Between Silver–Palladium Interconnections,'' *IEEE Trans. Elec. Insul.*, Vol. EI-10, 86 (1975).
49. J.J.P. Gagne, ''Silver Migration Model for Ag-Au-Pd Conductors,'' *IEEE Trans. Comp. Hybrids Manuf. Technol.*, Vol. CHMT-5, 402 (1982).
50. J.F. Graves, ''Thick Film Conductor Materials—Production Qualification

Requirements and Test Procedures,'' in *Proceedings of the International Microelectronics Symposium*, p. 155 (1977).

51. H.M. Naguib and B.K. MacLaurin, ''Silver Migration and the Reliability of Pd/Ag Conductors in Thick Film Dielectric Crossover Structures,'' *IEEE Trans. Comp. Hybrids Manuf. Technol.*, Vol. CHMT-2, 196 (1979).
52. W.M. Kane and C. Wood, ''Electrical Conduction in Thin Films of Silver Telluride,'' *J. Electrochem. Soc.*, Vol. 108, 101 (1961).
53. R.F. Diehl and N.A. Gifaldi, ''Elimination of Tin Whisker Growth on Interconnections,'' in *Proceedings of the 8th Annual Connector Symposium*, p. 328 (1975); also references cited therein.
54. J.R. Black, ''Physics of Electromigration,'' in *Proceedings of the Reliability Physics Symposium*, p. 142 (1974).
55. J.R. Black, ''Electromigration Failure Modes in Aluminum Metallization for Semiconductor Devices,'' *Proc. IEEE*, Vol. 57, 1587 (1969).
56. R.W. Pasco and J.A. Schwarz, ''The Application of a Dynamic Technique to the Study of Electromigration Kinetics,'' in *Proceedings of the Reliability Physics Symposium*, p. 10 (1983).
57. R.E. Hummel and R.M. Breitling, ''On the Direction of Electromigration in Thin Silver, Gold, and Copper Films,'' *Appl. Phys. Lett.*, Vol. 18, 373 (1971).
58. T.E. Hartman and J.C. Blair, ''Electromigration in Thin Gold Films,'' *IEEE Trans. Electron Devices*, Vol. ED-16, 407 (1969).
59. F.M. D'Heurle, A. Gangulee, C.F. Aliotta, and V.A. Ranierei, ''Electromigration of Ni in Al Thin-Film Conductors,'' *J. Appl. Phys.*, Vol. 46, 4845 (1975).
60. A. Mogro-Campero, ''Simple Estimate of Electromigration Failure in Metallic Thin Films,'' *J. Appl. Phys.*, Vol. 53, 1224 (1982).
61. D.J. LaCombe and Earl L. Parks, ''The Distribution of Electromigration Failures,'' in *Proceedings of the Reliability Physics Symposium*, p. 1 (1986).
62. J.M. Towner, ''Electromigration-Induced Short Circuit Failure,'' in *Proceedings of the Reliability Physics Symposium*, p. 81 (1985).
63. J.R. Lloyd, ''Electromigration,'' *J. Metals*, p. 54 (July 1984).
64. F.M. D'Heurle, ''Electromigration and Failure in Electronics: An Introduction,'' *Proc. IEEE*, Vol. 58, 1409 (1971).
65. R.R. Clinton, ''A Novel Method for Measuring Nonuniformities in Metallization Temperatures of an Operating Integrated Circuit,'' in *Proceedings of the Reliability Physics Symposium*, p. 19 (1986).
66. R.B. Marcus and M.H. Rottersman, ''Tin Whiskers and Filamentary Growths on a Thin Film Conductor in Response to Direct-Current Flow,'' *Electrochem. Technol.*, Vol. 5, 352 (1967).

6

THEORETICAL AND EXPERIMENTAL STUDY OF ELECTROMIGRATION

JIAN HUI ZHAO
Department of Electrical and Computer Engineering
Rutgers, The State University of New Jersey, Piscataway

6.1 INTRODUCTION

The continuous scaling down of integrated circuits has attracted greatly increased attention toward the investigation of integrated circuit reliability. It is known that integrated circuit reliability depends on a variety of factors and the most important one regarding VLSI and ULSI circuit reliability is that of electromigration. Electromigration describes the development of structural damage caused by metal ion transport as a result of a high-current stressing in thin metal films. In integrated circuits, electromigration occurs either in the metallization or at the metal–semiconductor contacts. The former causes an open circuit or short circuit and the latter results in poor performance or malfunction of semiconductor devices due to the degradation of ohmic and Schottky contacts.

Hundreds of papers have been published on electromigration in thin films since the first report by Blech and Sello in 1966 [1]. After more than 26 years of study, a complete understanding of this phenomenon is still lacking, largely because of its dependence on the complicated interaction among many different mechanisms involving the film composition, structure, geometry, deposition, electrical and mechanical stressing, and even environmental stressing conditions. Many tech-

Electromigration and Electronic Device Degradation, Edited by Aris Christou.
ISBN 0-471-58489-4

niques have been proposed and demonstrated for the characterization of electromigration, but none has been shown to be a candidate for generalized use in a production environment because each of them involves either a special sample design or a nonstandard experimental setup. Some of them are of particular interest for electromigration study at the early stage, which helps to understand the underlying mechanisms by avoiding catastrophic damage to the films. The others are of great value for determining information for practical applications. There are three methods that can be nondestructive and useful for the early stage electromigration study. These include the resistometric method [2–4], the low-frequency noise method [5–10], and the internal friction method [11]. Destructive methods include the lifetime test, which provides directly the practically useful information such as the mean time-to-failure (MTF) along with its standard deviation σ [12], and the drift velocity method [13], which measures directly the metal ion drift velocity. For each of the nondestructive and destructive methods mentioned, the application is very time consuming. Fast techniques that can provide useful electromigration information within a few seconds to a few hours have also been demonstrated. These include the Standard Wafer-level Electromigration Acceleration Test (SWEAT) [14], the Temperature-ramp Resistance Analysis to Characterize Electromigration (TRACE) [15], the Wafer-level Isothermal Joule-heated Electromigration Test (WIJET) [16], the Breakdown Energy of Metal (BEM) method [17], and the Resistance Ratio (RR) method [18]. Of all these fast techniques, SWEAT is the one that can generate useful information regarding the kinetics of the failure process in less than 15 seconds.

Along with the development of the experimental study of electromigration is an improved understanding of the basic physical mechanisms responsible for electromigration as compared to that of a quarter century ago. It is now understood that the driving force for electromigration includes not only the electrical force [19], that is, the net force due to the electron wind and the direct field, but also the thermal gradient induced force, the concentration gradient induced force, and the mechanical stress induced force. It is also known that healing effects affect electromigration rate considerably, and as a result a threshold current density exists [20]. The healing effects describe the impact to electromigration of the mass backflow due to electromigration-induced concentration nonuniformity and the increase of capillarity [19]. Several types of computer simulation models on electromigration have been proposed [21–34], and the simplest one is that of a probability model [35]. The advantage of a computer simulation as compared to that of experimentation is obvious. Computer simulation is powerful and efficient in that it is possible to study the dependence of the macroscopic quantities on all of the microscopic parameters and to reveal the underlying basic physics of electromigration.

In this chapter, a brief review of electromigration in metallization and in high-power devices (mainly at the metal–semiconductor contacts) will be presented. Concentration will then be focused on the description of the most important factors that need to be considered in the modeling of electromigration and the computer simulation implementation. Following the modeling and simulation is a detailed presentation of the representative experimental methods in each of the aforementioned categories, namely, the resistometric and the low-frequency noise methods that are suitable for the study of electromigration in the early stage, the drift velocity method that measures directly the migration velocity of metal ions, the lifetime test method that is currently most widely used and is the most informative for IC designs, and the SWEAT that is capable of providing valuable information in less than a minute. Current status on the studies of electromigration will then be presented to conclude and summarize this increasingly important research.

6.2 ELECTROMIGRATION IN INTEGRATED CIRCUITS

6.2.1 Electromigration in Metallization Systems

As one of the major failure phenomena in integrated circuits, especially in VLSI and USLI circuits where metallization linewidth falls into the submicrometer range, electromigration in metallization stripes is becoming increasingly involved due to the use of new compound materials such as silicones and polycides and new metallization schemes such as the multilayer metallizations.

The classical electromigration problem refers to the development of structural damage caused by ion transport in thin metal films as a result of a very high-current-density ($> 10^6$ A/cm^2) stressing. In the circuit metallization interconnects, such a high-current-density induced mass transport often manifests itself as voids or cracks, which could eventually lead to open circuits, or as hillocks, which may cause shorting between adjacent lines or layers as shown in Figure 6.1 [36]. Figure 6.1 is an SEM micrograph showing the typical appearance of electromigration damage. The conductor lines are e-gun deposited Al–0.5%Cu metallizations with a linewidth of 2.5 μm. The center line has been

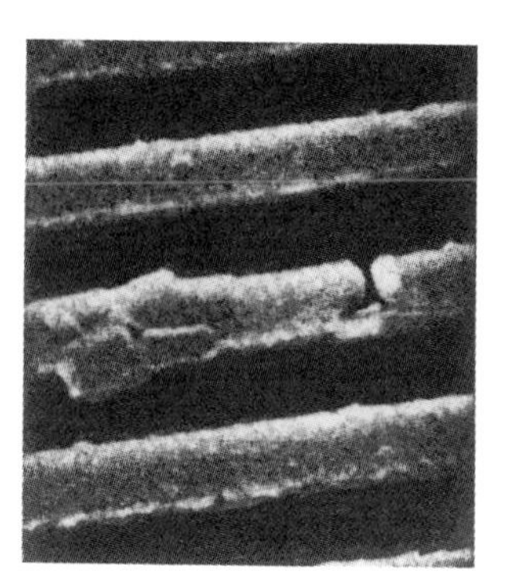

FIGURE 6.1. Scanning electron micrograph showing formation of cracks and hillocks. The center line has been stressed by a current density of 2×10^6 A/cm^2 at 250°C for 14 hours. After S. Vaidya et al., *Appl. Phys. Lett.*, Vol. 36, 464 (1980). Reprinted with permission.

subjected to a current density of 2×10^6 A/cm^2 for 14 hours at a temperature of 250°C. Both hillocks and voids have been developed due to electromigration, as shown clearly in the micrograph.

The key factor that determines the significance of the electromigration in a metallization line is the current density imposed on it. No electromigration has been detected for a current density below 10^5 A/cm^2, at least in the cases of gold and aluminum metallizations [21]. A concept of threshold current density has therefore been introduced and investigated [13, 20, 37–39], which represents a critical condition for electromigration damage to occur. It should be pointed out, however, that structural damage or change of surface topology in a metal line at high temperature and low current density may result from thermal grooving driven by temperature gradient rather than an electrical driving force.

It was found that the product of the threshold current (j_c) and the metal line length (l) is approximately a constant at a given temperature [38]. For example, for heat-treated Al films tested at 350°C, the product is $j_c l = 1260$ A/cm. Since current density is related to field E and the metal conductivity σ by $j_c = E_c \sigma$, the $j_c l$ product being a constant can be restated as

$$j_c l = E_c \sigma l = V_c \sigma = \text{constant} \tag{6.1}$$

where E_c has been replaced by the threshold voltage V_c across the metal line divided by its length l. Equation (6.1) suggests that as long as a critical voltage across a metal line is reached, electromigration is going to happen because σ in Equation (6.1) is a constant. The implication of Equation (6.1) is interesting because it states that as long as the voltage across a metal line reaches a critical constant electromigration will take place.

It is understood now that the physical origin of such a threshold current density is the counteracting mass flow induced by inhomogeneities such as gradient of mechanical stress, atomic concentration nonuniformity, and temperature gradient. Such a mass flow induced by structural inhomogeneities is always in the direction opposite to the electromigration mass transport and tends to cancel it. The value of the threshold current density therefore suggests the existence of a minimum energy barrier that the atoms must overcome to form a net mass flow. When the applied current density exceeds the threshold value, electromigration begins. As a result of the continuing scale-down of device dimensions and interconnection line sizes in the emerging technologies, such as in the ultra-large-scale integrated (ULSI) circuits, current densities of 1×10^6 A/cm^2 or higher exist in the interconnection lines, although the total current may be only a few tens of milliamperes. Therefore the effect of electromigration has become one of the major concerns for integrated circuit reliability.

The resistance degradation process of a conductor line shown in Figure 6.2 is also representative of the general failure process due to electromigration [2]. The fractional resistance of the conductor line first increases slowly with time due to the gradual development of structural

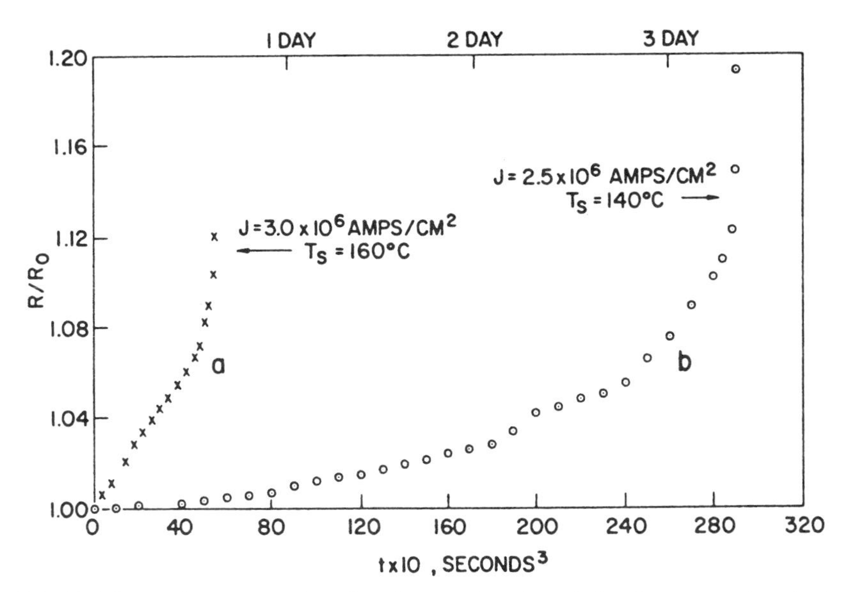

FIGURE 6.2. Resistance vs. time for electromigration of aluminum on mica. (a) j = 3×10^6 A/cm^2, T = 160°C; (b) j = 2.5×10^6 A/cm^2, T = 140°C in a silicone oil ambient. After R. Rosenburg et al., *Appl. Phys. Lett.*, Vol. 12, 201 (1968). Reprinted with permission.

damage, and then increases rapidly at a certain point indicating a catastrophic runaway [2, 4, 40]. The rate of resistance increase at the early stages has been found to be approximately constant [2, 4, 15, 40–43]. The very large or infinitely large value of resistance corresponds to an open circuit situation. This failure process can be viewed more directly in a microscopic way. At the early stage of electromigration, the voids or cracks formed at various locations are small in size in comparison with the film dimensions, and their existence does not significantly alter the overall current density and temperature distribution. This corresponds to the period of slow increase in the resistance. As the dimensions of those voids or cracks become comparable with the linewidth, the current densities around the voids are increased considerably. This local current density increase is referred to as current crowding. Current crowding causes the local temperature to rise since the Joule heating is proportional to the square of the current density. The elevated temperature accelerates the growth of the voids or cracks due to the exponential dependence of the atomic mobility on temperature, which in turn further increases the current crowding effect, eventually leading to the catastrophic degradation. This self-accelerated thermal runaway process can be summarized by the positive feedback loop shown in Figure 6.3. The initial slope on a resistance-versus-time curve is determined by the material properties, line geometry, and stressing conditions of the film, which characterize the failure rate of the early stages of electromigration. Thus the microscopic process of the early stage of electromigration can be macroscopically monitored by the fractional resistance change, provided the resistance measurement is of sufficient resolution. An in

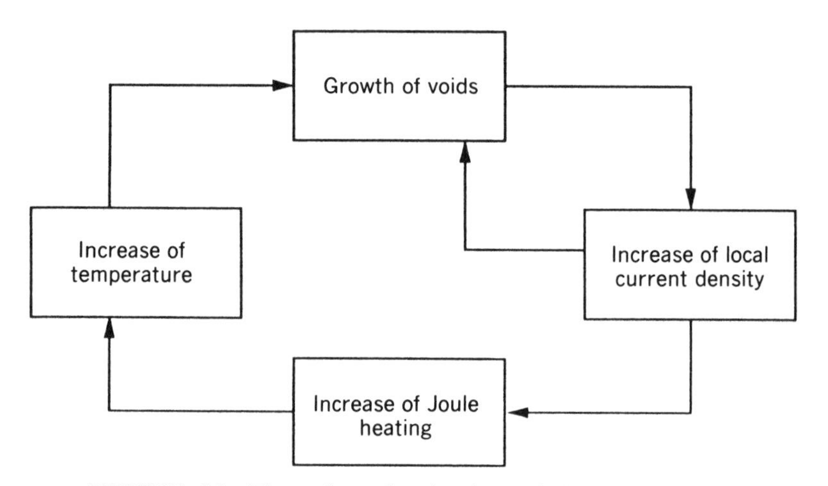

FIGURE 6.3. Thermal acceleration loop of electromigration.

situ microscopic examination of the development of voids and hillocks during stressing can also provide a direct measurement of the electromigration failure.

The empirical formula for the resistance change at the early stage of electromigration has been reported to be [4, 15]

$$R(T) = Cj^n te^{-Q/kT} \tag{6.2}$$

where C is a constant that depends on the material properties, temperature, and line geometry, n is an exponent of current density j, Q is the electromigration activation energy of the metal film, k is the Boltzmann constant, and T is the absolute temperature. Thus measurements of resistance change due to electromigration at different temperatures provide information on the activation energy, while results from different current densities yield the value of the exponent n. Much effort has been devoted to such a resistometric method in order to understand and characterize the early stage electromigration process. The interest in the early stage of electromigration is stimulated by the need to have a better understanding of the physics governing the phenomenon. At the early stage, the current density and temperature distribution have not been significantly distorted, and the system is still close to its initial state. There is little concomitant effects, such as thermal acceleration, involved at this stage. One may therefore gain more insight on the fundamental mechanisms of this phenomenon. Of course, correct information on early stages of electromigration depends on an accurate measurement of resistance, which sometimes represents a problem, and extra caution should be taken. The resistometric method will be discussed in the following sections.

The mean time-to-failure (MTF) of a metallization film has been well accepted to be

$$\mathrm{MTF} = C' j^{-n} e^{-Q/kT} \tag{6.3}$$

where C' is a constant depending on the film properties and testing conditions. Equation (6.3) is "Black's equation," derived by Black in 1969 [12]. Although a lifetime test is destructive and prohibits one from obtaining the physical origins of electromigration due to the involvement of other mechanisms, it has been the most commonly employed method to obtain information on metallization reliability. A prediction of circuit reliability would not be possible, at the present, with the nondestructive methods developed to study electromigration at the early stage. Therefore a complete lifetime test is still an important experiment for predicting circuit reliability. In addition, due to the randomness of the microstructure in polycrystalline metal films, the lifetime of an individual metallization line may appear to be random. Only the MTF, which is the median value of the lifetime distribution obtained from a large number of macroscopically identical metallization lines, is of interest from the IC reliability point of view, since it represents a statistically meaningful characteristic.

6.2.2 Electromigration in High-Power Devices

Electromigration may be especially important in high-power integrated circuits due to the large power dissipation and the high current density in high-power devices such as GaAs power MESFETs. In addition, this mass transport induced by a high current density takes place not only in the metallization lines of gates or ohmic contacts, but also at the interfaces of metal and semiconductor [44–59] since the interfaces represent discontinuity of structural properties. The large gate bias voltage and drain voltage for FETs at microwave frequencies result in a high current density in the gate metallizations on the order of 10^4–10^5 A/cm^2 [21, 48–51]. Although these values appear to be lower than the threshold current density commonly reported for electromigration, the higher current spikes during the AC signal plus structural inhomogeneities may still cause electromigration effects to arise. Beside, the relatively large DC bias current applied between source and drain may also result in electromigration failure, partially due to the elevated channel temperature caused by the high-power dissipation of the FETs. In power MESFETs, electromigration failure at both source/drain and gate have been observed [49, 51, 60–71]. Electromigration induces voids and openings in gate fingers, which result in a loss of control of device drain current and an increase in series resistance in ohmic contacts, which decreases I_{dss}, the device saturation current, and power gain. Taking into account the direction of the material transport induced by the electron wind, material depletion takes place at the end of drain contacts while accumulation can be observed on source metallizations. From the electrical performance point of view, the commonly reported degradations of GaAs FETs include (a) a decrease in the saturation current between source and drain, I_{dss}; (b) an increase in the leakage current, I_{dso}, associated with the impossibility of reaching the pinchoff conditions; (c) a decrease in gain accompanied by an increase in noise figure; and (d) an increase in the gate series resistance, R_g, which is partially responsible for the reduction of transconductance g_m. Almost all these degradations may be explained by electromigration in the ohmic and Schottky metallizations as well as at the metal–semiconductor interfaces, which will be discussed in the following examples.

Figures 6.4–6.6 show the typical evidences of electromigration-induced failure in the gate of GaAs power MESFETs. The devices under consideration are Al-gated X-band multigate transistors [44]. Each device has 20 gate fingers made of pure Al, 1 μm in length, 65 μm in width, and about 0.7 μm in thickness. A Ti/Pt interdiffusion barrier was used between Al-gate fingers and Au metallization pads. Ohmic contacts were obtained by Au-Ge-Ni/Au alloy and the surface was SiO_2 passivated. The devices were annealed with the source and drain short-circuited at 200°C and the maximum current density in the gate metallization was 5×10^5 A/cm^2. For both samples A and B shown in Figure 6.4, there is a large increase in the leakage current I_{dso} at a fixed drain-to-source voltage of 3 V after the annealing. The inserts give a comparison of the *I-V* characteristics of sample B before and after the test, which clearly shows the impossibility to pinch off the channel after the test. The mechanism for the above observed failure has been found to be the interruption of gate metallizations caused by electromigration.

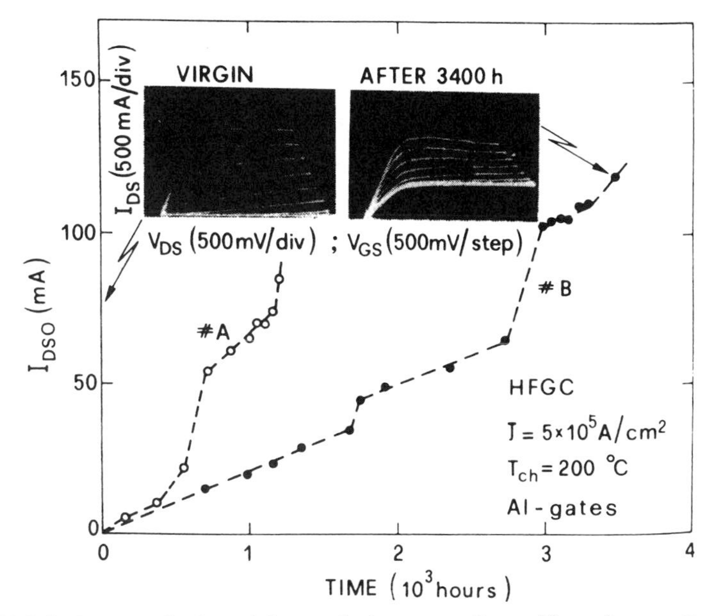

FIGURE 6.4. Increase in the minimum drain current I_{dso} at V_{ds} = 3v as a function of time for two samples after tested with gate current density of 5×10^5 A/cm^2 at 200°C. The inserts are the I-V characteristics of sample B before and after 3600 hours test. After C. Canali et al., *IEEE Trans. Elec. Dev.*, Vol. ED-34, 205 (1987). © IEEE 1987. Reprinted with permission.

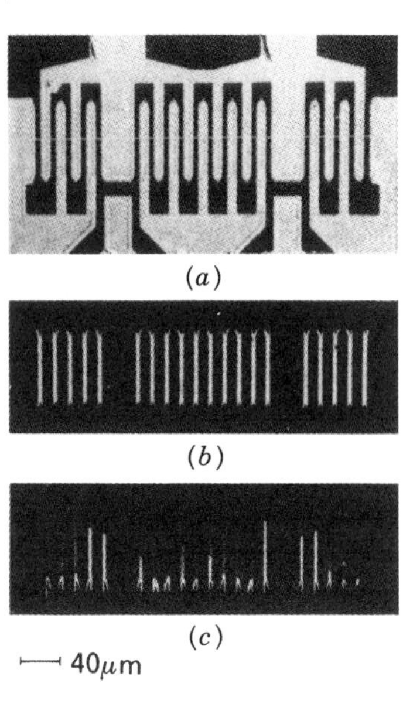

FIGURE 6.5. (*a*) SEM micrograph of the gate area of an Al multigate GaAs power MESFET; (*b*) EBIC image at the same magnification of an untested sample; (*c*) EBIC image of a degraded sample after 2000 test hours with a gate current density of 5×10^5 A/cm^2 at 200°C. Interruption of most of the gate fingers is evident. After C. Canali et al., *IEEE Trans. Elec. Dev.*, Vol. ED-34, 205 (1987). © 1987 IEEE. Reprinted with permission.

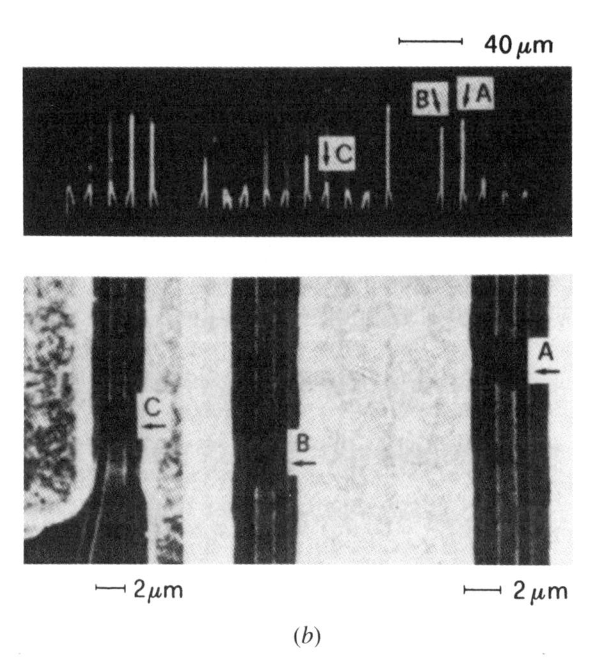

(*b*)

FIGURE 6.6. (*a*) EBIC image of a degraded sample after 2000 test hours with a gate current density of 5×10^5 A/cm^2 at 200°C; (*b*) magnified SEM micrograph showing points A, B, and C where the gate fingers are interrupted. After C. Canali et al., *IEEE Trans. Elec. Dev.*, Vol. ED-34, 205 (1987). © 1987 IEEE. Reprinted with permission.

Shown in Figure 6.5*a* is the SEM image of an untested sample, together with the EBIC image at the same magnification obtained by collecting the signal between the gate and the short-circuited source and drain shown in Figure 6.5*b*. Figure 6.5*c* depicts the EBIC image of a sample tested at 200°C for 2000 hours. The EBIC signal cannot be collected from those parts of the gate junction where the metal is depleted and therefore appears black in the figure. A more detailed SEM analysis of the breaking points is presented in Figure 6.6, where a higher magnification picture of the interrupted points, A, B, and C, is shown. The reduction in the effective area of gate contact explains the observed increase in pinchoff leakage current I_{dso} and the increase in gate series resistance R_g. The non-pinchoff condition can easily be understood since the channel can no longer be completely depleted due to the loss of the metallization in the gate area. Similarly, the increase in R_g is also a result of such a reduction of gate contact area. This failure mode has been found to be the most relevant for Al-metallized power MESFETs since it leads to catastrophic gate opening. Although this particular experiment is done for an aluminum gate, similar results may be expected in other gate metals such as gold.

Figure 6.7 shows the damaged drain metallization of a failed GaAs power transistor due to electromigration [51]. The transistor has the gate metallization of Ti/W/Au and the source and drain consisting of Au-Ge-Ni alloy covered by Ti/Pt/Au multilayer. The FET has been RF tested for 10,000 hours at a channel temperature of 184°C. The SEM micrograph shows the voids formed at drain metallization, as a result of electromigration. The enhanced view shown in Figure 6.7*b* indicates that the voids are essentially distributed along drain fingers and are approximately 1–3 μm in size. Figure 6.8 shows an example of the source metallization damage resulting from electromigration, which caused the device failure. The transistor has a gate metallization of Au-TiPt and Au-Ge-Ni ohmic contacts for the source and drain and has been subjected to RF tests. Electromigration-induced whisker formation has been observed on the source ohmic contact, which has formed a short-circuit bridge between the source and gate.

While the classical electromigration degrades the metallization reliability in the integrated circuits, the contact electromigration threatens the reliability of metal semiconductor contacts as shown in the previous paragraph. Both of them are critical in determining the circuit reliability.

A variety of techniques for the study of electromigration have been developed over the years and presently each of the techniques is good only for investigating some aspects of electromigration. No generalized

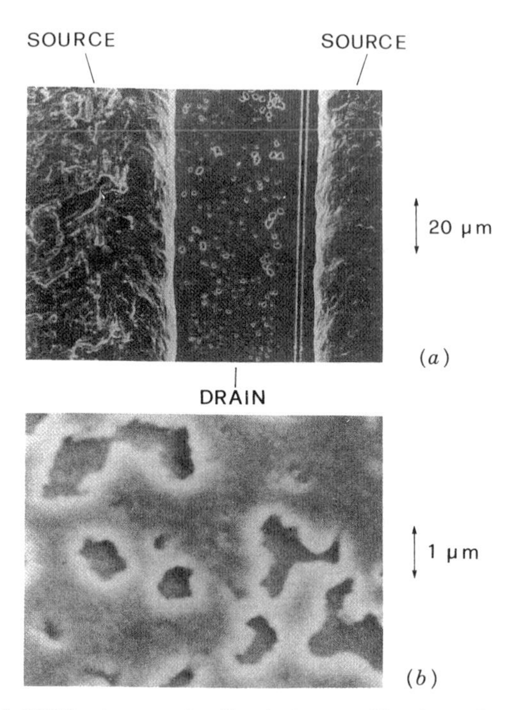

FIGURE 6.7. (*a*) SEM micrograph of a drain metallization of a GaAs power FET degraded after 10000 test hours at $T_{ch} = 185°C$ showing the presence of voids; (*b*) enhanced view of voids. After C. Canali et al., *Microelec. Rel.*, Vol. 24, 947 (1984). Reprinted with permission.

FIGURE 6.8. SEM micrograph of MMIC MESFET showing the whiskers developed on the source causing gate-source shorting. After A. Christou et al. [52].

approach has been demonstrated for use in a production environment. It should be pointed out, however, that a much better understanding of the basic mechanisms of electromigration has been obtained through the years of intensive research, although a relatively complete computer simulation has not been realized to include most of the important aspects of electromigration for which detailed models have been developed in the past. The next section is a review of those models and a presentation of the recent computer simulation results.

6.3 MODELING AND SIMULATION OF ELECTROMIGRATION

6.3.1 Modeling of Electromigration

Among various approaches developed to study the electromigration phenomenon, computer simulation is a powerful and efficient one. A computer simulation based on physical modeling makes it possible to take all the mechanisms associated with electromigration into consideration. It is therefore possible to study the dependence of the macroscopic quantities on all the related microscopic variables, which may not be measurable, to reveal the insights of this physical phenomenon. For example, it would be difficult to analytically or experimentally keep tracking along a conductor line the temperature and stress distribution, which are important factors affecting the degrading process as the electromigration-induced damage develops, whereas the instantaneous numerical solutions of the temperature and stress field equations provide such information; it would be impossible to know the grain boundary misorientation angles, θ's, and the inclinations with respect to electron flow, ϕ's, of all the grain boundaries in a conductor line, and it is therefore impossible to quantitize their effects on failure rate, while they can easily be found from the computer constructed film structures. Furthermore, only with the aid of a computer would it be possible to "conduct" a lifetime test of a large number of systems at the normal operating conditions to get statistically meaningful and physically realistic results of mean time-to-failure.

It is now understood that electromigration occurs by a diffusion-controlled mass transport process under a certain driving force mainly via grain boundaries [21, 28, 30, 72–75]. Such a mass transport therefore obeys the well established diffusion equation (Nernst–Einstein equation). The driving force, however, is more complicated than that in a pure diffusion process in which the concentration gradient of the moving species is the only component. There are a number of components in the electromigration driving force of which the predominant one is the "electron wind force." The electron wind force refers to the effect of momentum exchange between the moving electrons and the ionic atoms when an electrical current is applied through a conductor. When current density, which is equivalent to the electron flux density, is high enough, the effect of momentum exchange becomes significant, resulting in a noticeable mass transport: the electromigration. Nevertheless, while the atoms tend to move in the direction of the impulse during the momentum transfer, they also tend to move in the direction of the applied field since they are ionized. In the case of gold and aluminum, electron wind force dominates over the direct field force [38, 76]; the net electrical driving force is therefore in the direction of the electron wind. For simplicity, the term electron wind force is thus often used to refer to the net effect of these two electrical forces, as is the case for the following discussion. Besides the electron wind force, other components of the electromigration driving force result basically all from the inhomogeneities caused by the electron wind force induced damage, which will be discussed in the following sections. Most of the models reported in the literature include some components of these damage-induced driving forces, but none takes all of them into consideration due to the increased complexity.

Several types of computer simulation models on electromigration-induced degradation have been proposed [4, 19, 21–35]. One of the simplest is the probability model. In this kind of model, the statistical aspects of lifetime to failure such as the failure time distribution, the MTF, the standard deviation, and their dependence on line geometry are the main concerns and outputs [26, 30, 35]. The model is elaborated by a probability calculation. It assumes that a conductor line is composed of a number of unit elements connected in a certain manner, and the failure time of each unit element is an independent random variable following a lognormal distribution. With a predefined failure criterion, the failure time of each conductor film is calculated from the total failure probability of the assembly of all the unit elements. The probability model yields qualitative agreements on the line length and width dependence of MTF with experimental results and shows certain statistical features of conductor line failure in a simple way [26, 30, 35].

The three predominant mechanisms determining electromigration failure rate include (a) the metallurgical–statistical properties of the conducting film, (b) the thermal accelerating process, and (c) the healing effects.

The metallurgical–statistical properties refer basically to the microstructure parameters of the conductor film [19, 27, 34], such as grain size distribution or the number of grain boundaries, the distribution of grain boundary misorientation angle (mismatch angle) θ, and the grain boundary inclination angle ϕ with respect to the electron flow, as illustrated in Figure 6.9. The variation of all these microstructural inhomogeneities in a film causes a nonuniform atomic flow rate and the atomic flux divergence at the local regions where the number of atoms flowing in and out are not equal. Where a nonzero atomic flux divergence exists, there will be mass depletion or accumulation, leading to the formation of voids or hillocks, respectively. Microstructural inhomogeneities alone can cause nonzero flux divergence, regardless of the temperature distribution, concentration gradient, and so on. It is therefore the basic mechanism that any model should emphasize.

The thermal accelerating process refers to the acceleration of electromigration damage due to the damage-induced local temperature rising [22, 23, 27, 34]. A uniform temperature distribution along a conductor film is possible only before any electromigration damage occurs. Once a void is initiated, it causes current crowding because it reduces the cross-sectional area of the conductor. Since the Joule heating is proportional to the square of current density, the current crowding immediately leads to a local temperature rise around the void, which in turn increases the growth rate of the void due to the increase of both temperature and current density as discussed previously and illustrated in Figure 6.3. It should be pointed out, however, that temperature gradient may exist along a conductor line before any damage forms due to the existence of difference substrates or geometries. Care should therefore be paid to obtain an accurate temperature distribution along a conductor film in modeling the electromigration failure process.

The healing effects refer to all the mechanisms that contribute to the atomic flow in the direction opposite to electromigration. This backflow of mass begins to take place as soon as electromigration damage occurs. It tends to retard the electromigration rate when an electrical current is applied to the conductor and partially to heal the damage after the current is removed, as illustrated in Figure 6.10. The main reason for this backflow of mass is the inhomogeneities caused by electromigration damage, which moves the system away from its equilibrium state. Such a disturbed system tends to relax back to its original state to minimize its total system energy. One of the components in this mass backflow is the stress-induced mass flow discovered by Blech and Tai [39]. It was found that the mass accumulation and depletion induced by electromigration produce a stress gradient, which in turn induces a component of mass backflow. As a result, a threshold current density exists that cor-

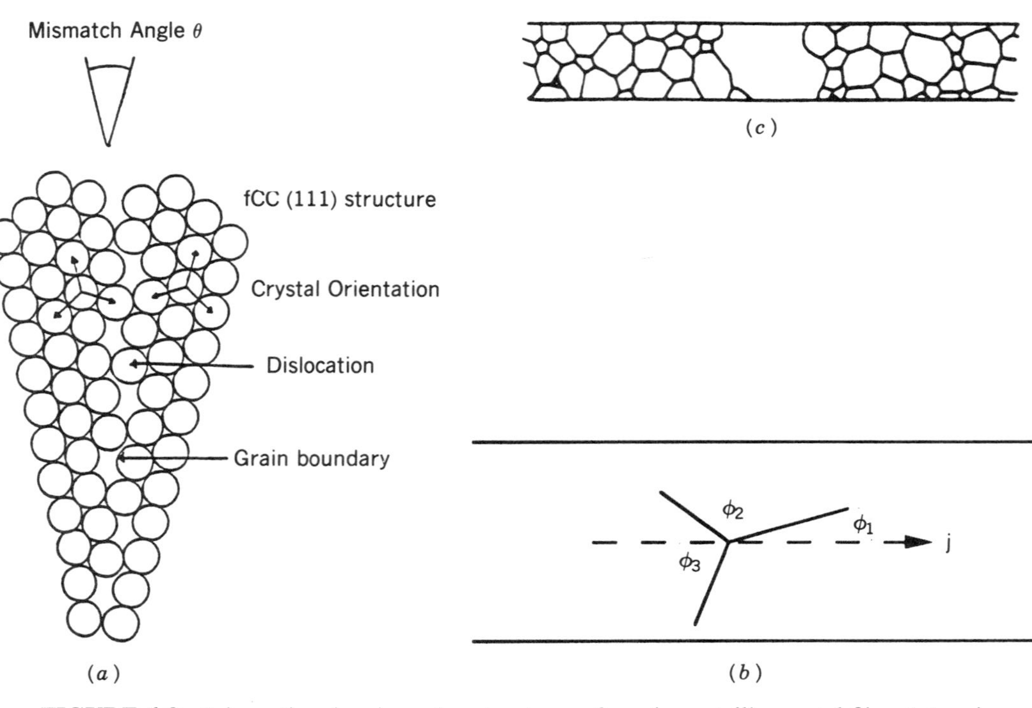

FIGURE 6.9. Schematics showing microstructures of a polycrystalline metal film: (*a*) grain boundary misorientation angle (mismatch angle) θ; (*b*) abrupt variation in grain size; (*c*) grain boundary inclination angle ϕ with respect to current flow.

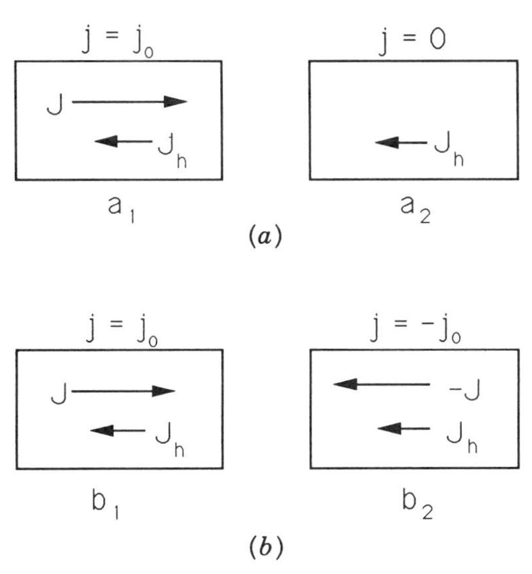

FIGURE 6.10. Schematical illustration of mass flows in (*a*) natural healing, (*b*) current-assisted healing. J and J_h are fluxes of electromigration and healing in a grain boundary, respectively.

responds to the minimum driving force required to overcome the effect of the stress gradient. Other examples of mass backflow include those driven by concentration gradient [72] and increase of capillarity [19]. Figure 6.10a_1 depicts electromigration flux J and the opposite healing flux J_h during the application of a current density $j = j_0$. Upon the determination of j, as shown in Figure 6.10a_2, flux J vanishes while the healing flux J_h may still exist for a time because of the existence of inhomogeneities caused by electromigration. Figure 6.10b illustrates another case where the stress current direction is changed to an opposite direction instead of being terminated. In this case, as shown in Figure 6.10b_2, the healing flux J_h may be assisting the electromigration flux J for a time. This has been confirmed experimentally [19]. Healing effect is important because a thorough understanding would make it possible to predict electromigration damage with the application of AC or pulsed DC current [31].

In the following sections, two-dimensional physical models will first be established based on the diffusion equation and only the dominant grain boundary diffusion will be considered. The driving force and the above-mentioned failure mechanisms will then be discussed in detail, which forms the basis of computer simulation. Statistical approaches will then be introduced to realize the models, and some typical results will be demonstrated.

Failure Sites of Nonzero Flux Divergence A two-dimensional conductor film can be considered as an assembly of grain boundaries and their intersections as illustrated in Figure 6.9*b*. Experimental observations have indicated that, in most cases, mass depletion and accumulation nucleate at grain boundary intersections such as triple junctions or stripe edges. The former would eventually lead to the formation of hillocks/whiskers or voids and the later hillocks/whiskers or cracks. The physical underline of such an observation is that grain boundary intersections are the locations where the mass flux would most easily diverge. There could be an abrupt change in grain size, which represents a change in the number of passes that mass moves through; there also could be a change in atomic diffusivity due to the different grain boundary structures. In both cases, the grain boundary would be the place where the amounts of mass moving into and out are not equal: a flux diverging site. Thus the failure sites, meaning the nucleation sites of mass accumulation or depletion, in this chapter are assumed to be at grain boundary intersections, in particular, in the triple junctions.

It is worth noting that whether a mass accumulation would result in a hillock or whisker mainly depends on the presence of a passivation layer on top of the metallization. When the passivation layer is present, the accumulated mass is under high mechanical pressure and therefore may extrude out, forming a whisker once a weak spot on the passivation layer is broken through. On the other hand, the mass tends to distribute itself in such a way as to minimize its total surface and therefore to form hillock or hump when the surface is free.

The atomic (ion) flux $\mathbf{J}$ is related to the total driving force $\mathbf{F}$ by the Nernst–Einstein equation [21]:

$$\mathbf{J} = \frac{ND}{kT}\mathbf{F} \tag{6.4}$$

where N is the atomic concentration, D is the diffusion coefficient, k is the Boltzmann constant, and T is the absolute temperature. Thus the atomic flux in the ith grain boundary is

$$J_i = \frac{N_i D_i}{kT_i} F_i \tag{6.5}$$

where the subscript i denotes the ith grain boundary.

At the beginning, the atomic concentration is assumed to be constant and is equal to that at the grain boundary—$N_i = N_{gb}$; and the temperature depends only on time and position in a conductor line—$T_i = T(x, y, t)$. For a given system and temperature, the diffusion coefficient in a grain boundary is determined by two factors: the grain boundary misorientation angle, θ, and the inclination angle of the grain boundary with respect to the electron flow, ϕ, as discussed in the previous section. A commonly accepted expression for D_i is

$$D_i = D_0' \sin \frac{\theta_i}{2} e^{-Q_0/kT} \qquad (\theta_i = 37°) \tag{6.6}$$

$$D_i = D_0' (\sin \theta_i) e^{-Q_0/kT} \qquad (37° < \theta_i < 60°) \tag{6.7}$$

where D_0' and Q_0 are the prefactor and the grain boundary activation energy of the film, respectively. In addition, the diffusion coefficient D_i can also be expressed as

$$D_i = D_0 e^{-(Q_0 + \Delta Q_i(\theta_i))/kT} \tag{6.8}$$

where D_0 is the prefactor, Q_0 is the average grain boundary activation energy, and $Q_i = Q_0 + \Delta Q_i(\theta_i)$ is the total activation energy of the ith grain boundary. The advantage of Equation (6.8) over Equations (6.6) and (6.7) is obvious. In both cases, the θ dependence can be represented by Θ:

$$\Theta = \sin\left(\frac{\theta}{2}\right) \qquad (\theta = 37°) \tag{6.9}$$

$$\Theta = \sin(\theta) \qquad (37° < \theta < 60°) \tag{6.10}$$

or

$$\Phi = e^{-\Delta Q_i(\theta_i)/kT} \tag{6.11}$$

Thus the diffusion coefficient D_i can be expressed as

$$D_i = D_0 \Phi_i e^{-Q_0/kT} \tag{6.12}$$

where Θ_i is defined by Equations (6.9) and (6.10) or (6.11), and it separates the structural difference of a particular grain boundary from the common part of the film diffusion coefficient. The constant D_0 depends on the material of the conductor film. The driving force due to the applied electrical field in the ith grain boundary is

$$F_i = Z^*qE(\mathbf{r}) \cos \phi_i = Z^*q\rho_0 j(\mathbf{r})[1 + \alpha(T(\mathbf{r}) - T_0(\mathbf{r}))] \cos \phi_i \tag{6.13}$$

where Z^*q is the effective charge of the ions that accounts for the net effect of electron wind force and the direct electrical field force, $E(\mathbf{r})$ is the applied field, T and T_0 are the temperature at time t and a reference temperature, respectively, ρ_0 is the resistivity of the film at the reference temperature, and α is the temperature coefficient of the resistivity.

Combining Equations (6.5), (6.12), and (6.13), the atomic flux in the ith grain boundary can be expressed as

$$J_i = \frac{N_{gb} D_0}{kT(\mathbf{r})} Z^*q\rho_0[1 + \alpha(T(\mathbf{r}) - T_0(\mathbf{r}))]\, \Theta_i j(\mathbf{r}) \cos \phi_i \, e^{-Q_0/kT} \tag{6.14}$$

By defining structural factor as

$$\Delta Y = \sum_{i=1}^{n_{gb}} \Theta_i \cos \phi_i \tag{6.15}$$

the flux divergence becomes

$$\nabla \cdot \mathbf{J} = \sum_{i=1}^{n_{gb}} J_i = \frac{N_{gb} D_0}{kT(\mathbf{r})} Z^*q\rho_0[1 + \alpha\Delta T(\mathbf{r})]\Delta Y j(\mathbf{r}) e^{-Q_0/kT} \tag{6.16}$$

Considering the stress effects by adding a component of counteracting current (i.e., the threshold current density j_c), Equation (6.16) is modified to

$$\nabla \cdot \mathbf{J} = \frac{N_{gb} D_0}{kT(\mathbf{r})} Z^*q\rho_0[1 + \alpha\Delta T(\mathbf{r})][j(\mathbf{r}) - j_c]\Delta Y e^{-Q_0/kT} \tag{6.17}$$

The number of the atoms accumulated in such an arbitrary intersection over a time period Δt is

$$\Delta N = -\delta h \, \nabla \cdot \mathbf{J} \, \Delta t \tag{6.18}$$

where δ is the grain boundary width and h is the film thickness. Thus the volume growth rate of the void formed at the grain boundary intersection under consideration is

$$\frac{\partial V}{\partial t} = \delta h \Omega_0 \nabla \cdot \mathbf{J} \tag{6.19}$$

where the Ω_0 is the atomic volume.

Current Crowding Effects The current density $j(\mathbf{r})$ in Equation (6.16) is usually not a constant. As soon as the electromigration-induced damage begins to form, current density becomes nonuniform due to the change of the line structure. For example, the current density around a void in a conductor line would be larger than that in the sections of the line where no damage has formed. As the damage grows, the whole field of current density keeps changing with time, and the nonuniformity of current density increases. Since the Joule heating is proportional to the square of the current density, the current crowding effect plays a double role in accelerating the degradation process. The instantaneous field of current density can be obtained by solving the two-dimensional Laplace equation for electrical potential $u = u(x, y)$ [34]:

$$\frac{\partial}{\partial x}\left(\sigma \frac{\partial u}{\partial x}\right) + \frac{\partial}{\partial y}\left(\sigma \frac{\partial u}{\partial y}\right) = 0 \tag{6.20}$$

where $\sigma = \sigma(x, y)$ is the electrical conductivity. In solving the above equation, the effects of the formed damage (voids) can be accounted for by assigning $\sigma = 0$ to those regions where material has been depleted and through which no current is being conducted. The current density field is related to potential by Ohm's law:

$$\mathbf{j}(x, y) = -\sigma\nabla + u(x, y) \tag{6.21}$$

Threshold Current Density In a drift velocity experiment, Blech and Tai [39] found that the threshold current density for electromigration, j_c, is proportional to the stress gradient created by mass depletion and accumulation at the two ends of the drifting saddle:

$$j_c \alpha \frac{\partial \sigma_{\mathrm{nn}}}{\partial x} \tag{6.22}$$

where σ_{nn} is the stress normal to the grain boundary. The threshold current density is also found to be inversely proportional to the stripe length, l:

$$j_c l = AT + B \tag{6.23}$$

where A and B are constants depending on film properties. Such an idea of stress-induced mass backflow can be extended into a broader scope. For a polycrystalline long metal line, mass accumulation and depletion do not appear only at the ends of the lines; rather, they occur at every vacancy source and sink such as grain boundary triple junctions. Thus it has been pointed out that length l in Equation (23) should be modified to be the ''dipole length'' of each pair of mass accumulation–depletion sites and it is therefore a variable along the conductor line. This idea has been suggested to explain the fact that not every triple junction in a conductor line, although there may be a nonzero flux divergence, would result in damage. The explanation is that each triple junction has its own value of j_c that may be quite different from that of the others. The applied electrical current may be high enough to overwrite some of the j_c's but not for the others, leading to the different responses in damage formation from all the triple junctions. However, the number of grain boundary intersections, mainly triple junctions, are usually quite large, and it is a good idea to work with a single average threshold current density for the whole conductor stripe.

Temperature Gradient Effects The variation of temperature with both time and location in a conductor line is one of the most significant factors in accelerating the electromigration degradation process because of the exponential dependence of the atomic diffusivity on temperature. Temperature gradient may exist well before any electromigration damage. For instance, closer to the bonding pads it is colder than at the center region of a line, and closer to a heat source, such as a gate junction or a resistor, it is hotter than in regions that are further away. As a result, flux divergence may occur solely because of the initial temperature gradient. As the damage starts to form, the additional Joule heat generated by current crowding increases local temperature rapidly. The increased temperature gradient in turn accelerates the growth rate of the voids/hillocks, as described in the positive feedback loop in Figure 6.3. One of the approaches simulating the evolution of temperature distribution is to solve the equation of temperature field assuming constant boundary conditions, that is, ambient temperature at the ends of the lines [34]:

$$\frac{\partial}{\partial x}\left(\tau \frac{\partial T}{\partial x}\right) + \frac{\partial}{\partial y}\left(\tau \frac{\partial T}{\partial y}\right) + j^2 \rho_0 (1 + \alpha \Delta T) = 0 \qquad (6.24)$$

where $\tau = \tau(x, y)$ is the thermal conductivity coefficient. This assumption may not be valid in many cases because there is no region on a conductor that can be kept at a fixed temperature. To avoid such a problem, the above equation needs to the modified so that the substrate temperature is considered constant rather than forcing the stripe ends to be at ambient temperature. Thus the temperature field equation becomes [27]

$$-\tau \frac{\partial^2 T}{\partial x^2} - \tau \frac{\partial^2 T}{\partial x^2} = Q_h - \frac{\lambda}{h}(T - T_s) \qquad (6.25)$$

where λ is the heat transfer coefficient between the film and the substrate, Q_h is the Joule heat generated per unit volume per unit time, and T_s is the substrate temperature obtained by

$$T_s = T_a + P_d R_{sa} \qquad (6.26)$$

where P_d is the power dissipated in the entire line, T_a is the ambient temperature, and R_{sa} is the thermal resistance between the substrate and the ambient conditions. The Joule heat Q_h can be obtained by

$$Q_h = j^2 \frac{Wh}{\lambda_0} R \qquad (6.27)$$

where λ_0 is the average grain size and R is the corresponding resistance, which will be discussed in a later section. The modification from Equation (6.24) to (6.25) can be extended further. In most cases, only the heat sink or package is at ambient temperature. The heat dissipation of the adhesive material between the chip in the package and the heat sink also need to be considered. The substrate temperature can now be obtained by

$$T_s = T_a + P_d \sum_{i=1}^{n_l} \theta_{R_i} \qquad (6.28)$$

where n_l is the number of layers between the chip and the substrate and θ_{R_i} is the thermal resistance of the ith layer.

Structural Factors of Triple Junctions The structural factor ΔY in Equation (6.16) reflects the effects of film microstructure on electromigration and plays an important role in determining the flux divergence. From the definition by Equation (6.15), the larger the variation in grain boundary diffusivity and misorientation angles, the greater the value of ΔY is. Thus the structural factor characterizes the grain boundary structural inhomogeneities. To analyze the structural factor in greater detail, it would be helpful to consider a general triple junction. A triple junction can be equivalently described in either way shown in Figures 6.11*a* and 6.11*b*. In Figure 6.11*a*, the triple junction is characterized by the misorientation angles and the three inclination angles with respect to the electron flow as discussed previously. In Figure 6.11*b*, the same triple junction can also be defined by the grain boundary misorientation angles, an orientation angle θ_0, and three relative angles ϕ_{ij}. With reference to Figure 6.11*b*, the structural factor can be reexpressed as

$$\Delta Y = \cos \theta_0 \, e^{-\Delta Q_1/kT} + \cos (\theta_0 + \phi_{12}) e^{-\Delta Q_2/kT} + \cos (\theta_0 + \phi_{12} + \phi_{31}) e^{-\Delta Q_3/kT} \tag{6.29}$$

where ϕ_{ij} is the relative angle between the ith and the jth grain boundary while $i, j = 1, 2, 3$. It can easily be shown that ΔY, and therefore the

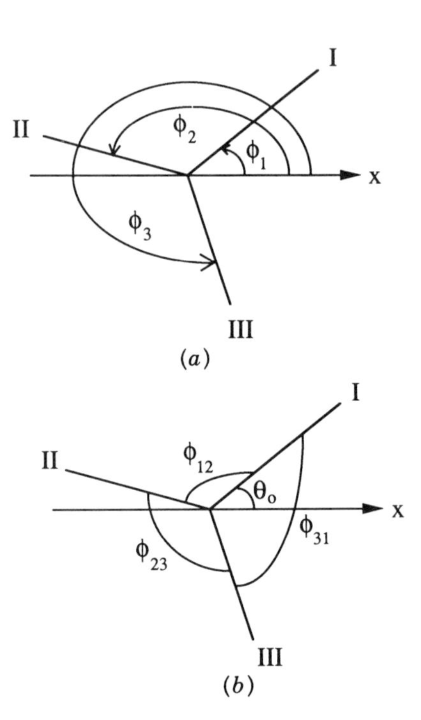

FIGURE 6.11. Top view of a triple junction described by two different sets of angular parameters.

flux divergence, vanishes for any value of θ_0 if the following conditions are met:

$$\Delta Q_1 = \Delta Q_2 = \Delta Q_3 = 0 \tag{6.30}$$

and

$$\phi_{12} = \phi_{23} = \phi_{31} = 120°\text{C} \tag{6.31}$$

Equation (6.30) represents a case of uniform diffusion coefficient for all the grain boundaries involved, while Equation (6.31) requires a perfect hexagonal grain structure. However, due to the variations in crystal orientation in a polycrystalline film, the activation energy and therefore the diffusion coefficient vary from one grain boundary to another. Also, the relative angles between two adjacent grain boundaries are not normally equal to 120°C due to the variation of grain sizes. Flux divergence is therefore likely to occur at a triple junction as a result of the existence of the structural inhomogeneities, leading to a mass depletion or accumulation. The knowledge of ΔY provides directions for one to improve electromigration reliability. For example, high-temperature annealing of the metallization helps to improve the uniformity of grain size and grain boundary diffusivity and consequently the flux divergence, which is a commonly employed method to enhance the lifetime of thin metallization films.

Concentration Gradient Effects Concentration gradient in grain boundaries results in another driving force component opposing further increase in concentration nonuniformity. The general expression for the concentration gradient induced force is therefore

$$F_c = -\frac{kT}{N} \nabla N \tag{6.32}$$

where N is the atomic concentration. The force due to the backdiffusion has also been proposed to take a value equivalent to the force exerted by a threshold electric field, E_{th}, defined by

$$F_c = Z^* q E_{\text{th}} = \frac{kT}{L} \ln \left[\frac{1 - F_g}{1 - F_l} \right] \tag{6.33}$$

where L is the length of the conductor line and F_g and F_l are the average fractional mass gain (negative porosity) and loss (positive porosity) along the line, respectively. Their values depend on the mass distribution along the conductor line, which is actually the solution of the mass transport equation determined by the total force including this component. Thus taking this force component into consideration, the mass transport equation would have to be solved iteratively [31]. An alternative approach is to include the effect of the concentration gradient together with that of the mechanical stress, and to build the total effect of the two into the threshold current density since they both induce backflow of mass and oppose the electron wind force.

It has been pointed out that the vacancy concentration reaches its supersaturation in a very short time period compared with the damaging process, usually within a few seconds [72]. After the supersaturation, vacancy condensation takes place, and the atomic concentration becomes a step function of position, that is, zero or a constant. The occurrence of the voids is taken care of by considering the zero electrical conductivity, the current crowding, and the associated temperature rise as well as those conditions discussed previously.

Resistance Change The evolution of structural damage can be monitored by the study of the volume of the accumulated or depleted mass using Equations (6.17) and (6.19). The resistance of a conductor line can be obtained immediately once the damage volume is known. In the calculation of resistance change caused by electromigration damage, it is assumed that voids increase the resistance of the conductor, whereas the hillocks have little effect on the value of resistance since very little current is likely to be diverted into a hillock. Strictly speaking, the resistance of the conductor line should be calculated by the integral of an elemental resistance over the entire line using the exact shape of every cross-sectional area. One of the simplified approaches, however, is to calculate the elemental resistance of each cell defined by a certain discretization approach and then add up the resistance of all the cells as if they are discrete resistors connected in analogy to an electrical circuit. The fractional resistance change of each single cell can be calculated based on the amount of mass that has been transported. For the simplest case where the void is assumed to be cylindrical in shape in the discretization of a conductor stripe of hexagonal grain structure shown in Figure 6.12*b*, the fractional resistance change of the cylindrical cell in Figure 6.12*a* can easily be shown to be

$$\left(\frac{\Delta R}{R}\right)_{ij} = \frac{w}{l}\left[\frac{2}{\sqrt{1-x^2}}\tan^{-1}\left(\frac{1+x}{1-x}\right)\right] - x - \frac{\pi}{2} \qquad (6.34)$$

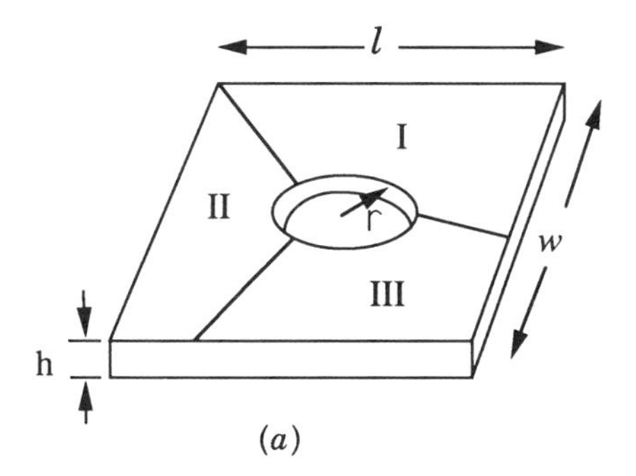

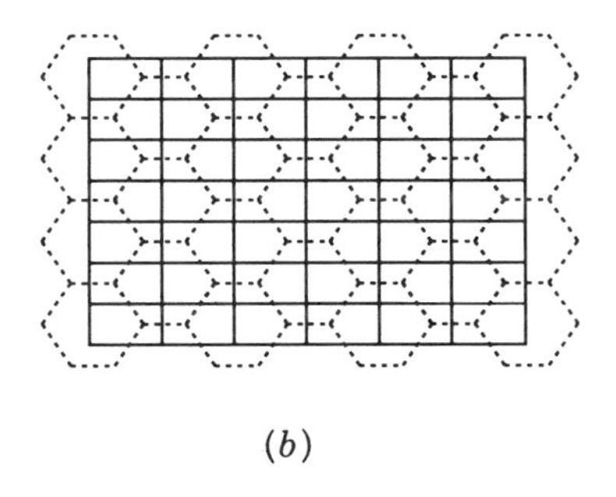

FIGURE 6.12. (*a*) A cylindrical hole of radius *r* formed at a triple junction; (*b*) discretization of a conductor stripe of hexagonal grain structure.

where w and l are the width and length of the triangular cell shown in Figure 6.12*a* and x is the diameter of the hole normalized by the cell width. To get the total resistance of the line, there are two possibilities of making the simplest electric circuit analogy: the "parallel of series" and "series of parallel" (PS and SP hereafter) models. In the SP model, the resistances of each cell along a column (from bottom to top) are first added up as resistors connected in parallel; the total resistance of the conductor stripe is then obtained by adding up the resistances of all the columns (from left to right). For the PS model, however, the two steps are reversed. Neither of the approaches is strictly correct since a metal film is made of a continuum material. When the number of grains is large and the discretization is fine enough, however, both represent a good approximation to the continuously distributed resistance of the conductor stripe. In the case where the length of the conductor line is much larger than the width, the SP circuit analogy should be employed. The total resistance in this case can be shown to take the following form:

$$R_T(t) = \frac{n_w R_T(0)}{n_l} \sum_{i=1}^{n_l} \left\{ \sum_{j=1}^{n_w} \left[1 + \left(\frac{\Delta R}{R_0} \right)_{ij} \right]^{-1} \right\}^{-1} \tag{6.35}$$

where R_0 is the initial value of the total resistance and $(\Delta R/R_0)_{ij}$ is given by Equation (6.34). n_w is the number of cells along the width and n_l is the number of cells along the length in the conductor stripe.

Failure Criterion The main goal of a computer simulation is the prediction of electromigration reliability. In order to calculate MTF, a properly defined failure criterion is needed. A traditional criterion defining failure is a complete opening of the conductor line under consideration. For example, when solving the electrical potential field, Equation (6.20), if the potential gradient is zero for two adjacent vertical grid lines used for the discretization, it signifies that no current is flowing through and therefore an open circuit has occurred [34]. Alternatively, the failure can be defined as when the maximum crack length reaches the width of the line or the maximum film temperature rises to the melting point of the conductor [27].

In most of the practical cases, the early stage of damage is of more concern. A complete open circuit is beyond interest and requires extensive computation time. The failure criterion may then be defined to correspond to a certain percentage increase in resistance [31], or a certain percentage of mass loss in the worst section along the conductor line [19]. As an example, the mass-loss criterion for the quasi-hexagonal grain network shown in Figure 6.12*b* is illustrated as follows. Defining a "column diameter" that is equal to the sum of all the void diameters along a column (perpendicular to the length of the line), the failure is then said to have occurred when the maximum column diameter reaches a certain percentage of the linewidth:

$$\text{Max}\left(\sum_{i=1}^{n_w} d_{ij}\right) = \frac{W}{f} \qquad j = 1, 2, \ldots, n_l \tag{6.36}$$

where d_{ij} is the diameter of the void in the cell located in the ith row and jth column and $1/f$ is the portion of the cross section lost in the column. The resistance failure criterion is straightforward and is more frequently followed since it makes the comparison with experimental results easier.

6.3.2 Computer Simulation of Electromigration

Computer Construction of Conductor Stripes The first thing to do in a computer simulation of electromigration is to generate a geometrical pattern that simulates the grain structure of a polycrystalline thin metal film. Since the mass transport in electromigration is mainly through grain boundaries and, in most of the applications, a thin film can be considered to be two-dimensional (neglect the variation across the thickness of the film), the problem is equivalent to the generation of an appropriate grain boundary network. The commonly accepted method is the "Voronoi polygon" method [77]. The method has been improved and modified by subsequent users [29, 34] and has become a powerful tool in the construction of polycrystalline films.

The method begins with laying down a random distribution of points on a discretized stripe surface at a prescribed cell density, namely, the number of points per cell. These points can be viewed as the crystal seed points (i.e., nucleating centers). The axes of the coordinate system are first moved to the central point. Rays joining adjacent points with this point and their respective perpendicular bisector lines are then calculated. The closest line is then realized. Next, the intersections of this line with all the others are calculated. The intersection point that has the shortest distance along this line (i.e., closest to the respective ray) forms the first corner of the polygon. Following a similar procedure, the rest of the corners of the polygon can be calculated. The entire edge of the first polygon is then traced out. The next polygon can be generated by moving the coordinate system to another point and then following the same steps so as to generate the rest of the polygons to cover the entire stripe. The points enclosing a polygon generated this way are the sets of points that have equal distances from the cell's seed as well as from the seeds of the neighboring cells. These cells then represent a carpet of metal grains. The way that the polygons are produced simulates the nucleating–growing–equilibrating process of a metal film during the deposition. Figure 6.13 shows a typical grain network generated using this approach [34].

When only the grain boundary intersections are of interest, for example, when the flux divergence is assumed to exist only a triple junctions and the linkage of all the grains is less important, the film construction procedure may be simplified so that only the grain boundary intersections are generated [4, 19]. In this approach, the conductor line is first defined and discretized into cells. Seeds are then randomly distributed into the cells according to a prescribed seed density. The seeds

FIGURE 6.13. A typical result of a simulated grain boundary network in a thin film conductor line constructed using Voronoi polygon technique. After P. J. Marcoux et al., *Hewlett-Packard J.*, 79 (1989). © Copyright 1989 Hewlett-Packard Company. Reprinted with permission.

now represent the grain boundary intersections. The inclinations of grain boundaries defining each intersection must then be specified, which will be discussed later. When the number of grains and therefore the grain boundary intersections are large, such a simplified approach may significantly reduce the computation time, while the results produced can still be statistically accurate [19]. This approach is termed the intersection-lattice method.

In both approaches discussed above, the grain boundary intersections are characterized by the defining boundaries. Once the microstructure of the grain boundaries is known, the flux divergences at the intersections are determined for a given stressing condition. Other ways of generating a grain boundary network have also been reported for special applications [26, 30].

Generation of Random Grain Boundary Structures Once the grain or grain boundary network is defined, the grain boundaries also need to be constructed, which would determine the structural factors at the intersections. For simplicity, the intersections again will be assumed to be triple junctions in the present section. As discussed earlier, a grain boundary is defined by its misorientation angle, θ, and the inclination angle with respect to electron flow, ϕ.

In a statistical computer simulation, these angles are randomly assigned to each grain boundary following certain distribution functions. The experimental evidence on the distribution functions of these parameters is not available; it is therefore often assumed to follow a uniform distribution [19, 34]. Other distribution functions may sometimes be required; for example, the relative angle between two adjacent grain boundaries is related to the grain size variation. It is therefore reasonable to define the ϕ's to follow a lognormal distribution since the grain size variation has been found experimentally to follow a lognormal distribution [26].

A triple junction is defined by three diffusivities for the three grain boundaries and another three inclination angles. The three diffusivities can be characterized by θ_1, θ_2, and θ_3 as the three grain boundary misorientation angles when diffusivity is defined by Equations (6.6) and (6.7), or ΔQ_1, ΔQ_2, and ΔQ_3 for the variation of grain boundary activation energies when the diffusivity is defined by Equation (6.8). When the intersection-lattice method is applied, the three inclination angles can be defined by ϕ_1, ϕ_2, and ϕ_3 in the case of Figure 6.11*a* or ϕ_{12}, ϕ_{23}, and θ_0 in the case of Figure 6.11*b*. However, there are certain limitations on the values of these parameters set by the microstructure physics. For example, the θ_i's ($i = 1, 2, 3$) should not exceed 60° due to the nature of (111) orientation, which confines the ΔQ's to be within a certain range; in the case of defining the three relative angles, the ϕ_{ij}'s ($i, j = 1, 2, 3$) should not be too far away from their equilibrium value, 120°C, because most of the conductor stripes are metallurgically stable. Furthermore, at each grain boundary intersection, a force balance is required. Denoting grain boundary surface tension for the three boundaries as γ_1, γ_2, and γ_3, the following condition represents the force balance between the three grain boundaries:

$$\frac{\sin \phi_{12}}{\gamma_1} = \frac{\sin \phi_{23}}{\gamma_2} = \frac{\sin \phi_{31}}{\gamma_3} \tag{6.37}$$

The value of γ has been reported generally to be in the range of 300–500 dyn/cm [78]. These values can be viewed as the limits when γ is randomly assigned in the intersection-lattice method. Thus the confinements of the microstructural parameters in defining grain boundaries associated with each triple junction can be summarized as follows:

$$120° - \Delta\phi_{\max} \le \phi_{12}, \phi_{23}, \phi_{31} \le 120° + \Delta\phi_{\max} \tag{6.38}$$

$$\sum_{ij=1}^{3} \phi_{ij} = 360° \qquad (\phi_{ij} = \phi_{ji}) \tag{6.39}$$

$$0° \le \theta_0 \le 360° \tag{6.40}$$

$$-\Delta Q_{\max}\ (\text{eV}) \le \Delta Q_i \le \Delta Q_{\max}\ (\text{eV}) \qquad (i = 1, 2, 3) \tag{6.41}$$

$$300\ \text{dyn/cm} \le \gamma_1 \le 500\ \text{dyn/cm} \tag{6.42}$$

$$\gamma_2 = \gamma_1 \frac{\sin \phi_2}{\sin \phi_1} \tag{6.43}$$

$$\gamma_3 = \gamma_1 \frac{\sin \phi_3}{\sin \phi_1} \tag{6.44}$$

Implementation of the Physical Model To implement the physical model, the grain structure of a conductor line is first constructed, followed by the construction of the grain boundaries in terms of the metallurgical statistics such as the diffusivities and grain boundary inclinations. The structure factor ΔY's are then calculated for all the grain boundary intersections over the entire line. At the beginning of each time interval, Δt, the current density distribution is first calculated following Equations (6.20) and (6.21). The temperature distribution is then calculated as discussed previously. The flux divergence and therefore the mass that has been transported in and/or out at each grain boundary intersection can then be obtained. The volume of all the voids and the line resistance are monitored at every step. The failure criterion is checked at the end of each time interval. If the line has failed, the time is recorded as the failure time of the line under consideration; otherwise, the above steps are repeated for the next time interval, Δt, until the line fails. The same calculation procedure will then be applied to the next line.

To obtain the mean time-to-failure, MTF, for a particular type of conductor film, failure time should be calculated for a large number of lines to obtain statistically meaningful results. These lines should be macroscopically identical; that is, they should have the same dimensions, same average activation energies, and the same prefactors for diffusivities, and should be stressed under the same conditions, such as the initial current density and temperature. The statistical properties should also be the same, meaning that the randomly generated quantities should follow the same distribution function. For specific tasks, certain approximations both in the physical models and line construction procedures can be utilized as discussed in the earlier sections, which usually greatly reduce the computation time.

Examples of Simulation Results Figure 6.14 illustrates a simulated result of the volume distribution of voids and hillocks [19] in a gold conductor film. The simulation is based on the simplified model for the early stage of electromigration damage. The line is constructed by the intersection-lattice method. The volume of the voids here is normalized by the cell volume, $w \cdot l$. The negative values represent hillocks, while the positive values are the voids. The vertical axis is the number of triple junctions normalized by the total number of triple junctions over the entire line. The result is for $t = 259$ hours at 250°C, with $\Delta\phi_{\max} = 10°$, $\Delta Q_{\max} = 0.02$ eV, and $j = 3$ MA/cm^2. The line dimensions are 20 cells across the linewidth, and 1400 cells along the length. For an average grain size of 1 μm, it corresponds to a line 8.7 μm wide and 1050 μm long. The figure shows that the distribution of voids and hillocks is symmetric about zero as expected due to mass conservation. Comparison of volume distribution with experimental results is possible

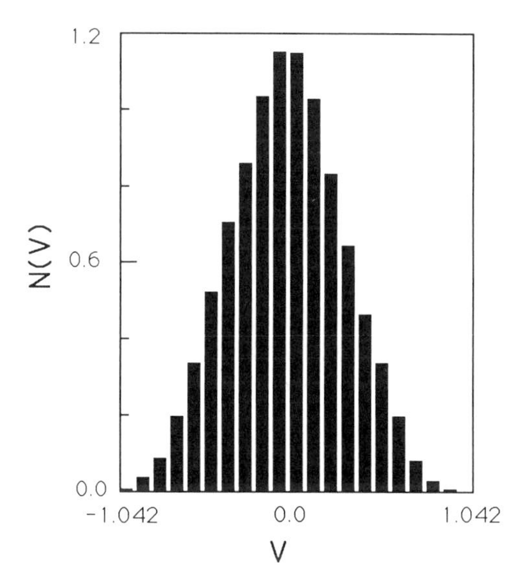

FIGURE 6.14. A simulated result of the volume distribution of voids and hillocks in a gold conductor film. After P. F. Tang, Ph.D. thesis, Carnegie–Mellon University, Pittsburgh, PA 1990.

with the aid of microscopic techniques, such as scanning electron microscope (SEM) or transmissional electron microscope (TEM) [19].

Figure 6.15 shows three simulation results of fractional resistance increase with time (solid lines) compared with the experimental data (discrete symbols) [4] based on the same conditions used for Figure

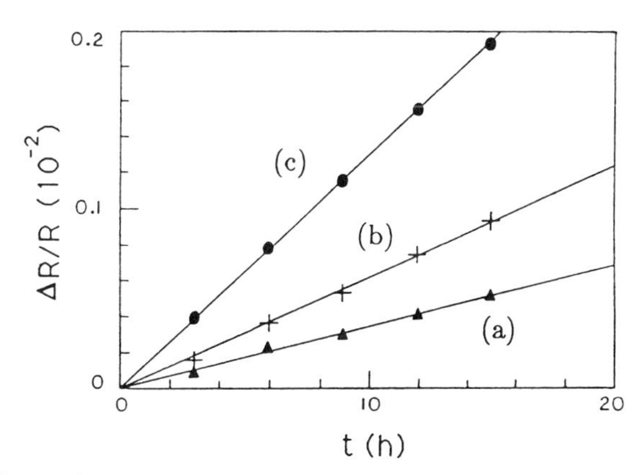

FIGURE 6.15. Typical computer simulation results of fractional resistance versus time compared with experimental data (solid dots). After P. F. Tang et al., *Proc. Mater. Res. Soc.*, Vol. 167, 341 (1989).

6.14 unless otherwise mentioned. The total resistance in this simulation has been obtained by the SP circuit analogy, as discussed earlier. The stressing current density in this simulation is 3 MA/cm^2, the temperature is 221°C for result (*a*), 240°C for (*b*), and 258°C for (*c*). The results show a quite linear relationship between the resistance and time at the early stage (i.e., up to 0.2%) of electromigration, and a good agreement when compared with the experimental data. The linearity has also been observed by other workers. However, because the result is from a simplified model that is only valid for the early stage, the resistance predicted may be overoptimistic when applied to later stages, that is, for more than 10% increase in fractional resistance change, due to the neglect of the current crowding and local heating effects.

The effects of variation of grain boundary diffusivity and grain size on the degradation rate can be illustrated by Figure 6.16, which shows the fractional resistance change for two different sets of microstructural parameters. The system simulated and the stress conditions are identical to that used in Figures 6.14 and 6.15, except for the grain boundary structures. The two groups of curves in Figure 6.16 show the resistance change versus time normalized by a time constant t_0, which is determined by the material and stress conditions. The use of t_0 is to subtract the effect due to stressing and show therefore the structural effects only. Group (*a*) is the result for $\Delta Q_{max} = 0.01$ eV, with the three curves corresponding to $\Delta\phi_{max} = 20°$, 15°, and 10° from top to bottom; group (*b*) represents the case for $\Delta Q_{max} = 0.02$ eV, with the same correspond-

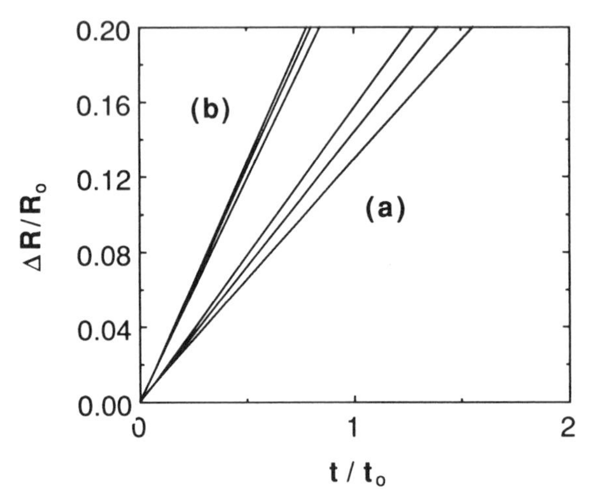

FIGURE 6.16. Simulated results of fractional resistance change for two different sets of microstructural parameters. Group (*a*) shows the result for $\Delta Q_{max} = 0.01$ eV and $\Delta\phi_{max} = 20°$, 15°, and 10°C for the three curves from top to bottom; Group (*b*) is for $\Delta Q_{max} = 0.02$ eV and the same corresponding $\Delta\phi_{max}$ as in group (*a*).

ing angle variations as in group (a). The positions of the two groups of curves are mainly determined by the ΔQ_{max} values, since the two groups are separated further than that within the curves in each group, which suggests that the variation in grain boundary diffusivity plays a more significant role in the structure inhomogeneities than that of the grain size variation.

The mean time-to-failure, MTF, was once believed to follow a lognormal distribution. Further careful investigations have shown that MTF does not strictly follow lognormal. Lognormal distribution, however, is still regarded as a good approximation in most cases. Figure 6.17 [19]

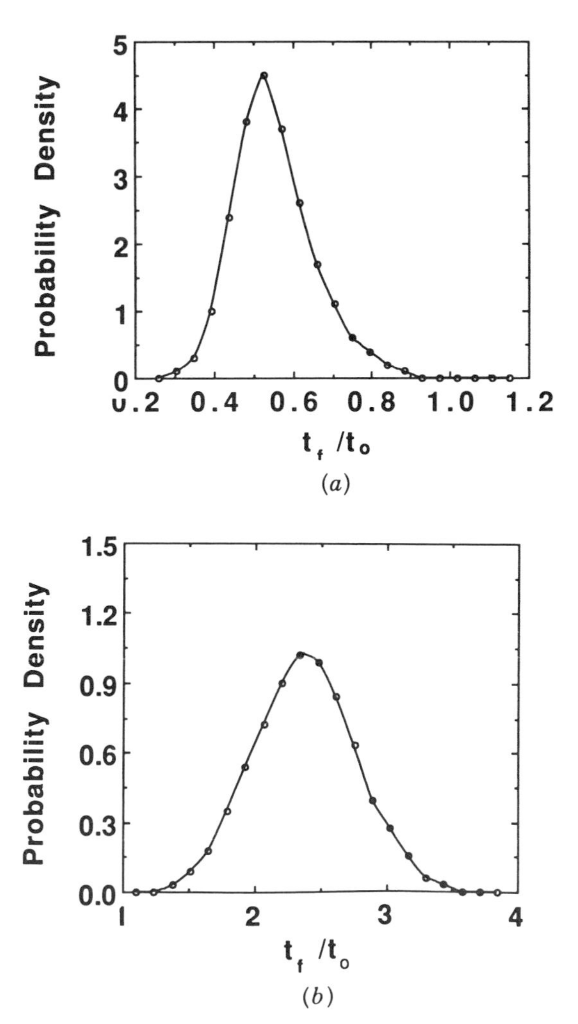

FIGURE 6.17. Simulated probability distributions of normalized failure time for lines of 1 × 50 segments (*a*), 15 × 50 segments (*b*). After P. F. Tang, Ph.D. thesis, Carnegie–Mellon University, Pittsburgh, PA 1990.

presents the results based on a simplified model for the early stage of electromigration, with the conductor line constructed by the intersection-lattice method. Figure 6.17*a* shows the normalized probability density versus the normalized time for a conductor line 1 cell wide and 50 cells long, while Figure 6.17*b* is for a line measuring 15 cells by 50 cells with all the other parameters and stressing conditions kept the same. Each distribution is obtained based on 3000 macroscopically identical lines to eliminate statistical errors. The probability density function curve appears to be closer to a normal distribution in the case of Figure 6.17*b*, while it is somewhere between lognormal and normal in the case of Figure 6.17*a*. The conclusion here is that the lifetime distribution is neither normal nor lognormal, which has been pointed out previously [86]. Nevertheless, when the cumulative probability is plotted, which is normally of greater interest, lognormal distribution often seems to be a good approximation, as can be seen in Figure 6.18 [4]. Figure 6.18 depicts the cumulative failure probability for three sets of lines plotted on a lognormal scale. The line sizes considered are (a) 40 × 50, (b) 15 × 50, and (c) 15 × 350 cells of width times length, respectively. Line (b) represents the same conductor line as in Figure 6.17 under the same stressing condition, but it fits a straight line fairly well especially in the range of 2–98%. In fact, the other two lines also fit approximately to straight lines in this range, although they all deviate somewhat more.

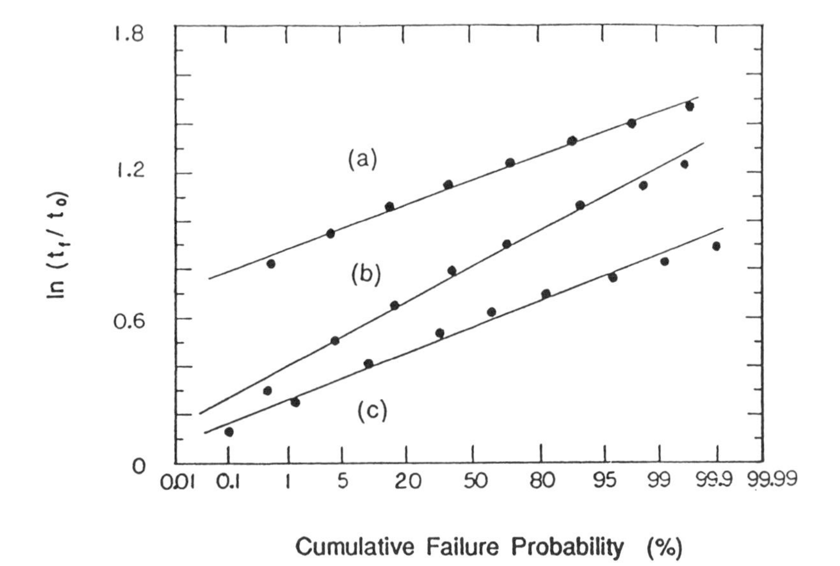

FIGURE 6.18. Simulated results of the normalized failure times shown by dots on a lognormal scale for lines of 40 × 50 (*a*), 15 × 50 (*b*), and 15 × 350 (*c*) segments. After P. F. Tang, Ph.D. thesis, Carnegie–Mellon University, Pittsburgh, PA, 1990.

Thus, if we are concerned only with MTF, which is defined as the 50% failure probability, the lognormal distribution may be considered as a good approximation. However, if the early failure is of central importance (i.e., less than 2%), more careful investigations are needed.

The cumulative probability plot shows both the MTFs and the standard deviations, σ's, while the latter can readily be determined from the line slopes. The line geometry dependence of MTF and σ can also be obtained by such plots. For instance, lines (a) and (b) shown in Figure 18 indicate that decrease of linewidth from 40 to 15 cells decreases MTF but increases σ; while lines (b) and (c) shown in Figure 6.18 suggest that increase of line length from 50 to 350 decreases both MTF and σ. The width and length dependence shown here are consistent with the experimental observations [35]. The dependence of MTF and σ on the line dimensions can easily be understood. As the line length increases, there is a larger chance for the line to contain a triple junction, where the largest possible flux divergence exists; the MTF is therefore decreased. As the linewidth is increased, on the other hand, the probability for those triple junctions of large flux divergence to all appear in the same column to form a large column diameter is lowered because the flux divergence tends to distribute more uniformly when the number of triple junctions is increased. Since the failure time actually depends on the maximum column diameter, or the total damage in the most-damaged column, MTF then increases. In either case, the larger the number of grains or triple junctions, the lower the statistical standard deviation is.

The linewidth dependence of MTF and σ changes drastically when the linewidth becomes smaller than the average grain size. Presented in Figure 6.19 is the simulation and experimental results showing this feature. The lines simulated are 100 μm long, 0.5 μm thick, and constructed by the stacking grains method [27]. The model accounts for both current crowding and local heating effects. The threshold current density is assumed to be of the form of Equation (6.23) with A and B being the same as the experimentally reported values for an Al conductor film [38]. Lines of width 1, 2, and 4 μm are simulated. All the results showed a steep increase as the width becomes less than the average grain size. These behaviors of MTF and σ are called the "bamboo effect" [36]. The name comes from the fact that when the linewidth becomes comparable with the average grain size, most of the grain boundaries are close to being perpendicular to the line length, and the grain structure of the line is similar to a piece of bamboo. The number of paths for atoms to move in the direction of the electron wind force is greatly reduced in this case; therefore the MTF increases. On the other hand, since the ratio of the total grain boundary area to the surface area

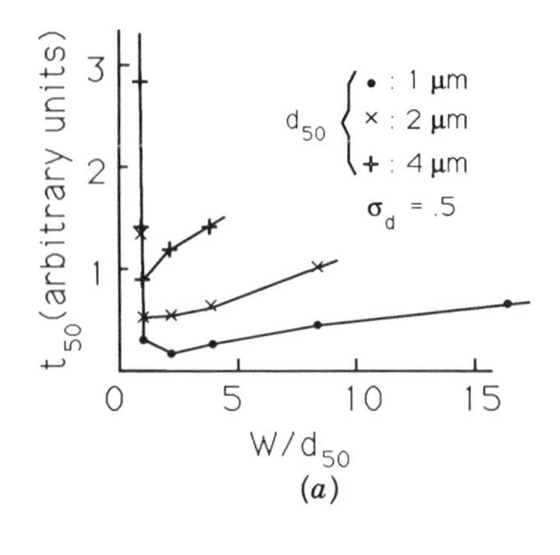

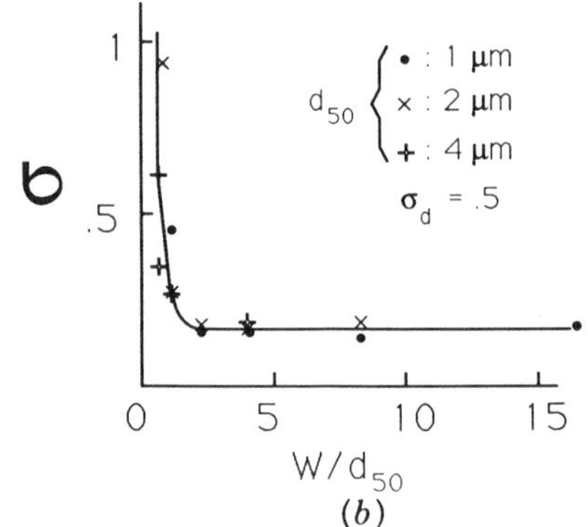

FIGURE 6.19. (*a*) Line width dependence of median time to failure (t_{50}) (calculations); (*b*) Line width dependence of a standard deviation of failure-time (σ) (calculations). After Nikawa et al., *IEEE IRPS,* 175 (1981). © 1981 IEEE. Reprinted with permission.

of the stripe is also reduced, the statistical error is increased. This type of dependence of MTF and σ on linewidth has been observed experimentally as shown in Figure 6.19*b*.

6.4 EXPERIMENTAL STUDY OF ELECTROMIGRATION

6.4.1 Resistometric Measurement

The resistance of a metal interconnect in integrated circuits is sensitive both to its microstructural and geometrical parameters. When electromigration creates voids, cracks, or hillocks within a metal line, the line microstructure and geometry are distorted and the line resistance changes. The Joule heating generated by the structural damage changes the resistivity of the metal, which results in a change in the total line resistance. Thus measurement of the resistance change is one way to monitor the electromigration failure process. As discussed in the previous section, the resistometric method is one of the techniques suitable for the study of the early stage of electromigration, where the conditions of uniform current density and temperature have not been disturbed significantly. Under the assumption of the early stage of electromigration—that is, the dimensions of the maximum voids are much less than the linewidth—it is straightforward to show that the line resistance of a thin film conductor has the following form [4]:

$$\frac{\Delta R}{R} = \frac{R - R_0}{R_0} = Cj^n te^{-Q/kT} \tag{6.45}$$

where R_0 is the line resistance at a reference temperature, C is a constant depending on the film geometry, the grain boundary structure, and the grain size, t is the time that the conductor has been stressed, and the other constants have been defined previously. The above expression has also been derived through different approaches [2, 3]. Fundamental analysis has shown that the exponent n is equal to unity [4]. However, experiments often result in larger values [2, 4]. This discrepancy is believed to be due to the Joule heating associated with the electromigration-induced structural change [23, 24].

As indicated by Equation (6.45), at a fixed current density j, which can be achieved reasonably well with a commercially available constant-current power supply, resistance measurements at different temperatures provide information on the value of the activation energy Q, and the measurements at different current density levels determine the value of the exponent n. Figure 6.20 shows a typical result of the tour-probe resistometric measurements [4]. The specimen tested in an Au thin film that is 7.5 μm wide, 1000 μm long, and 0.3 μm thick, deposited on a semi-insulating GaAs wafer. Such a test structure has been designed to simulate interconnection stripes between GaAs MESFETs on a semi-insulating substrate. The experiments have been conducted at various temperatures ranging from 210 to 280°C and at three different current levels, namely, 1×10^6, 2×10^6, and 3×10^6 A/cm^2. The slope of each line in Figure 6.20 is proportional to the activation energy in the following way:

$$\text{slope} = -\frac{Q}{1000} K \tag{6.46}$$

For a fixed temperature, the ratio of the slopes at different currents provides the value of the exponent n. From this experiment, the activation energy has been determined to be 0.73 eV with an n value ranging from

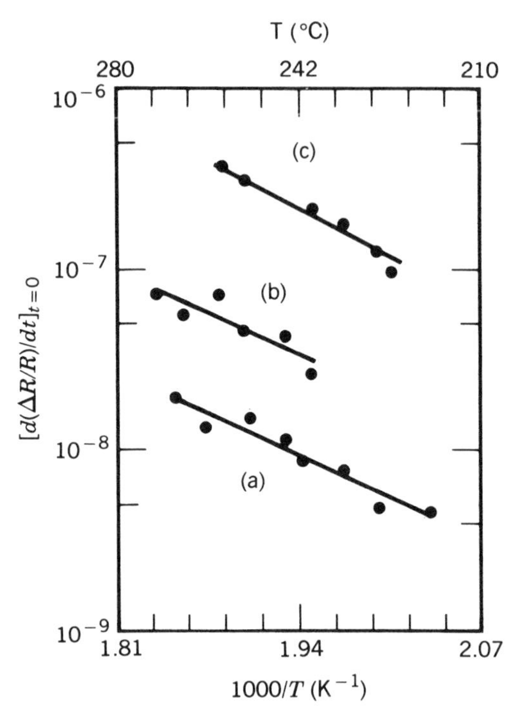

FIGURE 6.20. Initial changing rate of fractional resistance versus reciprocal of absolute temperature for current densities (*a*) 1, (*b*) 2, and (*c*) 3 × 10^6 A/cm^2. The resulted activation energy Q is 0.73eV. After P. F. Tang et al., *Proc. Mater. Res. Soc.*, 167, 341 (1989).

2 to 4. In the resistometric measurements, special care should be taken to ensure accurate temperature determination. The experimental accuracy and some practical approaches to improve the accuracy will be described next.

A typical test pattern designed for the four-probe measurement is shown in Figure 6.21. A constant current I is applied to the conductor line through contact pads I_1 and I_2 while the voltage drop across the conductor line, V, is tapped from pads V_1 and V_2. The line resistance, R, is therefore equal to the voltage drop divided by the current: $R = V/I$. In the case where the current I is measured by the voltage drop V_0 across a constant resistor R_0, the resistance R is then equal to R_0V/V_0. The accuracy of the resistance measurement is determined from the voltage measurement and the stability of the applied current, as described by

$$\left|\frac{dR}{R}\right| = \left|\frac{dV}{V}\right| + \left|\frac{dI}{I}\right| \tag{6.47}$$

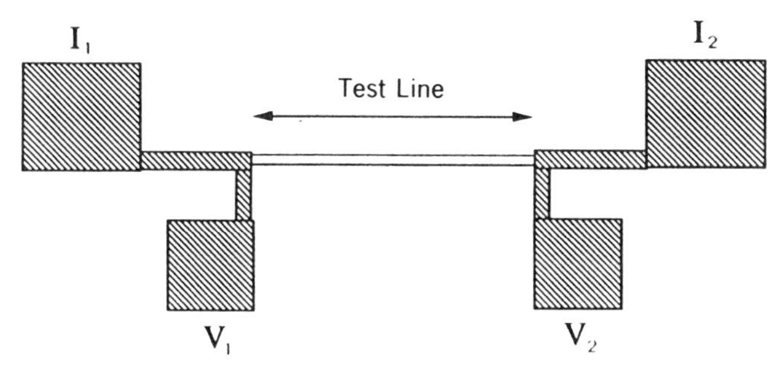

FIGURE 6.21. A typical sample pattern for 4-probe resistance measurement. The hatched areas are tapping pads.

or

$$\left|\frac{dR}{R}\right| = \left|\frac{dR_0}{R_0}\right| + \left|\frac{dV}{V}\right| + \left|\frac{dV_0}{V_0}\right| \tag{6.48}$$

Typically, it is possible to obtain the following precisions: $|dV/V| < 0.1\%$ and $|dR_0/R_0| < 0.001\%$. The accuracy of the resistance measurement can therefore be better than 0.1%.

Although the resistance can be measured to a high precision, the resistometric technique suffers from the resistance fluctuation caused by the fluctuation or inaccurate measurement of temperature. Temperature fluctuation is almost inevitable even in the "constant-temperature zone" of a furnace and the true value of the specimen temperature is sometimes difficult to obtain due to the small size of the conductor line in comparison with the size of the thermocouple used to monitor the temperature. As can be seen in Equation (6.45), the resistance is very sensitive to temperature. Assuming an activation energy of 0.7 eV and an inaccuracy or instability of temperature of $\pm 0.2°C$, the corresponding error in resistance is $\Delta R/R = 7 \times 10^{-3}$ at 200°C. This value is already comparable to resistance changes that may occur during the early stage of electromigration. Three approaches that can be used to solve this problem are introduced next.

In order to determine the real specimen temperature, the resistance calibration and thin film thermocouple measurements should be considered. In the temperature range of interest, the resistance is related to temperature by

$$R = R_0(1 + \alpha \, \Delta T) \tag{6.49}$$

where ΔT is the temperature rise from a reference temperature, R_0 is the resistance at the reference temperature, and α is the temperature coefficient of the conductor. The value of α may change from film to film since it depends on the microstructure of the polycrystalline thin film. Therefore α must be calibrated for each experiment. The calibration should be conducted under pulsed or very low current density in order to avoid any Joule heating of the conductor lines. The slope of R versus T yields α. With the known value of α, a true specimen temperature may be determined from its resistance value according to Equation (6.49). Note that the measurement of resistance has to be taken well before the electromigration-induced damage starts to happen. This is the simplest way of obtaining a true specimen temperature. However, it does not provide information on the instantaneous temperature of the sample when serious Joule heating and electromigration damage start. The approach is good only for the very early stage of electromigration, before the initial temperature distribution in the thin film is distorted.

A direct measurement of the true specimen temperature may also be realized by utilizing thin film thermocouples situated adjacent to the sample being tested [3]. The idea of a thin film thermocouple is schematically illustrated in Figure 6.22. Two thin metal stripes made of different materials, for example, Au and Ag, are so positioned that they overlap with each other at their tips. The overlapped tip area is located next to or on the specimen. The thermocouple should be fabricated by photolithography in order to achieve the small dimensions comparable to the sample. The voltage drop across the thermocouple is first calibrated at a series of known temperatures when the sample is uniformly heated without being electrically stressed. The calibration curve provides the instantaneous specimen temperature during any stage of electromigration testing. A limitation of this method is that temperatures

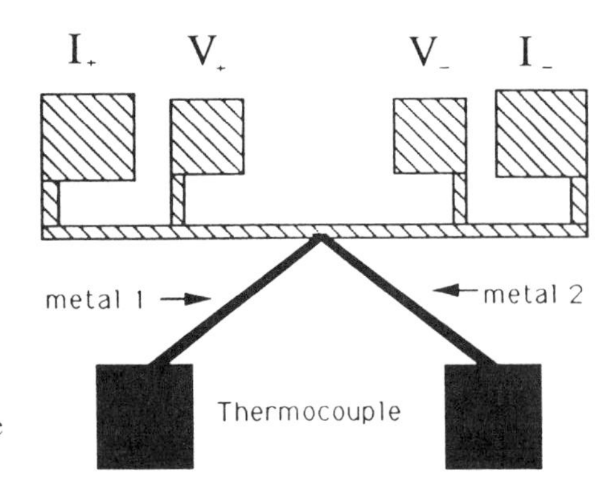

FIGURE 6.22. Thin film thermocouple in 4-probe resistance measurement.

only at locations near the thermocouple may be measured, which may not be sufficient in the presence of a large number of testing conductor lines or with large temperature nonuniformities.

In principle, a possible method for obtaining a high-resolution resistance is to compensate the temperature fluctuations using a second resistor, located very close to the stressed one, as a "thermometer" [43, 79, 80]. As an example, the work reported by Lloyd and Koch [43], the so-called AC Wheatstone bridge method, is described as follows. When an AC Wheatstone bridge network is used to measure the electromigration-induced resistance change, the conductor line to be tested forms one arm of the bridge. An adjacent, identical line on the same chip forms another arm of the bridge, which is used to monitor the pure temperature effect. Thus the two arms of the bridge are located together and undergo the same temperature fluctuation caused by the instability of the ambient or temperature controller. The conductor line being studied is stressed by a high DC current density, while the whole bridge is biased by a low AC current superimposed on the DC current. The AC signal is used to balance the bridge before DC stressing and to measure the voltage drop across the bridge caused by the resistance change of the stressed line during the DC stressing. Since the temperature fluctuation would be the same in both the monitoring and stressed lines, the measured voltage change would reflect the resistance change due to the applied current only. In the work described by Lloyd and Koch [43], the typical DC stressing current is 30 mA, while the AC signal is 0.5 mA and 320 Hz. The stressed and monitoring conductors are $0.5 \times 7.5 \times 275$ μm Al lines or $0.2 \times 7.5 \times 275$ μm Al–5%Cu lines. The monitoring line was DC stressed for 40 seconds while the resistance of the stressed line was being measured. This short-pulsed DC current application ensured that (a) the stressed and the monitoring lines were undergoing the same level of Joule heating when the measurements were taken, and (b) the structure of the monitoring line would not be damaged due to a long-time DC current stress. The system could easily measure current-induced resistance changes on the order of a few parts per million without significant temperature-induced error. It is noted that the reference resistor must be very stable with respect to the effects other than thermal fluctuations. A perfectly stable reference device may not be available, since annealing mechanisms are present during almost every thermal cycle [81, 82]. However, due to the high resolution of the bridge technique, electromigration can be detected at its very early stage when the fractional resistance is only a few parts per million. It is therefore possible to conduct the tests in a very short time interval while the resistance change in the monitoring line is essentially negligible.

This is possible if the film is annealed sufficiently after its deposition. It is also possible that a single line can withstand a number of electromigration tests at different temperatures so that one can extract the activation energy without having to change specimens.

6.4.2 Low-Frequency Noise Measurements

The low-frequency noise measurements provide a sensitive tool for the investigation of the structural changes induced in thin metal films by electromigration or by mechanical stresses [5–10]. When a constant noise-free DC current is applied to a metal thin film, noise is produced in the output current or voltage as a result of a generation–annihilation process of vacancies and dislocations within the material, that is, $1/f$ noise. When the structure of a thin film is changed, the $1/f$ noise characteristic changes as well. Thus one can investigate electromigration, using the $1/f$ noise method.

Considering the thermal noise generated in the output voltage, the noise voltage spectra can be represented by [10]

$$S_v(f) = 4kTR + \frac{KV^\beta}{f^\alpha} \tag{6.50}$$

where R represents the film resistance in ohms, V is the DC voltage applied across the film, f is the measurement frequency, and K, α, and β are constants used to characterize the current noise spectrum. The first term in Equation (6.50) represents the thermal noise. If the current distribution in a continuous metal thin film is uniform, the widely accepted equation for the current noise spectral density at room temperature can be used [83]:

$$S_c(f) = \frac{\gamma V^\beta}{N_c f^\alpha} = \frac{\gamma (IR)^\beta}{N_c f^\alpha} \tag{6.51}$$

where N_c is the total number of free charge carriers in the film, γ is a dimensionless constant equal approximately to 2×10^{-3}, and I is the DC current passing through the film. Comparing Equations (6.50) and (6.51), one can see that $K = \gamma/N_c$. It has been shown [10] that β is close to 2. The detailed studies on the parameters of α, β, and γ have previously been reported [5, 6, 84–86]. In metal films, the free-electron density is large (5×10^{22} cm^{-3}) compared with that of the semiconductors (10^{15}–10^{19} cm^{-3}). The only way to reduce the total number of electrons, N_c, is to reduce the film size. To ensure that the current or voltage noise is detectable in a $1/f$ noise experiment, one should increase the biasing current level and reduce the film size, as can be seen in Equations (6.50) and (6.51). This has been confirmed in $1/f$ noise electromigration studies. It has also been pointed out that the $1/f$ noise is affected by the current distribution in the conductor thin film [87]. For films of the same volume carrying the same current, the film that has a uniform current density generates the lowest $1/f$ noise. When a metal film or resistor has a nonuniform cross section that leads to a nonuniform current distribution, the magnitude of the $1/f$ noise changes. Generally, the current crowding effect will cause the $1/f$ noise to increase. This forms the basis of the application of current noise measurements for the study of electromigration.

In order to monitor frequency noise of the output voltage or current, a low-noise preamplifier and filter are needed. Unfortunately, the electromigration-generated defect-noise frequency in thin metal films is generally below 10 Hz, which is the frequency range where low-noise preamplifier performance becomes poor due to the generation of background noise (BN); namely, the BN may be comparable to the noise generated by electromigration. To deal with this difficulty, ultra-low-noise amplifiers (ULNAs) have been introduced [88, 89]. The ULNAs have been reported to allow the detection of a noise level significantly below the background noise of most preamplifiers. The main features of a ULNA designed for noise analysis in metals and other low-resistance samples are characterized by [88] (a) a bandwidth of 4 mHz to 230 kHz, (b) a gain of 80 dB, and (c) S_{BN}'s (the power spectral density of input equivalent voltage noise) of 0.6, 2.5, 10, and 70 $nV/(Hz)^{1/2}$ at 100, 1, 0.1, and 0.02 Hz, respectively. The ultra-low-noise measurement system used by Neri et al. [86] consists of a dual-channel cross-correlator circuit, which is shown schematically in Figure 6.23. The noise from the conductor film being tested is sent into two separate channels of amplifiers and filters. The output signals of the two filters are then sent into a cross-correlator circuit that gives an output proportional to the product of the instantaneous value of the two input signals. The output of the correlator is then averaged over a long period of time. This method eliminates the noise generated in the two channel amplifiers and filters, providing only the noise generated in the film [90].

The $1/f$ measurement setup is shown schematically in Figure 6.24. The measurement system consists of a power supply, usually a noise-free constant-current source, a low-frequency noise preamplifier or a ULNA, and a spectrum analyzer capable of performing power spectral density measurements in the frequency range of interest. The constant-

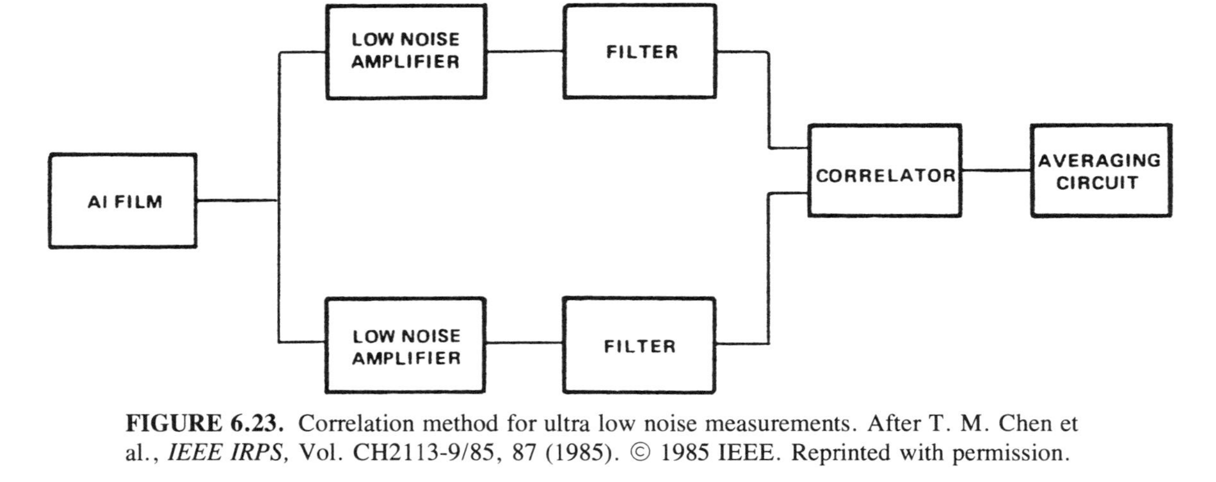

FIGURE 6.23. Correlation method for ultra low noise measurements. After T. M. Chen et al., *IEEE IRPS*, Vol. CH2113-9/85, 87 (1985). © 1985 IEEE. Reprinted with permission.

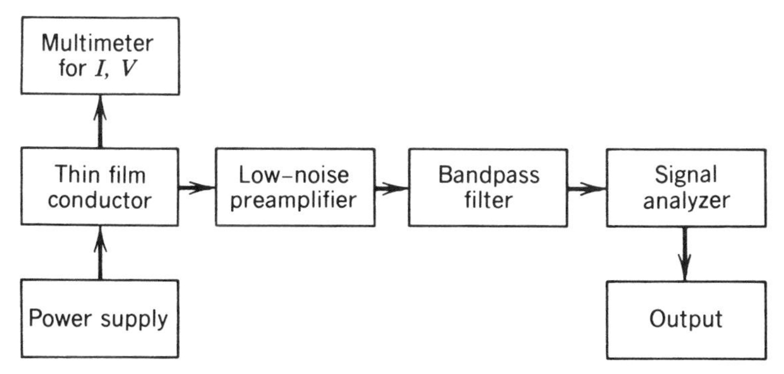

FIGURE 6.24. Schematic of the system assembly for low frequency noise measurements. After T. M. Chen et al., *IEEE IRPS*, Vol. CH2113-9/85, 87 (1985). © 1985 IEEE. Reprinted with permission.

current source could be rechargeable batteries [10]. The output voltage across the thin film conductor is sent to the preamplifier and filter, or a ULNA, and then to the analyzer. A typical output of the $1/f$ noise is shown in Figure 6.25 [10]. The sample tested in this work was a 3.5-μm wide pure Al thin film conductor sputter deposited on SiO_2 substrate and stressed under a current density of 2×10^6 A/cm^2 at room temperature. The time for the total resistance of the film to increase by 10% due to electromigration was found to be 38 minutes. It can be seen that the noise generated after the electromigration test is substantially

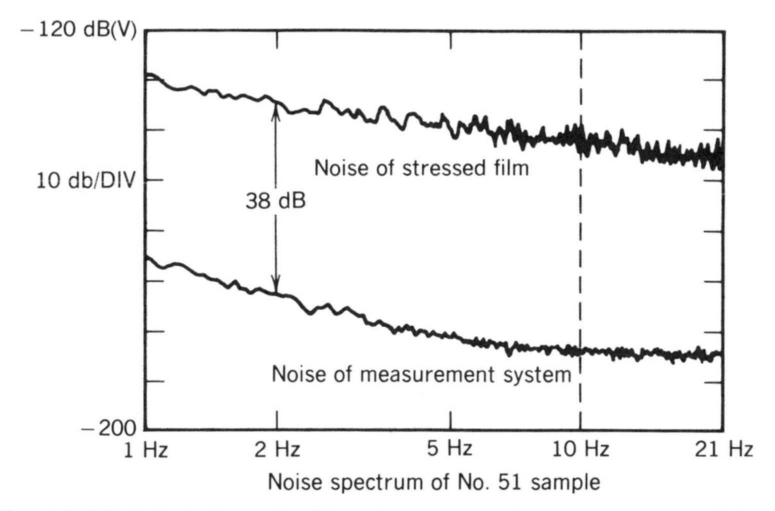

FIGURE 6.25. A typical result of low frequency noise measurements. After T. M. Chen et al., *IEEE IRPS*, Vol CH2113-9/85, 87 (1985). © 1985 IEEE. Reprinted with permission.

increased above the background noise measured before electromigration. For example, the noise level at a frequency of 2 Hz after stressing is increased by 38 dB above the background noise. The temperature effects on the noise measurement may also be eliminated by a proper calibration prior to the noise measurement.

The temperature characteristic of a metal film should first be measured using low current densities in order to keep the sample in thermal equilibrium with the sample-holder whose temperature varies slowly. After setting a current density suitable for the noise measurement, the sample temperature is allowed to be stabilized. Thermal fluctuations must be suppressed, since they can generate a low-frequency noise that hides the electromigration-generated noise. Electronic temperature controllers are not used in this type of experiment. The measurement is initiated immediately after the stabilization of the sample temperature. It has been noted that the noise measurement is possible only for current density above 10^6 A/cm^2 [5–10]. Lower current density does not result in detectable signals [10].

The power spectral density of the resistance fluctuation S_R has been proposed to be [84, 86]

$$S_R \propto \frac{j}{kT} e^{-Q/kT} \tag{6.52}$$

According to Equation (6.52), measurement of S_R at a given frequency f_0 as a function of T provides another method to measure the activation energy. The technique, named SARF (Spectral Analysis of Resistance Fluctuations), has been applied to the determination of activation energies of various thin film systems [40–92]. This technique is highly applicable to GaAs metallizations such as Au/Ti/Pt and Au-TiW systems, and additional results for typical systems are summarized in Table 6.1. Stressing conditions used to obtain the results by the SARF method are $j < 2 \times 10^6$ A/cm^2 and $T < 100°C$.

TABLE 6.1. Results of Activation Energy Measurement by SARF Method

Material	E_a (eV)
In	0.4
Au	0.94
Al unpassivated	0.64
Al passivated	0.88
Al–1%Si unpassivated	0.94
Al–1%Si passivated	1.07

The experimental results and subsequent analysis have shown that the percentage increase of $1/f$ noise is much larger than the corresponding percentage increase of the film resistance caused by electromigration [10]. The $1/f$ noise is much more sensitive to electromigration damage than the resistance change due to the fact that the noise spectral densities of voltage and current are proportional to the square of current density ($\beta \approx 2$), as described by Equations (6.50) and (6.51). Thus the current noise measurement method enables one to study electromigration at its early stage. It is possible to qualitatively relate a $1/f$ noise measurement to the MTF of a thin film conductor [10, 93, 94]. Generally, higher current noise during the electromigration test suggests a shorter lifetime. The quantitative correlation between a noise measurement and the corresponding MTF or the reliability of a given metallization system is not presently available. It appears that the $1/f$ method requires a much shorter stressing time, but the threshold current density is higher than that required by a regular resistometric method. The low-frequency noise method therefore forms the complementary part of the conventional resistometric technique for the study of electromigration phenomenon.

6.4.3 Drift Velocity Method

The mass transport of electromigration in a thin metal film can be investigated directly using the drift velocity method [13, 38, 95]. The experiment is also called the "saddle movement experiment." It was first presented by Blech in 1975 and has been widely adopted since then for the study of atomic drift velocity, the direction of the mass transport, and the activation energy, as well as other parameters.

A sample configuration for the drift velocity experiments is shown in Figure 6.26. Basically, a piece of metal thin film, the saddle, is deposited onto a less conducting metal track. For example, the metal film studied in the first drift velocity experiment [13] was Au while the metal track was Mo. A constant current is applied through the metal track, and most of the current is diverted into the highly conducting Au saddle in the overlapped region. As a result, the saddle experiences a contin-

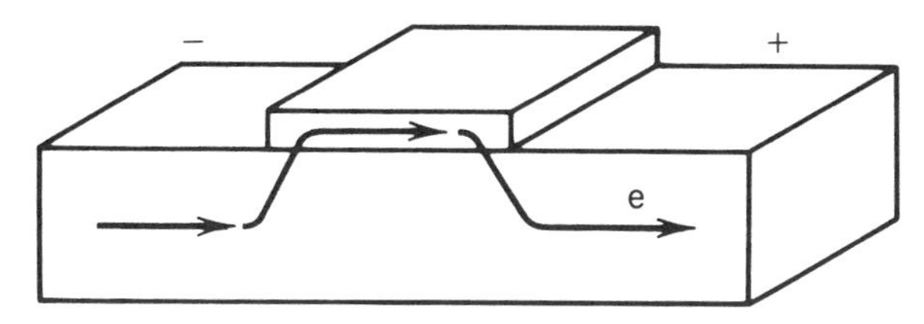

FIGURE 6.26. Schematic of sample configuration for drift velocity measurements.

uous impulse from the electron wind force, in the direction from the cathode to the anode edge. The mass in the saddle will therefore be transported in that direction, resulting in a depletion at the edge near the cathode and an accumulation at the edge near the anode. The average atomic drift velocity is assumed to be equal to the velocity of the mass transport, which is obtained from the effective movement of the saddle edge along the metal track. This average atomic drift velocity is given by

$$\nu = \frac{J}{N} = \frac{D_0}{kT} qZ^* \rho j e^{-Q/kT} \tag{6.53}$$

where ν is the drift velocity, J is the atomic flux, and ρ is the resistivity of the film. The other constants have been defined previously. Equation (6.53) can be rewritten as

$$\ln\left(\frac{\nu T}{j}\right) = \ln\left(\frac{D_0}{k} qZ^* \rho\right) - \frac{Q}{kT} \tag{6.54}$$

Thus measurement of drift velocity at various temperatures yields the value of activation energy. Equation (6.54) also indicates that measurement of the $D_0 Z^*$ product is possible if the drift velocity is measured as a function of the current density. Furthermore, the unique feature of this experiment is that it provides an experimental evidence for the direction of mass flow during electromigration.

Typical results from Blech's investigations [13] are shown in Figures 6.27 and 6.28. Figure 6.27 shows the SEM micrographs of the Au saddle having moved along the Mo track. The Au film is 20 μm wide, 175 μm long, and 1800 Å thick. The system has been stressed by a current density of 10^6 A/cm^2 at 430°C. Mass depletion at the cathode side and accumulation at the anode side are clearly seen. The direction of electromigration mass flow is therefore in the direction of electron wind for the system investigated. The same observation has been made for an aluminum on titanium nitride system [38]. The drift velocity of atomic movement in the Au saddle experiment has been reported to be proportional to the current density, as shown in Figure 6.28. The above results agree with the theoretical analysis [75, 96]. The activation energy was found to be between 0.6 and 0.9 eV for Au [13] and 0.55 eV for Al [38]. The value of $D_0 Z^*$ was determined to be between -0.4×10^{-3} and -1.9×10^{-3} cm^2/s. The advantage of the saddle movement experiment is that the effect of Joule heating may be reduced. By making

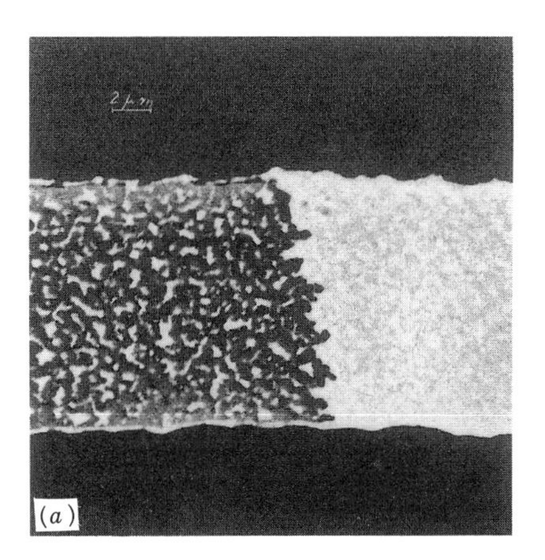

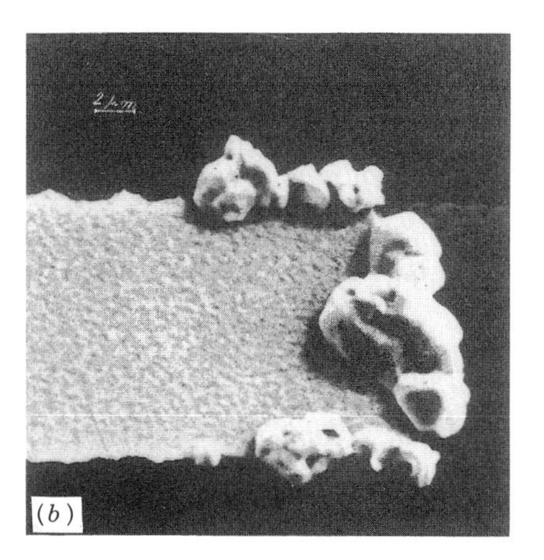

FIGURE 6.27. SEM micrograph of Au thin film edges after being stressed by a current density of 10^6 A/cm^2. (*a*) depletion of Au at cathode side; (*b*) accumulation of Au at anode side. After I. A. Blech et al., *Thin Solid Films,* Vol. 25, 327 (1975). Reprinted with permission.

the underlying metal track thick with respect to the saddle, it is possible to obtain a current density large enough to generate electromigration in the saddle, and yet to maintain the current density low enough in the track so that severe Joule heating in the track is avoided. The heat generated on the saddle itself is also limited due to its short length compared with the length required for a resistometric or lifetime measurement.

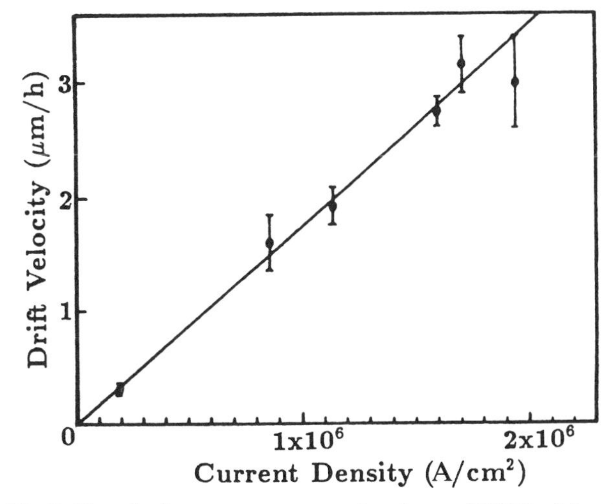

FIGURE 6.28. Drift velocity versus current density at 430°C. After I. A. Blech et al., *Thin Solid Films,* Vol. 25, 327 (1975). Reprinted with permission.

However, one must maintain caution in selecting the material systems so as to eliminate interdiffusion and intermetallic compound formation at the interface, which occurs at elevated temperatures. Such reactions will affect the temperature and the effective current diverted into the saddle and will therefore result in an inaccurate determination of the mass transport.

6.4.4 Lifetime Test Method

As discussed in the introduction, the measurement of MTF is still a widely employed technique in the study of electromigration reliability. The results on MTF for various systems have previously been published. In this section, the basic experimental approach and the typical results are reviewed.

In order to conduct a lifetime test, one must initially define the failure criterion. The failure criterion for a single conductor line is usually defined as an open circuit due to the formation of voids, or a short circuit to an adjacent line due to the formation of hillocks. In the early stage of electromigration, the failure criterion is often defined as a certain percentage of resistance increase. The lifetime for the individual conductor line is then defined as the time needed for the conductor to achieve the defined failure criterion. As discussed previously, the lifetime for an individual conductor line usually appears to be random regardless of the choice of the failure criterion. The randomness of a single conductor line is due to the randomness of the metallic film microstructure [26, 30, 35]. Since the lifetime of an individual conductor line is not representative of the system reliability, statistical measurements of lifetime, such as MTF and its standard deviation σ, become essential.

A common experimental configuration in conducting a lifetime test is to apply a constant voltage to a large number of macroscopically identical metal lines, connected in parallel. By monitoring the current flow in each line, the electromigration-induced degradation is monitored, since the line current provides a direct measurement of electromigration-induced line resistance change. The change in any line resistance or current does not affect the bias condition of the others since the lines are biased under a common constant voltage. The result of such an experiment is the distribution of lifetimes of all the lines tested. A cumulative failure probability can then be plotted against the failure time, which determines MTF. Figure 6.29 presents a set of the typical cumulative failure probability plots for electromigration [30]. The comparison between the four plots will be discussed later. The samples tested in this work are Al–2%Cu–0.3%Cr films of 2.2 μm in width deposited

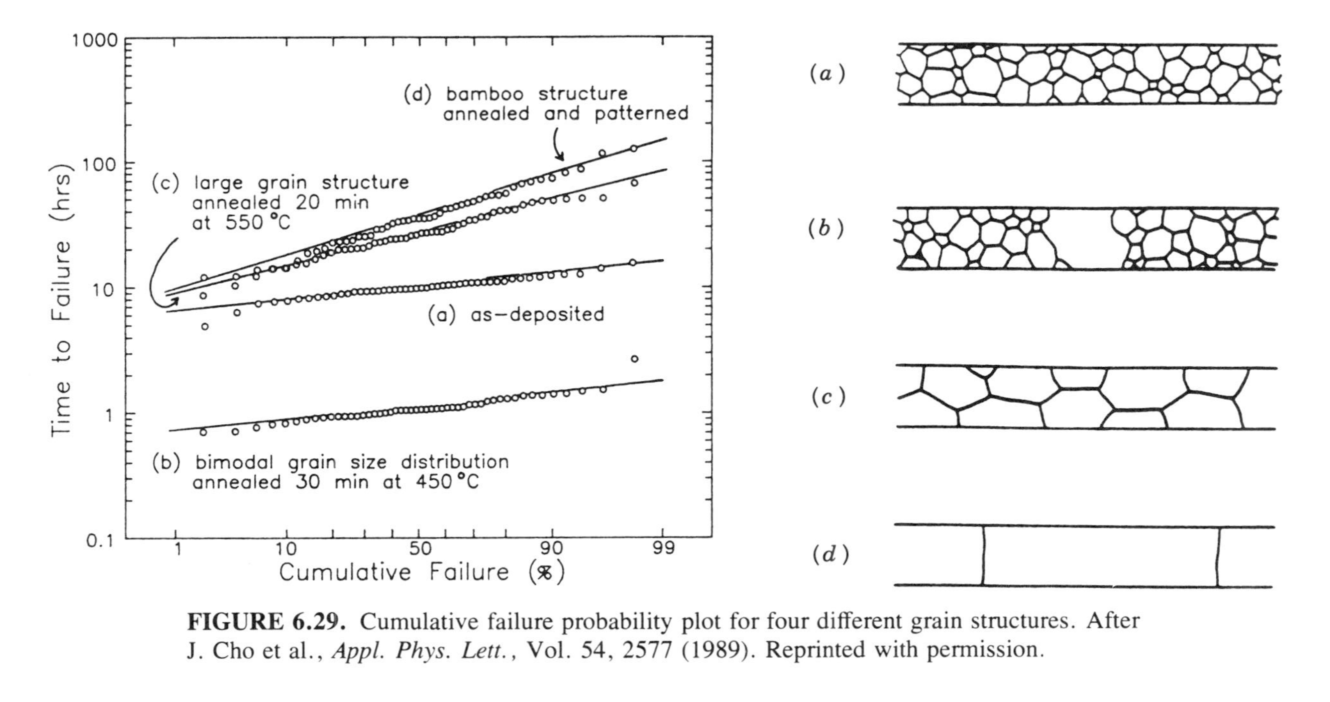

FIGURE 6.29. Cumulative failure probability plot for four different grain structures. After J. Cho et al., *Appl. Phys. Lett.*, Vol. 54, 2577 (1989). Reprinted with permission.

onto SiO_2 substrates. The strips were stressed under a current density of 1.23×10^6 A/cm^2 at a substrate temperature of 275°C. The MTF, defined as the time for 50% failure, is found to be 90 hours for the group (a) samples.

Important information that can be extracted from lifetime tests is the activation energy of the grain boundary diffusion. The normal procedure is to measure MTFs at different temperatures under a fixed initial current density and then plot them against $1/T$. The slope of such an Arrhenius plot is proportional to the activation energy of the grain boundary diffusion, according to Black's formula, Equation (6.3). In addition, the MTF at lower temperatures (i.e., at device operating conditions) may be extrapolated from the results obtained at the elevated temperatures. This yields the basis of the design rule formulation for metallization systems [97]. The underlying assumption here is that the failure mechanisms governing electromigration are not altered by the increase in temperature. The exponent factor of current density in Black's formula may also be determined from the MTF measurements, by plotting MTF versus current density at a fixed temperature. The value of n has been reported to be larger than unity [21, 41, 98], indicating the effect of Joule heating, which cannot be avoided in a lifetime test. The temperature determination of the conductor film during electromigration has always been a problem in the MTF technique, as discussed previously. When film structural changes occur, Joule heating becomes significant, causing the temperature to become nonuniform and varying with time. Since failure criteria in most of the lifetime tests approximate an open circuit (i.e., the last stage of electromigration damage), the difference between the initial and instantaneous temperatures may be substantial. A detailed analysis on the thermal response of the electromigration test structures to the applied stressing conditions has been carried out by Schafft [99]. However, the only approach capable of obtaining the instantaneous temperature distribution based on electromigration-induced Joule heating is that of computer-aided modeling of the electromigration failure process [27, 34], as discussed previously.

The dependence of MTF and σ on line geometry is of particular interest since they can be used to provide guidelines for metallization design rules [30, 35, 36, 100, 101]. Shown in Figure 6.30*a* are the same data presented in Figure 6.29 but normalized over its average grain diameter. It can be seen that MTF decreases with the linewidth, W, until it reaches the average grain diameter ($W/d \approx 1$) and then increases drastically as the linewidth becomes smaller than d. The standard deviation, σ, monotonically decreases as the linewidth increases. Figures 6.30*b* and 6.30*c* [35] show that both MTF and σ decrease as the line

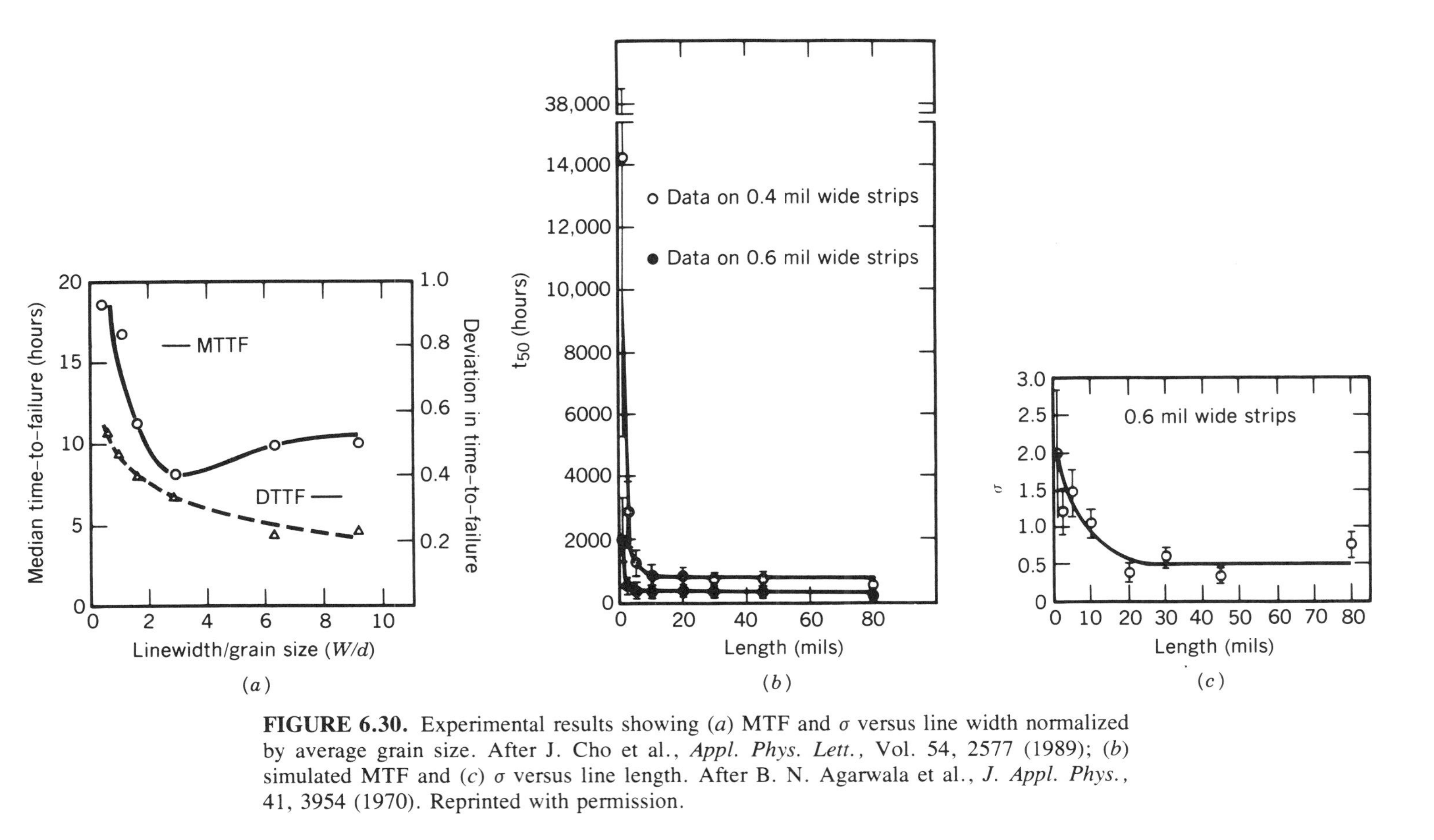

FIGURE 6.30. Experimental results showing (*a*) MTF and σ versus line width normalized by average grain size. After J. Cho et al., *Appl. Phys. Lett.*, Vol. 54, 2577 (1989); (*b*) simulated MTF and (*c*) σ versus line length. After B. N. Agarwala et al., *J. Appl. Phys.*, 41, 3954 (1970). Reprinted with permission.

length increases. Similar results have also been reported by other authors [36, 100, 101–104]. Such observed line geometry dependence of MTF and σ can be explained based on a statistical metallurgical model of electromigration in polycrystalline metal, proposed by Attardo et al. in 1971 [26]. As discussed previously, the mass transport of electromigration is mainly along grain boundaries, and the sites of nonzero flux divergence are mainly those of the grain boundary triple junctions in a polycrystalline metal thin film. The value of the flux divergence at a particular triple junction strongly depends on the microstructure of the three grain boundaries defining the triple junction. Since the microstructure of the grain boundaries is essentially random, the probability for a line to contain a triple junction that has the largest possible flux divergence increases as the line length increases; therefore MTF should decrease.

The experimental data shown in Figure 6.29 provide clear evidence supporting the aforementioned geometry dependence and the near lognormal distribution of MTF [30]. The films are manufactured to have four different grain structures: (a) as deposited, (b) annealed for 30 minutes at 450°C, (c) annealed for 20 minutes at 550°C, and (d) annealed for 20 minutes at 550°C before patterning. The corresponding sketches of the microstructures based on bright-field transmission electron micrographs of the conductor lines are shown to the right. The lines are 2.2 μm wide, tested under a current density of 1.2×10^6 A/cm^2 at 275°C [30]. For each of the grain structures, the lifetime approximately fits a straight line on the cumulative plot, indicating an approximate lognormal distribution. The results also show that a larger grain size with respect to linewidth yields a higher MTF.

It is important to predict the lifetime of a metal thin film system at the early stage of electromigration, prior to the occurrence of an open circuit. The lifetime technique produces concomitant problems as discussed above. On the other hand, the early stage resistance increase does not represent a real failure. A correlation between early failure and final failure time must therefore be developed [4, 19, 105–108]. Such correlations are especially important for multilayer interconnects [105], which do not open entirely when failed.

In developing such correlation, it has been observed that the product of the early stage resistance changing rate, dR/dt, and the failure time, t_f, appears to be a constant:

$$\frac{dR}{dt} t_f = C \qquad (6.55)$$

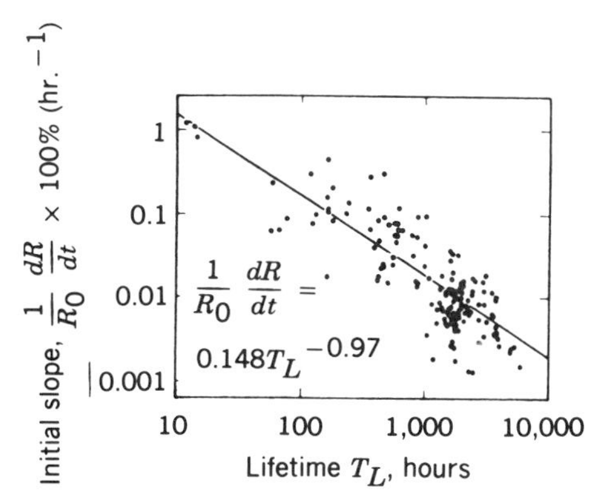

FIGURE 6.31. The relation between the initial slope of the *R–t* plot and lifetime. After A. T. English et al., *J. Appl. Phys.*, Vol. 45, 3757 (1974). Reprinted with permission.

where the constant C depends on the stressing conditions, such as the current density and temperature, the line geometry, and grain structures. Figure 6.31 shows an experimental result obtained from a system of Au on Ti deposited onto sapphire substrate [41]. The Au/Ti system is often employed for gate metallization in GaAs MESFETs, while sapphire is of high thermal conductivity, resulting in a minimum of Joule heating in the films. The samples are approximately 7 μm wide, 500 μm long, and 1000 Å thick. A total of 394 stripes were tested with constant current density of 2–8 $\times$ 10^6 A/cm^2 at 180°C. It can be seen that the initial slope of the resistance-versus-time curve is directly correlated with the failure time and is approximately inversely proportional to it. The same result has also been produced from a computer simulation of electromigration [4, 19], where Equation (6.55) holds true for metal lines of different dimensions, tested at different conditions. The results reported so far, however, are far from being conclusive, and additional investigations are needed in order to develop a criterion for predicting electromigration reliability based on early stage failure behavior.

6.4.5 SWEAT Method

Several techniques have been developed for monitoring electromigration rapidly so that a reliability evaluation of a metal line within a very short time such as a few seconds to a few minutes is possible. The common features of these techniques are (a) a fast response (within a few seconds and never beyond 10 hours), (b) the possibility for a wafer level setup, and (c) a ramping stress (current density or temperature) rather than a constant stressing condition. A fast response is necessary not only for an efficient reliability prediction but also for the possibility of wafer level tests. A ramping stress reveals the kinetics of the failure process in a dynamic way. One of the most important fast ramping techniques, the Standard Wafer-level Electromigration Acceleration Test (SWEAT), is reviewed in this section.

SWEAT was developed to monitor electromigration susceptibility at the wafer level [14]. The test is extremely fast, allowing for the qualitative evaluation of a metal line reliability in less than 15 seconds. The main idea in this method is that the temperature increase that is necessary to accelerate the test is achieved solely by means of Joule heating. In a production line environment, external heating elements, such as a hot stage or a temperature chamber, are not practical. First, heating the entire wafer to a high temperature, for example, 200°C, may cause a number of side effects, such as mechanical stresses or reactions between different materials; second, it takes a much longer time to have the entire system reach thermal equilibrium. On the other hand, it takes only a few hundreds of milliseconds or less to achieve a thermal equilibrium on the film being tested after the high current density is applied, due to its high thermal conductivity and small size. The measurement procedure in a SWEAT technique can be summarized as follows [14]:

1. Calibrating thermal coefficient of resistivity α; choosing a reference resistance R_0 at a reference temperature, such as room temperature.
2. Calibrating the power P dissipated inside the metal line versus the corresponding temperature increase using short pulsed currents.
3. Defining a minimum acceptable lifetime $t_{\min}$ of the metallization system according to Black's equation at the device normal operating conditions including T and j.
4. Defining the accelerating stress factor S_F from $t_{\min}$ divided by the required testing time t_s.
5. Gradually increasing the current density j and continuously monitoring the power $P = IV$. Obtaining the instantaneous temperature on the line using the $P - T$ calibration described in step 2. Stopping the current ramping when the time-to-failure, TTF, calculated according to Black's equation with the instantaneous values of j_0 and T_0, is equal to t_s.
6. Stressing the sample at j_0 and T_0 until failure occurs.
7. Accepting the wafer if the measurement failure time is greater than TTF $= t_{\min}/S_F$, otherwise rejecting the wafer.

Except for the initial current density ramping process, the technique essentially follows the lifetime measurement approach and suffers from the same disadvantage as the MTF method.

6.5 SUMMARY

Theoretical and experimental studies of electromigration, a major failure mechanism for the metallizations and contacts in integrated circuits, have been reviewed. Detailed physical models have been discussed and illustrated through computer simulations. Representative electromigration characterization techniques, both destructive versus nondestructive and slow versus fast, have been presented along with typical experimental results. While the destructive techniques such as the MTF method are widely used, the nondestructive ones are still at the experimental stage. The latter, however, are capable of providing valuable information for a better understanding of the basic physics of electromigration. At the present time, only qualitative correlation has been obtained between these two techniques and much effort is obviously needed to develop quantitative relationships between the quantities measured by these two techniques. Such a quantitative relationship would make it possible to eliminate the need to extrapolate reliability of an integrated circuit from high stress conditions to the circuit normal operating conditions because extrapolation may involve large error.

Electromigration in integrated circuits at the present time is not under good control and new models need to be developed to study electromigration in the latest integrated circuit technologies that utilize multilayer metallization schemes and new compound materials for metallizations. The specific role that electromigration plays in contact degradation is far from being well understood and detailed studies of metal–semiconductor contact electromigration are needed, especially for those new devices that have junction depths less than 0.25 μm.

REFERENCES

1. I.A. Blech and H. Sello, "A Study of Failure Mechanisms in Silicon Planar Epitaxial Transistor," in *Physics of Failure in Electronics*, Vol. 5, T.S. Shilliday and J. Vaccaro, eds., Rome Air Development Center, 1966, p. 496.
2. R. Rosenburg and L. Berenbaum, "Resistance Monitoring and Effects

of Nonadhesion During Electromigration in Aluminum Films,'' *Appl. Phys. Lett.*, Vol. 12, No. 5, 201 (1968).

3. R.E. Hummel, R.T. Dehoff, and H.J. Geier, ''Activation Energy for Electrotransport in Thin Aluminum Films by Resistance Measurements,'' *J. Phys. Chem. Solids*, Vol. 37, 73 (1976).
4. P.F. Tang, A.G. Milnes, C.L. Bauer, and S. Mahajan, ''Electromigration in Thin Films of Au on GaAs,'' *Proc. Mater. Res. Soc.*, Vol. 167, 341 (1989).
5. F.N. Hooge and A.M.H. Hoppenbroouwers, ''$1/f$ Noise in Continuous Thin Gold Films,'' *Physica*, Vol. 45, 393 (1969).
6. J.W. Eberland and P.M. Horn, ''Excess ($1/f$) Noise in Metals,'' *Phys. Rev. B*, Vol. 18, 6681 (1978).
7. K.P. Rodbell and P.J. Ficalora, ''Effect of Hydrogen on Electromigration and $1/f$ Noise in Gold Films,'' *Appl. Phys. Lett.*, Vol. 50, 1415 (1987).
8. R.H. Koch, J.R. Lloyd, and J. Cronin, ''$1/f$ Noise and Gain Boundary Diffusion in Aluminum and Aluminum Alloys,'' *Phys. Rev. Lett.*, Vol. 55, 2487 (1985).
9. J.G. Gottle and T.M. Chen, ''Activation Energies Associated with Current Noise of Thin Metal Films,'' *J. Electron. Mater.*, Vol. 17, 467 (1988).
10. T.M. Chen, T.P. Djeu, and R.D. Moore, ''Electromigration and $1/f$ Noise of Aluminum Thin Films,'' *IEEE IRPS*, Vol. CH2113-9/85, 87 (1985).
11. F. Vollkommer, H.G. Bohn, K.-H. Robrock, and W. Schilling, ''Internal Friction: A Fast Technique for Electromigration Failure Analysis,'' in *Proceedings of the 28th IEEE International Reliability Physics Symposium*, p. 51 (1990).
12. J.R. Black, ''Electromigration Failure Modes in Aluminum Metallization for Semiconductor Devices,'' *Proc. IEEE*, Vol. 57, 1587 (1969).
13. I.A. Blech and E. Kinsbron, ''Electromigration in Thin Gold Films on Molybdenum Surfaces,'' *Thin Solid Films*, Vol. 25, 327 (1975).
14. B.J. Root and T. Turner, ''Wafer Level Electromigration Test for Production Monitoring,'' in *Proceedings of the 23rd IEEE International Reliability Physics Symposium*, pp. 100–107 (1985).
15. R.W. Pasco and J.A. Schwartz, ''Temperature-Ramp Resistance Analysis to Characterize Electromigration,'' *Solid State Electron.*, Vol. 26, No. 5, 445 (1983).
16. R.E. Jones and L.D. Smith, ''A New Wafer-Level Isothermal Joule-Heated Electromigration Test for Rapid Testing of Integrated-Circuit Interconnect,'' *J. Appl. Phys.*, Vol. 61, 4670–4678 (1987).
17. C.C. Hong and D.L. Crook, ''Breakdown Energy of Metal (BEM)—A New Technique for Monitoring Metallization Reliability of Wafer

Level,'' in *Proceedings of the 23rd IEEE International Reliability Physics Symposium*, pp. 108–114 (1985).

18. W. Baerg, K. Wu, P. Davies, G. Dao, and D. Fraser, ''The Electrical Resistance Ratio (RR) as a Thin Film Metal Monitor,'' in *Proceedings of the 28th IEEE International Reliability Physics Symposium*, pp. 119–123 (1990).
19. P.F. Tang, *Modeling of Electromigration Phenomena with Application to GaAs on Au*, Ph.D. thesis, Carnegie–Mellon University, Pittsburgh, PA, 1990.
20. I.A. Blech and Conyers Herring, ''Stress Generation by Electromigration,'' *Appl. Phys. Lett.*, Vol. 29, 131 (1976).
21. F.M. d'Heurle and P.S. Ho, ''Electromigration in Thin Films,'' in *Thin Films: Interdiffusion and Reactions*, J.M. Poate, K.N. Tu, and J.W. Mayer, eds., Wiley, New York, 1978, pp. 243–303.
22. R.A. Sigsbee, ''Failure Model for Electromigration,'' in *Proceedings of the 11th IEEE International Reliability Physics Symposium*, pp. 301–305 (1973).
23. R.A. Sigsbee, ''Electromigration and Metallization Lifetimes,'' *J. Appl. Phys.*, Vol. 44, No. 6, 2533–2540 (1973).
24. J.D. Venables and R.G. Lye, ''A Statistical Model for Electromigration Induced Failure in Thin Film Conductors,'' in *Proceedings of the 10th IEEE International Reliability Physics Symposium*, pp. 159–164 (1972).
25. J.M. Schoen, ''Monte Carlo Calculations of Structure-Induced Electromigration Failure,'' *J. Appl. Phys.*, Vol. 51, No. 1, 513–521 (1980).
26. M.J. Attardo, R. Rutledge, and R.C. Jack, ''Statistical Metallurgical Model for Electromigration Failure in Aluminum Thin-Film Conductors,'' *J. Appl. Phys.*, Vol. 42, No. 11, 4343–4349 (1971).
27. K. Nikawa, ''Monte Carlo Calculations Based on the Generalized Electromigration Failure Model,'' in *Proceedings of the International Reliability Physics Symposium*, pp. 175–181 (1981).
28. M.J. Attardo and R. Rosenburg, ''Electromigration Damage in Aluminum Film Conductors,'' *J. Appl. Phys.*, Vol. 41, No. 6, 2381 (1970).
29. P.P. Meng, *Computer Simulation of Electromigration in Thin Films*, M.S. thesis, Rensselaer Polytechnic Institute, Troy, NY, May 1988.
30. J. Cho and C.V. Thompson, ''The Grain Size Dependence of Electromigration Induced Failures in Narrow Interconnects,'' *Appl. Phys. Lett.*, Vol. 54, No. 25, 2577 (1989).
31. J.W. Harrison, Jr., ''A Simulation Model for Electromigration in Fine-Line Metallization of Integrated Circuits Due to Repetitive Pulsed Currents,'' *IEEE Trans. Electron. Dev.*, Vol. 35, No. 12, 2170 (1988).
32. C.A. Ross and J.E. Evetts, ''A Model for Electromigration Behavior in Terms of Flux Divergences,'' *Scripta Metall.*, Vol. 21, 1077–1082 (1987).

33. A.P. Schwarzenberger, C.A. Ross, J.E. Evetts, and A.L. Greer, "Electromigration in Presence of a Temperature Gradient: Experimental Study and Modeling," *J. Electron. Mater.*, Vol. 17, No. 5, 473–478 (1988).
34. P.J. Marcoux, P.P. Merchant, V. Naroditsky, and W.D. Rehder, "A New 2D Simulation Model of Electromigration," *Hewlett-Packard J.*, pp. 79–84 (June 1998).
35. B.N. Agarwala, M.J. Attardo, and A.P. Ingraham, "Dependence of Electromigration Induced Failure Time on Length and Width of Aluminum Thin-Film Conductors," *J. Appl. Phys.*, Vol. 41, No. 10, 3954 (1970).
36. S. Vaidya, T.T. Sheng, and A.K. Sinha, "Linewidth Dependence of Electromigration in Evaporated Al–0.5%Cu," *Appl. Phys. Lett.*, Vol. 36, No. 6, 464 (1980).
37. E. Kinsbron, I.A. Blech, and Y. Komem, "The Threshold Current Density and Incubation Time to Electromigration in Gold Films," *Thin Solid Films*, Vol. 46, 139 (1977).
38. I.A. Blech, "Electromigration in Thin Aluminum Films on Titanium Nitride," *J. Appl. Phys.*, Vol. 47, No. 4, 1203 (1976).
39. I.A. Blech and K.L. Tai, "Measurement of Stress Gradients Generated by Electromigration," *Appl. Phys. Lett.*, Vol. 30, No. 8, 387 (1977).
40. Kenji Hinode and Yoshio Homma, "Improvement of Electromigration Resistance of Layered Aluminum Conductors," *IEEE IRPS*, Vol. CH2787-0, 25 (1990).
41. A.T. English, K.L. Tai, and P.A. Turner, "Electromigration of Ti-Au Thin Film Conductors at 180°C," *J. Appl. Phys.*, Vol. 45, No. 9, 3757 (1974).
42. Donald J. LaCombe and Earl Parks, "A Study of Resistance Variation During Electromigration," *IEEE IRPS.*, Vol. CH2113-9, 74 (1985).
43. J.R. Lloyd and R.H. Koch, "Study of Electromigration-Induced Resistance and Resistance Decay in Al Thin Film Conductors," *Appl. Phys. Lett.*, Vol. 52, No. 3, 194 (1988).
44. C. Canali, F. Fantini, A. Scorzoni, L. Vmena, and E. Zanoni, "Degradation Mechanisms Induced by High Current Density in Al-Gate GaAs MESFETs," *IEEE Trans. Electron Devices*, Vol. ED-34, No. 2, 205 (1987).
45. C. Canali, L. Vmena, and F. Fantini, "Increase in Barrier Height of Al/n-GaAs Contacts Induced by High Current," *IEEE Electron Device Lett.*, Vol. EDL-7, No. 5, 291 (1986).
46. C. Canali, F. Castaldo, F. Fantini, D. Ogliari, L. Vmena, and E. Zanoni, "Gate Metallization "Sinking" into the Active Channel in Ti/W/Au Metallized Power MESFETs," *IEEE Electron Device Lett.*, Vol. EDL-7, No. 3, 185 (1986).

47. C. Canali, F. Fantini, E. Zanoni, A. Giovannetti, and P. Brambilla, "Failure Induced by Electromigration in ECL 100k Devices," *Microelectron Rel.*, Vol. 24, No. 1, 77 (1984).
48. H. Fukui, S.H. Wemple, J.C. Irvin, W.C. Niehaus, J.C.M. Hwany, H.M. Cox, W.O. Schlosser, and J.V. Dilorenzo, "Reliability of Power GaAs Field Effect Transistors," *IEEE Trans. Electron Devices*, Vol. ED-29, 395 (1982).
49. P.M. White, C.G. Rogers, and B.S. Hewitt, "Reliability of Ku-band GaAs Power FETs Under Highly Stressed RF Operation," in *Proceedings of the 21st IEEE International Reliability Physics Symposium*, p. 297 (1983).
50. A.C. Macpherson, K.R. Gleason, and A. Christou, "Voids in Aluminum Gate Power GaAs FETs Under Microwave Testing," *Proc. ATFA*, p. 155 (1978).
51. C. Canali, F. Castaldo, F. Fantini, D. Ogliari, M. Vanzi, M. Zicolillo, and E. Zanoni, "Power GaAs MESFET: Reliability Aspects and Failure Mechanisms," *Microelectron Rel.*, Vol. 24, No. 5, 947 (1984).
52. A. Christou, P. Tang, and J.M. Hu, "Dual Functional Distribution of Failures in GaAs Microwave Monolithic Integrated Circuits (MMICs)," *IEEE Trans. Electron Devices*, Vol. 39, 2229 (1991).
53. R. Esfandiari, T.J. O'Neill, T.S. Lin, and R.K. Kono, "Accelerated Aging and Long-Term Reliability Study of Ion-Implanted GaAs MMIC IF Amplifier," *IEEE Trans. Electron Devices.*, Vol. 37, 1174 (1990).
54. R. Esfandiari, T. Sato, J. Furuya, L. Pawlowicz, and L.J. Lee, "Reliability and Failure Analysis of MMIC Amplifier Fabricated on Various GaAs Substrates," *IEEE GaAs IC Symp.*, p. 325 (1990).
55. P. Ersland and J.P. Lanteri, "GaAs FET MMIC Switch Reliability," *IEEE GaAs IC Symp.*, p. 57 (1988).
56. W.J. Roesch and M.F. Peters, "Depletion Mode GaAs IC Reliability," *IEEE GaAs IC Symp.*, p. 27 (1987).
57. M. Spector and G.A. Dodson, "Reliability Evaluation of a GaAs IC Preamplifier HIC," *IEEE GaAs IC Symp.*, p. 19 (1987).
58. K. Katsukawa, T. Kimura, K. Ueda, and T. Noguchi, "Reliability Investigation On S-Band GaAs MMIC," *IEEE Microwave Monolithic Circuits Symp.*, p. 57 (1987).
59. J.C. Irvin, "The Reliability of GaAs FETs," in *GaAs FET Principles and Technology*, J.W. Dilorenzo and D.D. Khandelwal, eds., Artech House, Dedham, MA, 1982, Chap. 6.
60. E. Zanoni, A. Callegari, C. Canali, F. Fantini, H.L. Hartnagel, F. Magistrali, A. Paccagella, and M. Vanzi, *Metal–GaAs Interaction and Contact Degradation in Microwave MESFETs*, Wiley, New York, 1990.
61. A. Christou, "Reliability Problems in State-of-the-Art GaAs Devices and Circuits," *Quality Reliab. Eng. Int.*, Vol. 5, 37–46 (1989).

62. A. Christou and B.A. Unger, eds., *Semiconductor Device Reliability*, NATO ASI Series, Series E: Applied Science, Vol. 175, Kluwer Academic Publishers, Norwell, MA, 1990.
63. I. Drukier and J.F. Silcox, "On the Reliability of Power GaAs FETs," in *Proceedings of the IEEE International Reliability Physics Symposium*, p. 150 (1979).
64. W.J. Slusark, G.L. Schnable, V.R. Monshaw, and M. Fukuta, "Reliability of Aluminum-Gate Metallization in GaAs Power FETs," in *Proceedings of the IEEE International Reliability Physics Symposium*, p. 211 (1983).
65. E.D. Cohen and A.C. Macpherson. "Reliability of Gold-Metallized Commercially Available Power GaAs FETs," in *Proceedings of the IEEE International Reliability Physics Symposium*, p. 156 (1979).
66. B. Dornan, W. Slusark, Y.S. Wu, P. Pelka, R. Barton, H. Wolksterin, and H. Huang, "A 4 GHz GaAs FET Power Amplifier: An Advanced Transmitter for Satellite Down-link Communication Systems," *RCA Rev.*, Vol. 41, 474 (1980).
67. M. Otsubo, Y. Mitsui, M. Nakatani, and H. Wataze, "Degradation of GaAs Power MESFETs Due to Light Emission," *IEEE Int. Electron Devices Meet.*, p. 114 (1980).
68. V. Singh and P. Swarup, "Early Gate Failures in Biased GaAs Superimposed on Kirkendall Voids," *Thin Solid Films*, Vol. 97, 277 (1982).
69. K. Katsukawa, Y. Kose, M. Kanamori, and S. Sando, "Reliability of Gate Metallization in Power GaAs MESFETs," in *Proceedings of the IEEE International Reliability Physics Symposium*, p. 59 (1984).
70. A. Christou, E. Cohen, and A.C. Macpherson, "Failure Modes in GaAs Power FETs: Ohmic Contact Electromigration and Formation of Refractory Oxides," in *Proceedings of the IEEE International Reliability Physics Symposium*, p. 182 (1981).
71. F. Wihelmsen and I. Zee, "Effect of Electromigration on GaAs FET Reliability," *Proc. ISTFA*, p. 163 (1984).
72. R. Rosenberg and M. Ohring, "Void Formation and Growth During Electromigration in Thin Films," *J. Appl. Phys.*, Vol. 42, No. 13, 5671 (1971).
73. K.L. Tai and M. Ohring, "Grain-Boundary Electromigration in Thin Films II, Tracer Measurement in Pure Au," *J. Appl. Phys.*, Vol. 48, No. 1, 36 (1977).
74. M. Ohring, "Electromigration Damage in Thin Films Due to Grain Boundary Grooving Processes," *J. Appl. Phys.*, Vol. 42, No. 7, 2653 (1971).
75. H.B. Huntington and A.R. Grone, "Current-Induced Marker Motion in Gold Wires," *J. Phys. Chem. Solids*, Vol. 20, Nos. 1/2, 76 (1961).

76. I.A. Blech and R. Rosenberg, "On the Direction of Electromigration in Gold Thin Films," *J. Appl. Phys.*, Vol. 46, No. 2, 579 (1975).
77. J.K. Crain, *Comput. Geophys.*, Vol. 4, 131 (1978).
78. H. Hu, ed., *The Nature and Behavior of Grain Boundaries*, Plenum Press, New York, 1972.
79. G.L. Baldini and A. Scorzoni, "A New Wafer Level Resistometric Technique for Electromigration," in *Proceedings of the 1st European Symposium on Reliability of Electron Devices, Failure Physics and Analysis* (ESREF-90), Bari, p. 245 (1990).
80. J.A. Maiz and I. Segura, "A Resistance Change Methodology for the Study of Electromigration in Al-Si Interconnects," in *Proceedings of the 26th IEEE International Reliability Physics Symposium*, p. 209 (1988).
81. A.P. Dorey and J. Knight, "The Variation of Resistance of Gold Films with Time and Annealing Procedure," *Thin Solid Films*, Vol. 4, 445 (1969).
82. A. Kinbara and Y. Sawatari, "The Decay in Electrical Resistance in Evaporated Gold Films," *Jpn. J. Appl. Phys.*, Vol. 4, 161 (1965).
83. F.N. Hooge, *Physica*, Vol. 60, 130 (1972).
84. T.M. Chen, P. Fang, and J.G. Cottle, "Electromigration and $1/f\ \alpha$ Noise in Al-Based Thin Films," in *Noise in Physical Systems*, A. Ambrozy, ed., Akademiai Kiado, Budapest, 1989, p. 515.
85. P. Dutta and P.M. Horn, "Low-Frequency Fluctuations in Solids: $1/f$ Noise," *Rev. Mod. Phys.*, Vol. 53, 497 (1981).
86. B. Neri, A. Dilligenti, and P.E. Bagnoli, "Electromigration and Low-Frequency Resistance Fluctuations in Aluminum Thin-Film Interconnections," *IEEE Trans. Electron Devices*, Vol. ED-34, 2317 (1987).
87. T.M. Chen and J.G. Rhee, *Solid State Technol.*, Vol. 30, 49 (1977).
88. T.M. Chen and A van der Ziel, *Proc. IEEE*, Vol. 53, 395 (1965).
89. B. Neri, B. Pellegrini, and R. Saletti, "Ultra Low-Noise Preamplifier for Low-Frequency Noise Measurements in Electronic Devices," *IEEE Trans. Instrum. Meas.*, Feb. 1991.
90. T.M. Chen, Ph.D. thesis, University of Minnesota, 1964.
91. A. Diligenti, P.E. Bagnoli, B. Neri, S. Bea, and L. Mantellassi, "A Study of Electromigration in Aluminum and Aluminum–Silicon Thin Film Resistors Using Noise Technique," *Solid State Electron.*, Vol. 32, 11 (1989).
92. B. Neri, A. Diligenti, P. Aloe, and V.A. Fine, "Electromigration in Thin Metal Films: Activation Energy Evaluation by Means of Noise Technique: Results and Open Problems for Indium and Gold," *Vuoto (Sci. Technol.)*, No. 4, 219 (1989).
93. M.I. Sun, J.G. Cottle, and T.M. Chen, "Determination of Al-Based

Thin Film Lifetimes Using Excess Noise/Measurements,'' in *Noise in Physical Systems*, A. Ambrozy, ed., Akademiai Kiado, Budapest, 1989, p. 519.

94. A. Diligenti, B. Neri, P.E. Bagnoli, A. Barsanti, and M. Rizzo, ''Electromigration Detection by Means of Low-Frequency Noise Measurements in Thin Film Interconnections,'' *IEEE Electron Dev. Lett.*, Vol. ED-6, 606 (1985).
95. A.P. Schwarzenberger, C.A. Ross, J.E. Evetts, and A.L. Greer, ''Electromigration in the Presence of a Temperature Gradient: Experimental Study and Modeling,'' *J. Electron. Mater.*, Vol. 17, No. 5, 473 (1988).
96. H.B. Huntington, ''Effect of Driving Forces on Atom Motion,'' *Thin Solid Films*, Vol. 25, 265 (1975).
97. S.P. Sim, ''Procurement Specification Requirements for Protection Against Electromigration Failures in Aluminum Metallizations,'' *Microelectron. Reliab.*, Vol. 19, 207 (1979).
98. C.A. Martin, J.C. Ondrusek, and J.W. McPherson, ''Electromigration Performance of CVD-W/Al-Alloy Multilayered Metallization,'' *IEEE IRPS*, Vol. CH2787-0, 31 (1990).
99. H.A. Schafft, ''Thermal Analysis of Electromigration Test Structures,'' *IEEE Trans. Electron Devices*, Vol. ED-34, 664 (1987).
100. T. Kwok, ''Effect of Metal Line Geometry on Electromigration Lifetime in Al-Cu Submicron Interconnects,'' in *Proceedings of the 26th IEEE International Reliability Physics Symposium*, p. 185 (1988).
101. G.A. Scoggan, B.N. Agarwala, P.P. Peressini, and A. Brouillard, ''Width Dependence of Electromigration Life in Al-Cu, Al-Cu-Si and Ag Conductors,'' in *Proceedings of the 13th IEEE International Reliability Physics Symposium*, p. 151 (1975).
102. Janet M. Towner, ''Are Electromigration Failures Lognormally Distributed?'' in *Proceedings of the IEEE International Reliability Physics Symposium*, p. 100 (1990).
103. H.H. Hoang, E.L. Nikkel, J.M. McDavid, and R.B. MacNaughton, ''Electromigration Early-Failure Distribution,'' *J. Appl. Phys.*, Vol. 65, 1044 (1989).
104. D.J. LaCombe and E.L. Parks, ''The Distribution of Electromigration Failures,'' in *Proceedings of the 24th IEEE International Reliability Physics Symposium*, p. 1 (1986).
105. J.C. Ondrusek, A. Nishimura, H.H. Hoang, T. Sugiura, R. Blumenthal, H. Kitagawa, and J.W. McPherson, ''Effective Kinetic Variations with Stress Duration for Multilayered Metallizations,'' in *Proceedings of the 26th IEEE International Reliability Physics Symposium*, p. 179 (1988).
106. A. Bobbio and O. Saracco, ''A Modified Reliability Expression for the

Electromigration Time-to-Failure,'' *Microelectron. Reliab.*, Vol. 14, 431 (1975).

107. K.P. Rodbell and S.R. Shatynski, ''Electromigration in Sputtered Al-Cu Thin Films,'' *Thin Solid Films*, Vol. 108, 95 (1983).
108. K.P. Rodbell and S.R. Shatynski, ''A New Method for Detecting Electromigration Failure in VLSI Metallization,'' *IEEE J. Solid State Circuits*, Vol. SC-19, No. 1, 98 (1984).

7

GaAs ON SILICON PERFORMANCE AND RELIABILITY

P. PANAYOTATOS
Department of Electrical and Computer Engineering, Rutgers, The State University of New Jersey, Piscataway

A. GEORGAKILAS and N. KORNILIOS
Foundation for Research and Technology–Hellas, Research Center of Crete, Institute of Electronic Structure and Lasers, Heraklion, Crete, Greece

7.1 INTRODUCTION

High bandwidth computing introduces the requirement of massively parallel interconnetions that need to be fast and, at several levels, reconfigurable. Such realization of fast interconnects is not possible with the use of solely electrical interconnets and electronic components. Optical interconnects, with no line capacitance and mutual coupling, appear as the alternative of choice. The need for the utilization of optical interconnects in combination with optoelectronic components has become apparent for such applications as clock and data distribution or even power distribution [1]. Such utilization can be visualized at various levels such as board-to-board, chip-to-chip, or intra-chip. The crossover in the plot of the figure of merit versus interconnect length depends on

Electromigration and Electronic Device Degradation, Edited by Aris Christou.
ISBN 0-471-58489-4

the application and architecture, but clear trends can be observed and values determined for each case separately. Thus comparison of the interconnections between an NMOS source and a lumped capacitance load indicates that optical interconnects become the preferred approach for board-to-board applications for line lengths over 25 cm and for chip-to-chip applications for lengths between 1 and 25 cm [2].

Interest in optical interconnections has been one of the reasons for investigations in the growth of GaAs on silicon substrates, and the main interest of the writers. Comprehensive reviews on the subject of growth of GaAs/Si and applications in general have appeared in the literature [3]. On the subject of optical interconnects, significant progress has been made in the hybrid integration of compound semiconductor optoelectronic components with the Si substrate. The old technique of self-aligning solder bump connections [4] (''flip-chip'', or C-4) was revitalized in work out of Plessey, Caswell (currently GEC–Marconi) [5, 6] in alignment of optical fibers with integrated optics. Compared to hybrid, monolithic cointegration of the optoelectronic components on the same chip as the signal processing silicon circuitry has not seen the same development. A complete study of the cost–performance trade-offs in the application of monolithic versus hybrid integration is not available yet. Nevertheless, it is generally expected that the improved performance and lack of need for separate alignment of the optoelectronic component, which is present in the hybrid approach, will justify the compound semiconductor thin film growth cost, which is the major cost driver in the monolithic approach. Monolithic cointegration hinges on the availability of high-quality GaAs layers grown in windows opened in silicon substrates with prefabricated circuitry.

Early cointegration results were reported from MIT/Lincoln Labs, where Si MOSFETs were integrated with GaAs MESFETs [7]. It was followed by the integration of a double heterostructure (DH) GaAs/AlGaAs LED with the driving 5-μm-gate Si MOSFET. The LED was modulated to 27 Mbit/s, limited by the silicon device [8]. An improved version with 10 MOSFETs achieved modulation of over 100 MHz [9]. Lifetime measurements at room temperature for 144 hours indicated a drop of LED output by about 30%. The silicon circuitry was reported relatively unaffected, even though the process seems to involve an 850°C oxide desorption step [7].

The results from Texas Instruments [10, 11] broke new ground on the cointegration of GaAs with silicon circuitry. They demonstrated the circuit level cointegration of GaAs and Si on a coplanar structure. MBE window growth of GaAs in prefabricated misorientated silicon substrates yielded layers that were successfully used for the definition of a

GaAs 1-μm-gate MESFET buffered FET logic (BFL) ring oscillator. A composite ring oscillator was demonstrated consisting of 35 Si CMOS inverters and 12 GaAs MESFET BFL inverters and level shifters. A complete process flow was developed in the second publication [11]. TiW/Au was used for metallization for both types of circuits. Practical problems, such as low yield due to misalignment in lithography patterns because of the use of different DSW steppers, were stated as one of the difficulties encountered. Changes in the threshold voltage of the silicon CMOS devices were reported, which could only partially be rectified by annealing in forming gas. Window growth included a substrate oxide desorption step at 950°C for 5 minutes.

The concerns about aluminum spiking in first-level metallized CMOS substrates led to the adoption of lower temperature processing by several laboratories. Work at Lawrence Livermore on the cointegration of photoconductors (PCs) on silicon [12, 13] on layers grown by both MBE and MOCVD resulted in PCs with rise times of 50 ps (optical pulse-width limited) and responsivity of 1.6 A/W for PCs defined on layers that had been grown at low temperature (400°C). Work at CNM–Madrid, for the cointegration of asymmetric Fabry–Perot MQW modulators with fully Al metallized CMOS circuitry on the substrate by low-temperature atomic layer MBE (ALMBE) demonstrated the feasibility of the process [14]. Shifts in CMOS characteristics could be totally attributable to damage during e-gun contact deposition.

7.2 PROBLEMS STEMMING FROM THE GaAs/Si HETEROEPITAXY

In terms of reliability, there are two questions to be addressed. The first deals with the failure mechanisms of the CMOS substrate, which are generally recognized to be primarily due to aluminum spiking, interdiffusion, and contamination by GaAs. This part of the problem is of interest, in the context of this book, only to the extent in which attempts to address these silicon failure mechanisms introduce failure mechanisms in the GaAs layer. The second question deals with failure in the GaAs layer itself, for which novel mechanisms could be introduced by aspects of the heteroepitaxial growth of GaAs on silicon.

There are fundamental differences between GaAs grown on GaAs and GaAs grown on silicon. They are the result of a number of Si–GaAs incompatibilities: there is a 4.1% lattice parameter mismatch between GaAs and Si and a difference of more than a factor of 2 in the thermal expansion coefficients. The former results in interfacial misfit dislocations and crystal defects that can reach the active layer, affecting device performance. The latter leads to stress when the structure is cooled down from growth temperature to room temperature, especially in blanket growth, resulting in wafer bowing and, for thicker layers, in microcracks. It may also be one of the causes [15, 16] of threading dislocations. There are additional fundamental difficulties stemming from the growth of a polar compound semiconductor (GaAs) on a nonpolar substrate (Si), leading to the problem of antiphase domains and boundaries (APDs and APBs) [17, 18]. APDs are areas of reverse Ga and As positions so that APBs are planes of Ga–Ga or As–As bonds. APBs can be highly charged, can degrade overall mobilities, and are generally deleterious to device performance. Contributions from several laboratories have provided different techniques for responding to both problems. In terms of diminishing crystal defects, the techniques mostly used are the two-step growth method [19, 20], strained-layer superlattices (SLSs) [21] and thermal treatments [22, 23]. The two-step growth method [19, 20] consists in first growing a thin GaAs layer at low temperatures, then increasing to normal growth temperatures in order to grow the required layer thickness. In such a process, a localization of misfit dislocations is believed to occur near the interface. Another method utilizes SLSs, which, when used as buffer layers, are very efficient in reorienting dislocation axes parallel to the growth plane, suppressing their propagation toward the active region (dislocation bending, anchoring) [21]. The typical remaining dislocation density is approximately 10^8 cm^{-2} when observed by transmission electron microscopy (TEM) [24–26]. It should be noted here that selective etching to reveal dislocations by KOH solution underestimates the real density of dislocations [27] by up to three orders of magnitude and is the origin of the early optimistic claims in the field and, we believe [26], recent spectacular values [28]. The remaining stress at room temperature is approximately 10^9 dyn/cm^2, much too high for several optoelectronic devices [29]. From past studies of GaAs/AlGaAs double heterostructure lasers, it is known that the maximum acceptable residual stress is 2×10^8 dyn/cm^2, otherwise dislocations tend to glide under optical injection.

In terms of performance, the ultimate target for GaAs/Si thin films is the electrical performance of homoepitaxial GaAs/GaAs. A necessary condition is the elimination of APBs and the reduction of threading dislocations, stacking faults, microtwins, and oval defects. A study was performed [26] that was geared toward optimizing the preparation of the substrate, the growth time parameters, and postgrowth treatment for best active layer properties. In particular, the study of growth involved optimization of substrate preparation, study of silicon substrate orientation, ex situ treatment, as well as multilayer and silicon buffer layers. For determination and quantification of film quality, a number of characterization methods were used both in situ (RHEED) and ex situ (electrical (*I-V*, *C-V*, DLTS, Hall), Nomarski interference optical microscopy, electron microscopy (TEM and SEM), electron channeling patterns (ECP), and x-ray double crystal diffractometry (DDX)). Several device structures such as Schottky diodes, *p-n* heterojunctions, MESFETs, and HEMTs were fabricated on these films and used for quantification of layer quality. Metal–semiconductor–metal photoconductors/photodetectors (MSM PC/PDs) were chosen to test the quality of the grown films in a simple optoelectronic application [30].

In terms of reliability, changes in the thin film induced by the schemes for improvement were observed through structural (mainly TEM) and electrical (mainly Hall) characterization. For devices, reliability studies were carried out by a comparative failure analysis [31] of GaAs MESFETs on GaAs / GaAs and GaAs / Si as well as by an accelerated aging study of GaAs / Si MSM PC / PDs.

7.3 MATERIAL PROBLEMS AND SCHEMES FOR IMPROVEMENT

The presence of APBs is a major consideration. Although the problem appears solved in practice, the solutions generate constraints, such as temperature treatment steps. Surface morphology, surface defects, and smoothness correlate with the presence and density of APBs and were observed by Nomarski and scanning electron microscopy. Cross-sectional TEM (XTEM) reveals APB self-annihilation with distance away from the interface. APBs were studied by utilization of a chemical stain [32] (Figure 7.1), by chemical lift-off [33], and by conventional TEM and XTEM observations [34]. Results were obtained using an $HF:HNO_3$ stain solution for APBs proposed before [32]. This etch reveals the APBs as trenches (Figure 7.1). APB density at different depths was examined by the use of an isotropic GaAs etch before application of the APB stain. Following APB stain, the samples were examined by SEM. APB density was quantified by the number of APBs encountered along 1 cm of sample surface. An average was considered along various directions. Investigation of the effect of the following specific variations were carried out: Si substrate orientation, temperature and thickness variation, prelayer type, and deposition temperature [26].

Electron microscopy observations in various samples grown at dif-

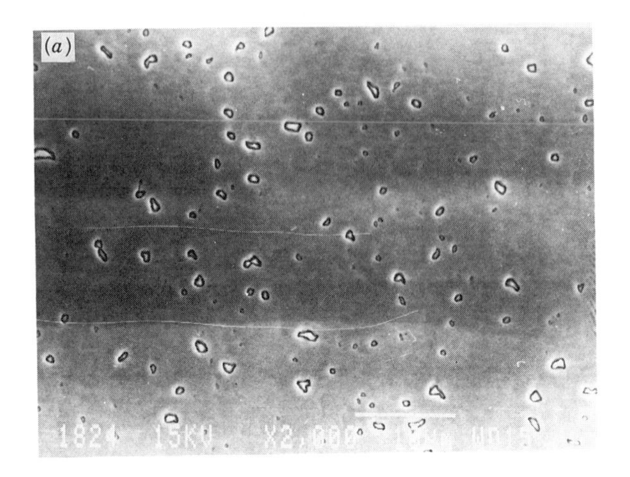

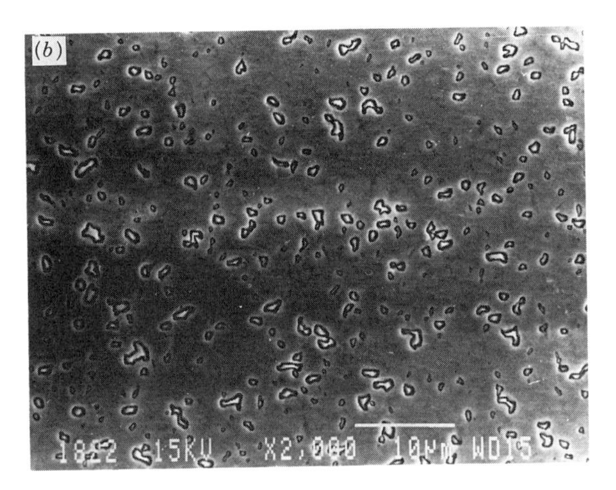

FIGURE 7.1. SEM micrographs of stained APBs on the surface of 2 μm GaAs samples on Si(001) misoriented by 3.5° towards ⟨100⟩ and grown at 575°C. Nucleation temperature: (*a*) T_N = 190°C (*b*) T_N = 565°C.

ferent growth temperatures and orientations have indicated that, in GaAs on Si, high growth temperatures improve the GaAs crystal. Low-temperature growth always resulted in the appearance of significant densities of stacking faults and microtwins, while a high-temperature annealing following the nucleation GaAs layer was found to assist in the reduction or elimination of planar faults in the overall epilayers. Si technology has matured on (001)Si wafers. This and the early observations of Kawabe and Ueda [35] resulted in the enhanced interest in Si(001) substrates misoriented toward ⟨110⟩. This misorientation was found

[26] to result in the best material under the optimum two-step growth conditions proceeding as follows: uniform As prelayer at low temperature (350°C) on Si, thin nucleation layer at a low growth rate, annealing of nucleation layer, high growth temperature. Oval defects were hillocks with or without core particulate of a typical density of 10^2–10^4 cm^{-2}. Thus, for APB-free layer formation, one finds that misorientation by at least 2° toward ⟨110⟩ is indispensable. Uniform preexposure of the Si surface is also necessary; without it, even this misorientation resulted in APBs.

APB density was not found to depend significantly on the growth temperature for variations between 500 and 600°C. However, nucleation temperature is strongly related to APB density [34]. This is also apparent by comparison of the SEM micrographs of Figure 7.1, where the sample with the lower nucleation temperature is shown to have the lower APB density. Low nucleation temperature ensures that small GaAs islands coalesce [36, 37]; thus APB annihilation is accomplished more efficiently. APB annihilation is achieved during growth, as indicated by the comparison of APB density at, say, 1.5 μm of a 3.3-μm film with the density on the surface of a 1.5-μm film (Figure 7.2).

FWHM correlated to film thickness and growth temperature is shown in Figure 7.3. DDX FWHM was found to decrease significantly for as small an increase in growth temperature as 20°C (Figure 7.3*a*). Correlation with thickness exhibits values as low as 140 arc seconds for

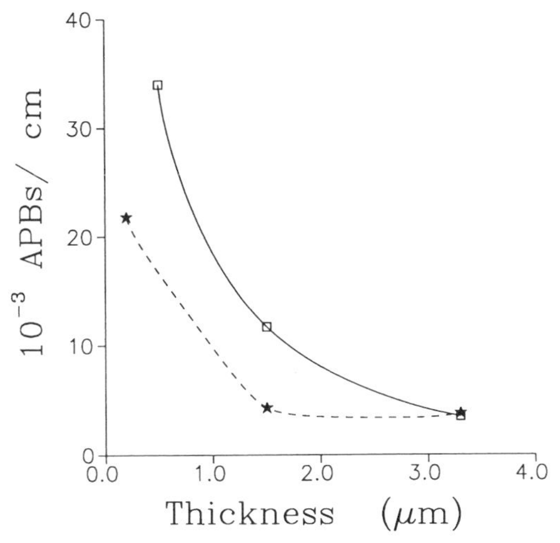

FIGURE 7.2. Comparison of APB density depth profile (dashed line) of a 3.3 μm GaAs/Si film to surface APB densities of GaAs/Si samples with various thicknesses (solid line).

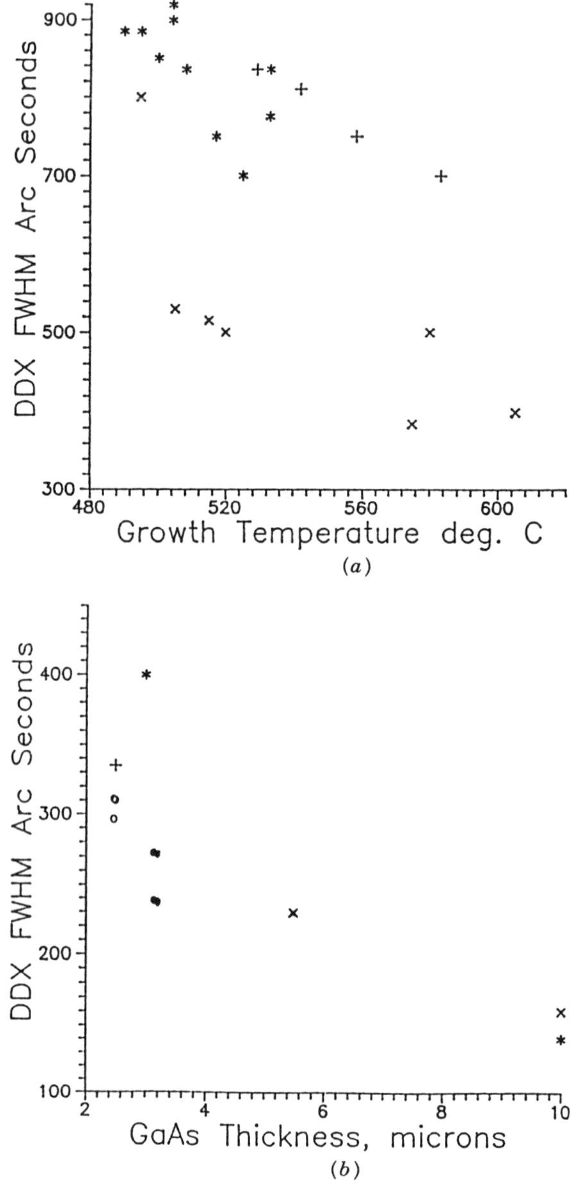

FIGURE 7.3. (*a*) DDX FWHM correlated with growth temperature for GaAs/Si(001) 4° → ⟨100⟩, with thickness as a parameter x: 3.0–3.5 μm, *: 1.5 μm, +: 1.5 μm (*b*) DDX FWHM correlated with GaAs thickness x: GaAs/Si (001) *: GaAs/Si (001) 3° → ⟨110⟩, T_G = 580°C open circles: GaAs/Si (001) 3° → ⟨110⟩, T_G = 600°C +: AlGaAs/Si (001) 2° → ⟨110⟩, solid circles: GaAs/Si (001) 3° → ⟨110⟩ with MBE silicon buffer.

thicker films of 10 μm (Figure 7.3*b*). There is, of course, a limit on how thick a GaAs one can grow, which is set by the appearance of cracks for film thicknesses above about 5 μm. DDX FWHM was found to be relatively insensitive to low densities of APBs; thus GaAs grown on exact (001)Si exhibit only slightly higher FWHM (Figure 7.3*b*). Since even low densities of APBs degrade mobilities significantly and are deleterious to device performance, this also is a measure of the limit of usefulness of DDX measurements.

In the above study [26] we did not rely solely on RHEED observations during growth for the determination of GaAs orientation and sublattice occupation [38–41] but correlated the GaAs orientation with the Si misorientation direction after growth. The Si $\langle 110 \rangle$ misorientation direction was deduced easily by ECPs [42] observed in the SEM. For such Si substrates where the (001) planes are inclined 3°–4°, the macroscopic GaAs/Si surface ECP center is shifted along the misorientation direction. GaAs layer orientation was deduced by the utilization of a preferential $HCl:H_2O_2:H_2O$ (1:4:40) etch [43] to open trenches parallel to the $\langle 110 \rangle$ GaAs directions. Inspection of the slope of the trench walls indicated the GaAs orientation, since the walls coincide with the $\{111\}$A planes. GaAs orientation can also be correlated to the longitudinal axis of the GaAs oval defects [44]. Selective etching showed that the GaAs textured morphology is oriented along the $[1\bar{1}0]$ direction. Also parallel to the $[1\bar{1}0]$ direction were the longitudinal axes of the hillock-type oval defects, while the hole-type oval defects were elongated in the [110] direction.

Hall mobility measurements were the major means of quantification of electrical quality of these layers. For a substrate of Si(001) misoriented by 3° toward $\langle 110 \rangle$, films grown with 1.0 μm buffer and 1.5 μm *n*-GaAs at 2×10^{17} cm^{-3} exhibited Hall mobility at 300°C (μ_{RT}) of 3800 and 4100 cm^2/Vs at 77°C (μ_{77K}) as compared to the homoepitaxial control sample with values of $\mu_{RT} = 4000$ and $\mu_{77K} = 4400$ cm^2/Vs. The DDX FWHM ranged between 295 and 310 arc seconds. Samples on nonmisoriented substrates consisting of 1.1 μm buffer and 1.9 μm *n*-GaAs at 2×10^{16} cm^{-3} exhibited maximum Hall mobilities of $\mu_{RT} = 4500$ and $\mu_{77K} = 11000$ cm^2/Vs, not uniformly distributed on the wafer due to the APBs resulting from the nonmisoriented substrate. These should be compared with the homoepitaxial control with $\mu_{RT} = 5300$ and $\mu_{77K} = 13000$ cm^2/Vs. The DDX FWHM ranged from 240 to 270 arc seconds. The results indicate [26] that it is essential to completely eliminate APBs from the early stages of growth in order to obtain device quality GaAs films on Si. The effect of APB density on Hall mobility is shown in Figure 7.4 where it is clearly shown that

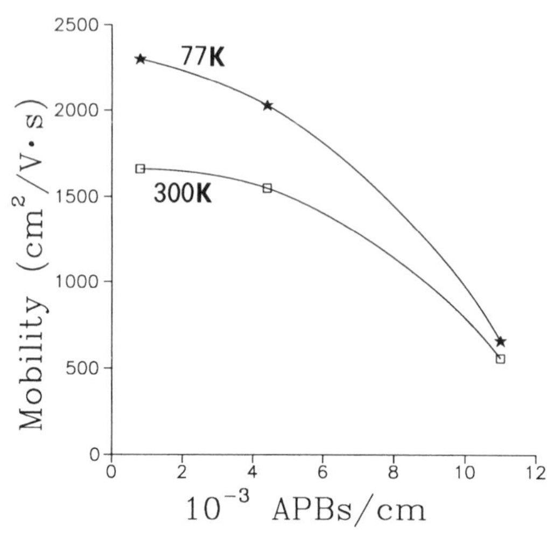

Figure 7.4. Electron Hall mobility at 77 and 300 K, as a function of APB density for n-GaAs (10^{17} cm^{-3}) grown on (001)Si substrates 3° off towards ⟨100⟩.

mobility rapidly decreases with APB density. Samples consisted of 1.0 μm of Si-doped GaAs (10^{17} cm^{-3}) grown on 1.0 μm of GaAs buffer grown on (001)Si substrates 3° off toward ⟨100⟩. A significant mobility degradation was also observed for GaAs grown on nonmisoriented (001)Si, or misoriented toward ⟨110⟩. Spatial variation of the mobility on the same wafer by a factor of 3 (from 4600 to 1500 cm^2/Vs) was observed for extremely low APB densities depending on the choice of areas with or without APBs [45]. Thus APBs seem to scatter carriers very efficiently, which is consistent both with the presence of charges at APBs and with the lattice distortion introduced. Strong Si compensation was observed for high APB density in addition to mobility degradation for Si(001) misoriented toward ⟨100⟩. Combined gradual etching with Hall measurements indicated a carrier concentration and mobility change from surface values of $N_s = 2.0\text{–}2.2 \times 10^{12}$ cm^{-2} (and μ_{RT} = 2200–2500 cm^2/Vs) to 1.4×10^{11} cm^{-2} (and μ_{RT} = 900 cm^2/Vs) at 0.4 μm away from the interface. In contrast, samples on (001)Si misoriented toward ⟨110⟩ with APBs showed an additional carrier concentration near the interface (from $N_s = 1.0\text{–}1.2 \times 10^{13}$ cm^{-2} and μ_{RT} = 1400–1700 cm^2/Vs to values of $N_s = 1.3 \times 10^{13}$ cm^{-2} and μ_{RT} = 800–900 cm^2/Vs at 0.2 μm away from the GaAs/Si interface). This increase may be attributed to same-charged APBs with a nonzero average charge introduced through favoring by the anisotropy of the misorientation. Such a two-dimensional concentration of charges near the interface has also been reported by Chand et al. [24].

Additional approaches for material improvement consisted of Si and ternary strained layers and strained layer superlattices (SLSs) included in the grown film and the nucleation layer, laser assisted MBE (LAMBE), postgrowth treatment by rapid thermal annealing (RTA), and the use of MBE-grown Si buffer layers between the substrate and the GaAs nucleation layer. In a brief comparative study of Si-, AlGaAs-, and InGaAs-, based SLSs, the lowest dislocation density of 4×10^8 cm^{-2} was found when, following a 30-nm GaAs nucleation layer, a structure consisting of {0.38 μm of GaAs + 3 periods (20 nm $In_{0.10}Ga_{0.90}As$/20 nm GaAs)} was repeated twice. This, however, represented a small difference in dislocation density, which could be due to the generation rather than the propagation of threading dislocations and could be a result of deviation of the GaAs nucleation on Si. Nevertheless, XTEM evidence indicates that these superlattice buffers do inhibit dislocation propagation and that the effect is stronger for increased SLS buffer thicknesses [46].

A III-V growth MBE chamber was used, which shares the same preparation chamber with a silicon MBE, so that samples can be taken from one chamber to the next without breaking the UHV. The Si MBE is equipped with a Si e-gun source. GaAs on Si with an MBE Si buffer (GaAs/Si/Si) were grown in the following structure: 1–3 μm of GaAs grown in the III-V growth chamber on 0.2–1.0 μm of Si grown in the Si growth chamber on a substrate of p-Si(001) misoriented by 4° toward ⟨110⟩. The standard Si process [26] was used for the preparation of the substrates. Oxide desorption, however, took place in the Si MBE. The procedure includes 20 minutes of heating at a temperature estimated as less than 750°C, 10 minutes under a low Si flux [47] at the same temperature, and 10 minutes while heating the Si source for higher deposition rate. This resulted in a silicon oxide desorption temperature at least 100°C lower than without the use of the Si beam, so that the high-temperature treatment process step previously included for step doubling was eliminated. As a result, the lower thermal stress had the effect of the complete elimination of the slip lines that are normally observed at the edges of the Si wafer after oxide desorption in the high-temperature process. Si buffer and GaAs thicknesses were used as parameters. The growth rate was 1 μm/h for GaAs and 0.2–0.4 μm/h for Si and the growth temperatures were 600°C for GaAs and 680–750°C for Si. XTEM observations indicated that the crystal quality of the GaAs layers was improved overall. A typical XTEM micrograph for a sample with 2 μm GaAs and a Si buffer thickness of 0.73 μm is shown in Figure 7.5. A high dislocation density is present in a thin interfacial region, which drops rapidly with distance away from the interface to about 10^9

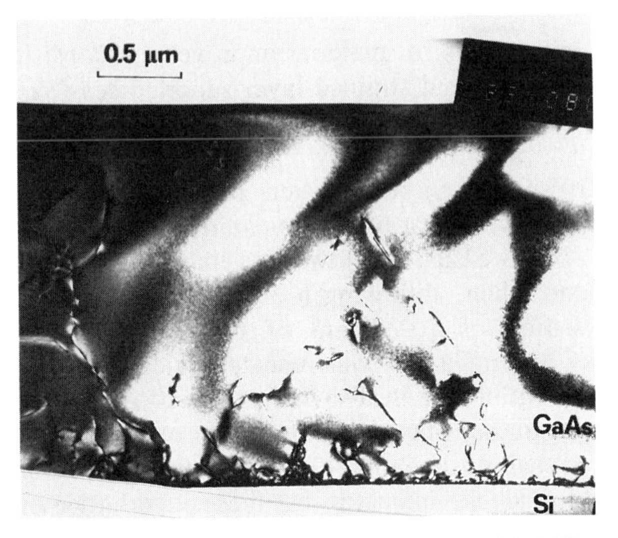

Figure 7.5. XTEM micrograph for 2 μm of GaAs on Si with 0.73 μm of MBE Si buffer. The electron beam was parallel to the misoriented Si step edges.

cm^{-2} midway and 10^8 cm^{-2} at the GaAs surface. No planar faults, such as stacking faults, microtwins, and APBs, were observed. This indicates that utilizing such an improved Si surface ensures a two-dimensional growth mode [48], from the early stages of growth, which follows a homogeneous three-dimensional nucleation despite the low-temperature (<750°C, rather than 850–1000°C) treatment used to clean the Si substrate. A two-dimensional growth mode is also suggested by the absolutely smooth GaAs surfaces observed by SEM and by the regular Moire pattern of the GaAs/Si interface [26].

In the micrograph of Figure 7.5 there is no dislocation crossing the GaAs surface along the 4-μm length of the specimen. This seems to be one of the lowest densities ever obtained in GaAs on Si. Even so, this does not guarantee dislocation densities lower than 10^8 cm^{-2}. In order to be transparent, the cross-sectional specimen thickness is about 0.2 μm. Thus a dislocation density of 10^8 cm^{-2} would approximately correspond, after a geometrical correction, to the appearance of one dislocation per 10 μm of specimen length. Therefore, even in our improved material, the dislocation density for 2-μm-thick GaAs/Si/Si samples is estimated to be in the low 10^8 cm^{-2} range. From several XTEM micrographs an average dislocation density of 2×10^8 was estimated for the sample shown in Figure 7.5 with a 0.73 μm Si buffer layer while an average of 4×10^8 cm^{-2} was estimated for a sample with a 0.34 μm Si buffer layer. More accurate estimates of the dislocation density were obtained using TEM observations in planar specimens prepared from the top 0.3 μm of the samples, indicating dislocation densities of 3.5×10^8 and 5.5×10^8 cm^{-2}, respectively. Although the dislocation density remains in the 10^8 cm^{-2} range, we believe that these results are probably the best values ever obtained for 2-μm GaAs/Si samples without annealing treatment. Dislocation densities orders of magnitude lower have indeed appeared in the literature, but they are invariably determined by the etch pit density technique, which is unreliable [27]. X-ray techniques (DDX FWHM) are somewhat more reliable. DDX results are very encouraging for GaAs/Si/Si as compared to similar thicknesses of GaAs/Si without a Si buffer. The value of 255 arc seconds FWHM for a 1-μm film was achieved, which is the best value ever reported. For our 2-μm GaAs/Si/Si samples, DDX FWHM had a value of about 215 arc seconds, which is one of the best values ever reported for samples of such thickness. In addition, it seems that there is a much weaker correlation between DDX FWHM value and thickness when a Si buffer is present, which suggests that thinner GaAs layers may be sufficient with the use of GaAs/Si/Si. We found [26], however, that one needs to use the "upper limit" or "cap" calculated from DDX FWHM values with caution since the formula $D(\text{cm}^{-2}) \approx 1630(\text{FWHM})^2$ (with FWHM expressed in arc seconds) underestimates the dislocation density in GaAs on Si by about a factor of 5. TEM measurements, which are the ones producing the highest numbers for D, should be the ones that do leave the fewest dislocations uncounted. Indeed, other researchers [24] who obtained the best x-ray FWHM ever reported for 3 μm of GaAs/Si (about 150 arc seconds) also found their dislocation density in the low 10^8 cm^{-2} range.

A LAMBE parameter study was carried out on the structure of 1.4 μm of n-GaAs (3.8×10^{17} cm^{-3} as determined from C-V profiler) on 0.5 μm of undoped GaAs grown on substrates of n-type Si(001) misoriented by 2°–3° toward ⟨110⟩. Both MBE and LAMBE structures were grown at a nucleation temperature of $T_N = 350°\text{C}$ and a growth temperature of $T_G = 600°\text{C}$, for a growth rate of 0.8 mm/h. Some laser parameters are determined by other than optimization considerations. There is a constraint imposed by the absorption of the arsenic-coated quartz window on the repetition rates. Experiments performed at different repetition rates determined that 15 Hz is probably the highest one should attempt without endangering the quartz window from overheating and risking MBE implosion. The other parameters were a wavelength of 308 nm, a rectangular 2×1 cm^2 beam of 10–15 ns pulse with 40 mJ/cm^2 impinging on the quartz window. The resulting structures were characterized by photoluminescence and it was found that the FWHM of PL light hole exciton peak was lower at 10 meV for the

LAMBE structure as compared with 15 meV of the MBE structure. It was also found that the shift from homoepitaxial control of the peak at 1.488 eV was the same, at about 26 meV, in both cases indicating similar tensile thermal strain; that is, laser processing (at this energy density) does not affect strain. Surface morphology was of similar high quality in both cases. Thus LAMBE is judged to result in a (small) improvement of the otherwise optimized structure.

We have examined the effect of RTA on the crystalline quality of the relatively optimized GaAs/Si/Si sample. Two RTA treatments were involved. In both cases the sample was treated at 870°C for approximately 3 seconds. However, in one case the sample was allowed to cool to room temperature, while in the other case cooling was achieved in steps with the sample kept at 350°C for 10 minutes and at 250°C for another 10 minutes before it was allowed to reach room temperature. These low-temperature steps were included in order to provide the time for dislocations to move while they are still mobile. Indeed, residual strain measurements [49] have shown that the terminal strain present in these GaAs layers agrees with a calculated value based on the thermal expansion coefficient difference consistent with a growth temperature of 250°C rather than the actual 580°C. In other words, dislocations freeze to where they are when the sample goes through 250°C on cooling, suggesting dislocation motion at temperatures higher than 250°C.

The XTEM micrographs of Figure 7.6 show the overall GaAs quality with RTA either followed by the low-temperature steps (Figure 7.6a) or not (Figure 7.6b). Cracks appeared for the directly cooled sample (Figure 7.6b) with a density of approximately one crack per 10 μm in XTEM specimens. In both cases, there was an apparent dislocation density reduction in the GaAs surface of about a factor of 3 to values around 1.5×10^{18} cm^{-2}. XTEM micrographs indicate that dislocation movement (glide or climb) and reactions between dislocations occur during RTA. Thus perfect dislocation loops are formed (Figure 7.6a and 7.6b). Close to the cracks the dislocation density is lower, probably because dislocations escape at cracks, resulting in the formation of denuded zones around them (Figure 7.6b). Straight defects, which appear inclined 55° from the interface, are perfect dislocations laid on the $\{111\}$ planes at different depths in the specimen and projected as straight lines on the plane of observation. No SFs were observed. The XTEM micrograph in Figure 7.7 shows the GaAs/Si interface of the sample of Figure 7.6*a*. The 220 reflection was strong and by somewhat tilting the specimen, dislocation contrast appeared for both of the two orthogonal rows (along $\langle 110 \rangle$) of misfit dislocations. These reflections show an ideally periodic contrast with periodicity of 9 nm in both directions.

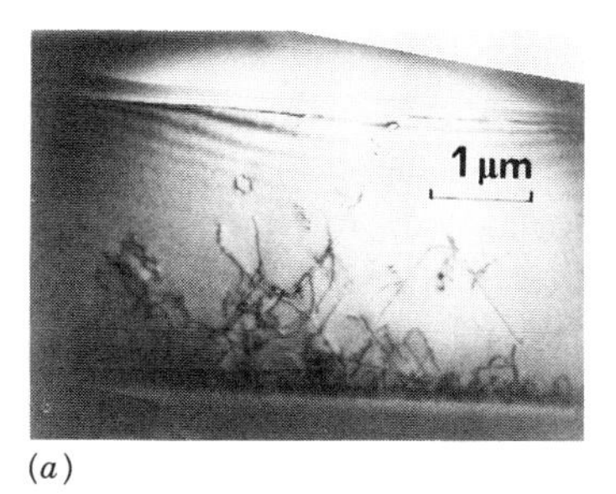

(*a*)

(*b*)

Figure 7.6. XTEM micrographs after RTA of 870°C, 3 sec, for a sample of 2 μm of GaAs on Si with 0.34 μm of MBE Si. Cooling to room temperature (*a*) with cooling steps of 10 min at 350°C and 250°C, and (*b*) immediately (cracks).

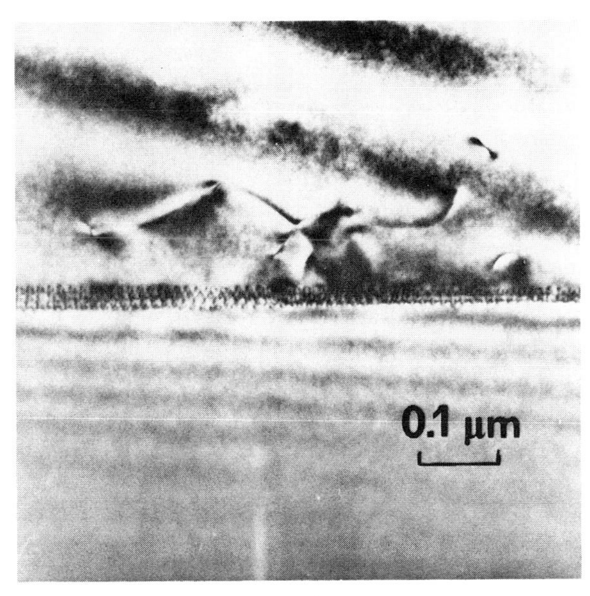

Figure 7.7. XTEM micrograph showing the ideally periodic row of misfit dislocations in the GaAs/Si interface for the Rapid Thermal Annealed sample of Figure 7.6*a*.

Thus they form an ideal square grid of perfect edge-type dislocations [50] to accommodate the misfit between GaAs and Si. Although XTEM observations indicated a significant dislocation density reduction, TEM observations in planar (001) specimens near the GaAs surface have shown that only a dislocation redistribution actually occurs. Thus a higher density appears in local, relatively limited area regions, while a less dislocated crystal appears everywhere else.

7.4 DEVICE PERFORMANCE AND RELIABILITY

7.4.1 Diodes

P-n heterojunction diodes were used to characterize the GaAs–Si interface and Schottky diodes were used to characterize the surface of the active layer and compare with that on homoepitaxial layers. The *p-n* heterojunctions were formed between the n^-- or n^+-GaAs epilayer on *p*- or p^--Si with resistivities of up to 2000 Ωcm. The band bending value obtained from the intercept voltage is not significantly different from the one calculated from the measured doping concentration of the n^--GaAs (1×10^{15} cm^{-3}) and the *p*- substrate (5×10^{14} to 1×10^{15} cm^{-3}). From such comparisons we may conclude that interface states do not play a major role in determining the intercept voltage and that these interfaces between GaAs and silicon are not completely dominated by interface states.

Schottky diodes were formed routinely on the grown films [26]. The performance of these diodes was compared to the characteristics of similar devices formed on GaAs/GaAs homoepitaxial material. The figures of merit examined were leakage current, ideality factor (*n*-value), barrier height, and deep trap density and location. The diode characteristics were chosen as a measure of the density of electrically active surface defects, and deep level transient spectroscopy (DLTS) was used to determine the location and concentration of traps deeper in the film. *I-V* results indicated that *n*-values are typically 1.2 for Al Schottky diodes and 2.0 for TiW/Si contacts before annealing. The comparison between devices on GaAs/Si and on GaAs/GaAs films in term of breakdown voltage shows that for a doping level of 2×10^{16} cm^{-3} the GaAs/Si devices exhibit breakdown at about 22 V, approximately the same as the homoepitaxial device. For a doping level of 2×10^{17} cm^{-3}, the breakdown voltage is about 7 V for the GaAs/Si devices as compared to the homoepitaxial device, which breaks down at about 8 V. Doping was determined from *C-V* curves, which were typically well behaved with a good straight line character for C^{-2} versus V, indicating constant doping level in the thin film. Barrier heights determined from these C-V measurements were typically in the range of 0.7–0.8 eV.

7.4.2 MESFETs

The FET structure consisted of 45-nm 2×10^{18} cm^{-3}/0.13-μm 3×10^{17} cm^{-3}/2-μm undoped GaAs/0.8-μm $p(6 \times 10^{16}$ cm$^{-3})$ MBE Si/p-Si 2000 Ωcm. MESFET characterization for devices with a gate length of 1.5 μm and width of 250 μm resulted in maximum extrinsic transconductance values of $g_m \simeq 230$ mS/mm at $V_{GS} = 0.6$ V and $V_{DS} = 2.5$ V [31], when Si buffer layers were used, which is actually higher than the value exhibited on GaAs/GaAs homoepitaxial layers characterized for comparison. In addition, GaAs/Si/Si MESFETs exhibited negligible light sensitivity, indicating a very low trap density in the active layer. The gate–source Schottky diodes in GaAs/Si/Si exhibited low leakage and high breakdown voltages. The value for g_m/g_D was 45–70 and leakage currents were low, albeit higher than for devices on homoepitaxial layers. HEMT characterization for 2DEG mobility at liquid nitrogen temperature resulted in a value of 31,000 cm^2/Vs with processed transistors exhibiting transconductance on the order of 200 mS/mm. These values are in the same range with the best results reported in the literature [24].

A comparative reliability and failure analysis was performed [31] in order to assess differences in GaAs MESFET performance and reliability between devices defined on GaAs grown on GaAs versus devices on layers grown on Si. MBE-grown 1-μm gate GaAs MESFETs on GaAs substrates (group A) were compared to GaAs MESFETs on Si substrates (group C). In the latter case a 2-μm GaAs buffer layer was grown on a substrate of (001)Si 2° misoriented toward ⟨110⟩. Group A had AuGe/Au drain and source metallizations and TiWSi/Au for the gate contact. Group C utilized AuGe/Ni/Au for the ohmic contacts and Al/Au gate metallizations. Closed ceramic ovens regulated within 1°C to the specified value were used for thermally accelerated aging in the range of 200–250°C. No passivating layers were used for the devices, which were not assembled into packages.

The parameters that were judged most sensitive to monitoring performance degradation were the drain–source current and the transconductance g_m at different gate biases. The characteristics of the FET, gate diode, and the pinchoff voltage (V_p) were monitored. The resistance of the active channel and the drain–source resistances ($R_D + R_S$) were also measured in the ohmic region of the FET. R_S and R_D serve to monitor the degradation of ohmic contacts and the active channel resistance was used to characterize the interaction between the gate and the semiconductor in the channel region. The DC characteristics before aging of the GaAs/GaAs and GaAs/Si FETs were similar.

During aging all devices exhibited a decrease in both drain–source saturation current at zero bias (I_{dss}) and g_m. The results for the degradation of I_{dss} with time t with different aging temperatures as a parameter for device groups A and C are shown in Figure 7.8. Each point represents an average over ten devices. Standard deviations remained about 10% throughout aging. Comparing the curves of 200°C and 250°C in the GaAs/GaAs (A) and GaAs/Si (C) groups, one observes that there seem to be no substantial differences between the two groups. Similar differences for different storage temperatures are observed for the two groups with I_{dss} degradation gradual at 200 and 220°C and more rapid for 240 and 250°C. In addition, it appears that, for the higher temperature, the degradation is sharp for the initial period of aging to become subsequently less pronounced. These two different regions were associated [31] with two different degradation mechanisms. The first, occurring at 240 and 250°C, is due to "gate sinking." This degradation was expected from previous results [51] and has been shown to be negligible below this range of temperature [31]. Gate sinking was assumed, rather than doping by metallization atoms in the channel, because of the stability of the I–V characteristics of the gate diode before and after annealing. The other type of degradation is due to the failure of the ohmic contacts. It was revealed by scanning electron microscopy, energy dispersive x-ray analysis, transmission electron microscopy examinations, and by the increase of the ohmic contact resistance together with the decrease of the source–drain saturation current during aging. At the higher temperatures and for the initial period of aging, both degradation mechanisms were present. The similarity of the I_{dss} aging curves for GaAs/GaAs and GaAs/Si suggest that both degradation mechanisms exist for both types of samples. Small differences can be attributed to metallization differences.

The second degradation mechanism, which is the only one present at lower temperatures and was experimentally confirmed to be an increase in the source and drain contact resistance, can provide an additional basis for quantitative comparison. Expressing the variation rate of the contact resistance ($R_S + R_D$) during aging as $R = R_0 e^{at}$, with R_0 the value before aging, resulted in experimentally determined coefficients $a(\mathrm{h}^{-1})$ ranging from 1.0×10^{-3} to $2.0 \times 10^{-4}\ \mathrm{h}^{-1}$ with comparable values for both groups A and C.

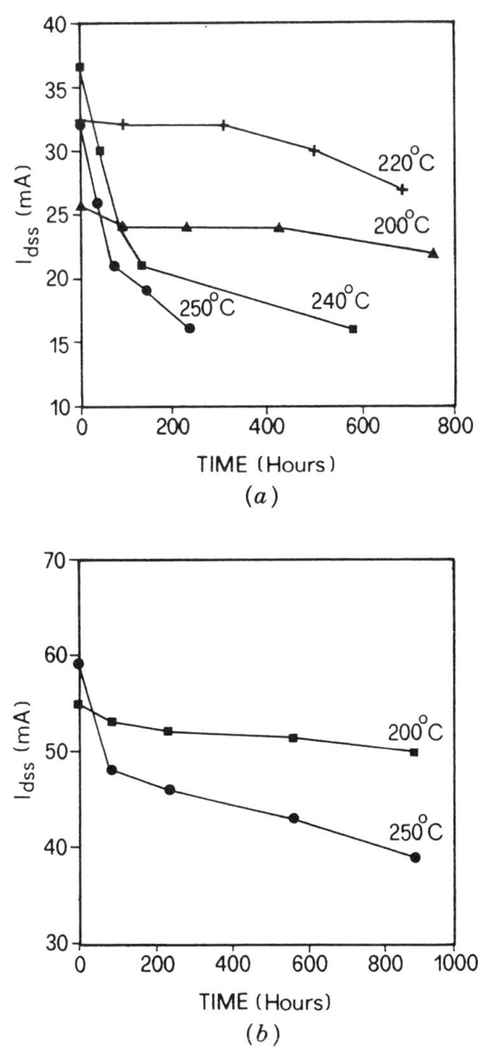

Figure 7.8. Average value of I_{dss} over 10 GaAs MESFETS taken at $V_{gs} = 0$ V and $V_{ds} = 3$ V as a function of aging at several storage temperatures. (*a*) group A, GaAs/GaAs; (*b*) group C, GaAs/Si.

Mean time-to-failure (MTTF) was defined based on a 20% decrease in drain–source saturation current. This current was assumed to obey an Arrhenius type relationship of $I_{\mathrm{ds}} = I_{\mathrm{ds0}}\,e^{-at}$ due to the increase of ohmic contact resistance. Then group A with activation energy of 1.5 eV and MTTF at 110°C of 1×10^7 h is found to be very similar to group C with values of 1.4 eV and 9×10^6 h, respectively. Failure analysis by SEM indicated in both device groups A and C the creation of interelectrode short-circuiting bridges [31] appearing as fingers extending from the source and drain areas toward the gate, resulting in a catastrophic failure. The conclusion drawn is that MESFETs on both GaAs and GaAs/Si exhibit the same reliability and the same failure mechanisms.

7.4.3 Photoconductors/Photodetectors (PCs/PDs)

PCs/PDs were chosen as simple devices that can be used to test the optoelectronic quality of the films grown. The parameters chosen as most sensitive for DC characterization were photocurrent at a specific light level and dark leakage current in 10-V bias. In order to establish the best thermal treatment parameters, in addition to photosensitivity optimization (I_{ph}) and dark (leakage) current (I_d) minimization, the parameters of Schottky diodes were chosen to be optimized. Since these PCs/PDs were defined on undoped GaAs/Si, additional Schottky diodes were defined on n-GaAs/Si, fabricated on the same runs. Then the barrier height (φ_b), and the ideality factor (n-value) were also measured as a function of annealing temperature (T_a).

Annealing is necessary in order to remove sputtering damage and decrease the n-value of the Schottky contacts. Thus annealing is effectively the first step of the accelerated aging process. Figure 7.9a shows I_d and I_{ph} versus T_a. The initial loss of photocurrent may be related to optical effects such as increased reflectivity. Otherwise, I_{ph} remains relatively unaffected. On the contrary, I_d decreases with T_a up to 250°C and remains relatively constant up to 300°C, after which it increases again. This is consistent with previous [52] DLTS investigations, which indicated that a majority carrier trap at 0.624 eV can be annealed out after heat treatment at 200°C for 4 hours, and that a novel trap appears at 300°C, possibly due to interdiffusion mechanisms. On the other hand, Schottky diode parameters indicate that sputter damage is annealed out better at 300°C. Indeed, the barrier height increases with T_a, and at 300°C the n-value is practically ideal (1.0), and increases again with increasing T_a (Figure 7.9b). The optimum temperature compromise was determined to be at 280°C. At this temperature the barrier height established from I-V agrees best with the value determined from $1/C^2$ versus V lines, which is an additional indication of near-ideal n-values, and thus of minimum interface states. Such optimally annealed TiW/Si contacted devices performed better in terms of both I_{ph} and I_d than Pd-based test structures. In terms of other processing, it was found that etching all the way to the silicon substrate guarantees reduction by an order of magnitude in the leakage current, and that an additional improvement of a factor of 5 can be picked up by the utilization of MBE Si buffers.

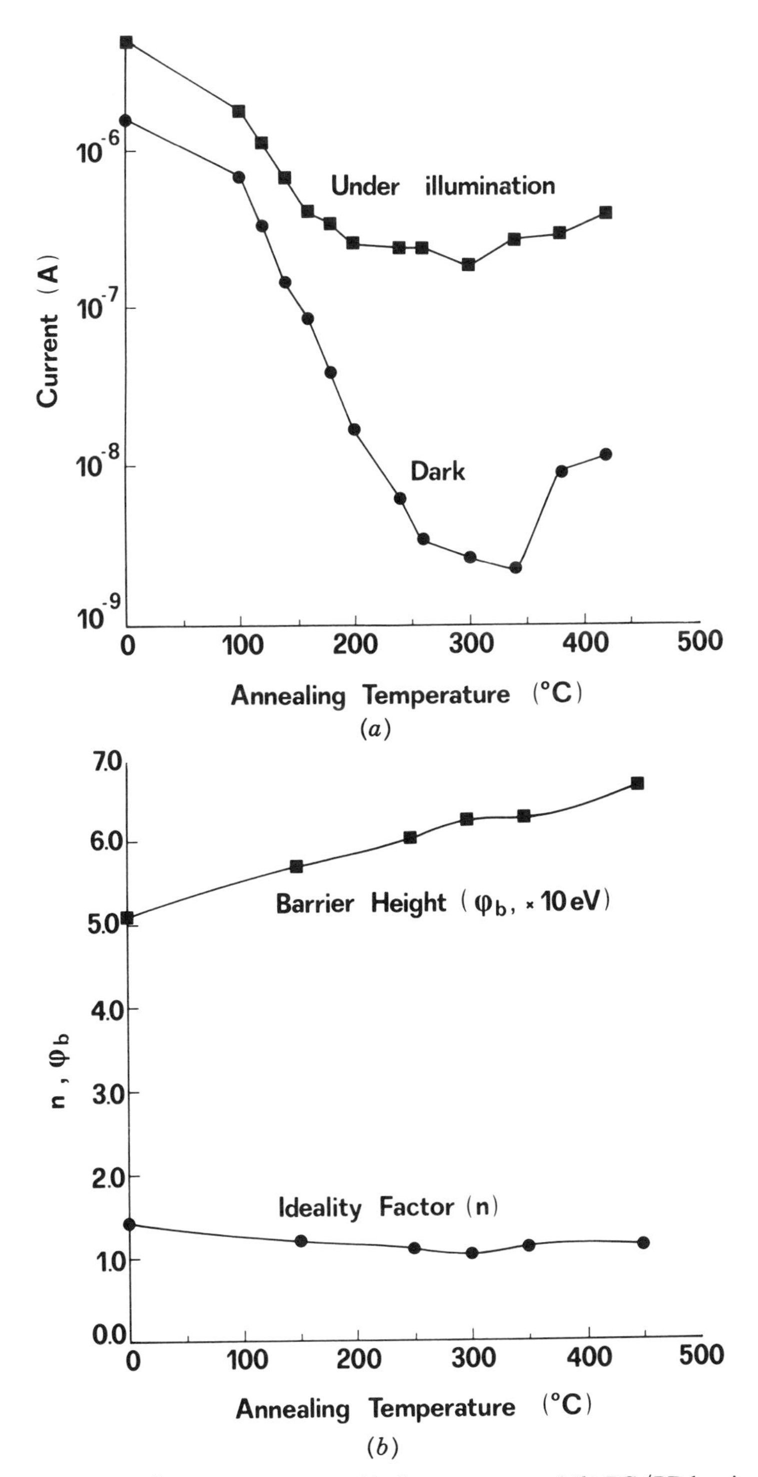

Figure 7.9. (*a*) PC/PD photocurrent and leakage current and (*b*) PC/PD barrier height and ideality factor as a function of annealing temperature.

The comparison of such optimized GaAs/Si devices with devices fabricated and treated at the same time on GaAs/GaAs homoepitaxially grown material shows that, in terms of DC, devices on GaAs/Si have comparable leakage currents at or below 1 nA, and slightly lower photoresponse (Figure 7.10). Transient response was also found to be comparable [53].

Accelerated aging of 20 interdigitated PCs/PDs, ten samples each at temperatures of 200 and 240°C, was carried out. The dark current and photocurrent at three illumination levels for a 10-V bias, judged as the most sensitive parameters, were used as the figures of merit. Approximately every three days, the samples were removed to room temperature and the dark current and photocurrent were recorded. A microscope illuminator was used for the measurements under illumination at three intensity levels. Figure 7.11 shows the average dark current (I_d) and photocurrents (I_{ph}) at three different intensity levels for ten devices with values exhibiting a standard deviation of about 5% in all measurements over three months. There was a consistent initial spectacular increase in photocurrent during the first 30 hours for the 240°C group or the first

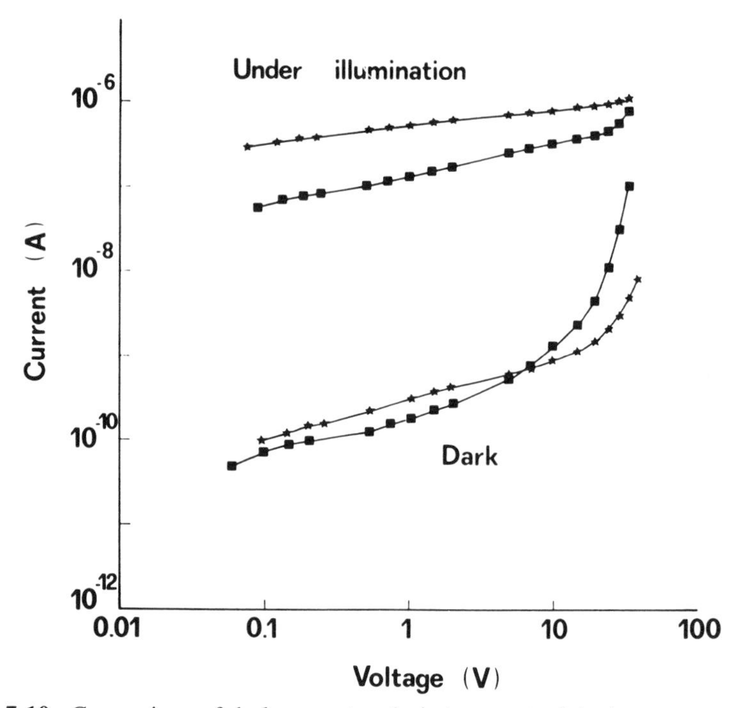

Figure 7.10. Comparison of dark current and photocurrent of devices grown on heteroepitaxial GaAs/Si (squares) as compared to homoepitaxial devices (asterisks).

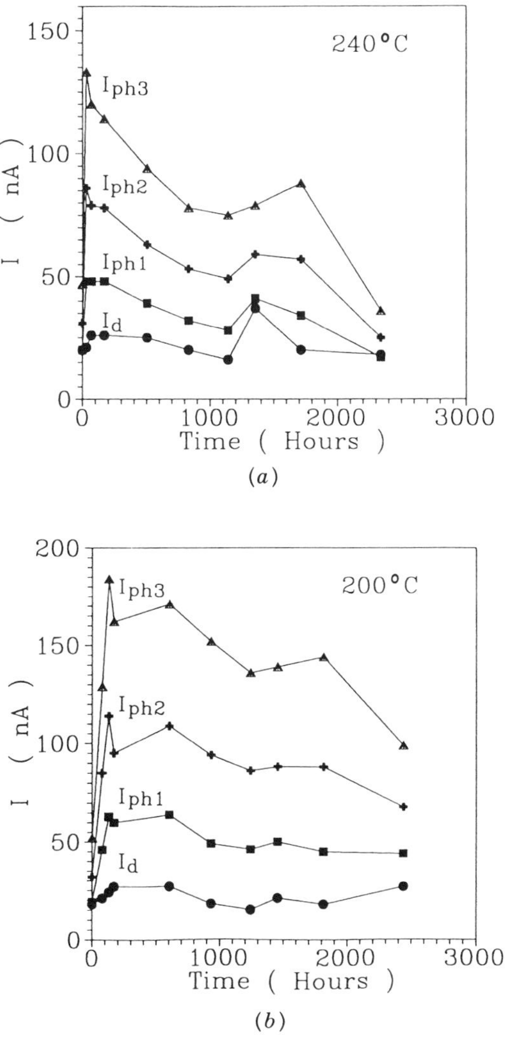

Figure 7.11. Average value of I_d and I_{ph} at three different illumination intensities over 10 photoconductor/photodetectors defined on GaAs grown on Si for a storage temperature of (*a*) 240°C and (*b*) 200°C.

78 hours for the 200°C group, while the dark current remained relatively unaffected. This suggests an initial annealing effect, possibly due to annealing out of traps or the coexistence of stabilization and degradation during this period. Loss of this improvement, possibly by the introduction of new traps, is observed beyond 2000 hours at 200°C and much sooner at 240°C (Figure 7.11). This is consistent with the discussion above on annealing and new trap introduction. Using a 100%

increase of the dark current as the failure criterion, the median lives are deduced for 200 and 240°C as 3000 and 1350 hours, respectively. The activation energy is calculated at 1.2 eV, and the extrapolated median life at an operating temperature of 110°C exceeds 3×10^6 hours.

7.5 CONCLUSIONS

GaAs growth on silicon introduces concerns that are unique to the system and qualitatively different than for homoepitaxial layers. The optimized two-step growth method on (001)Si substrates misoriented by at least 2° toward ⟨110⟩, following uniform preexposure to As, can result in complete APB annihilation near the interface. Further improvement of heteroepitaxial layer quality can be achieved by the introduction of InGaAs-based strained layer superlattices, LAMBE, and, especially, by the use of MBE grown Si buffer layers. Also, high-temperature processing for oxide desorption and step doubling can be avoided by the use of silicon beams. Rapid thermal annealing is shown to simply redistribute dislocations, rather than reduce them, and RTA with fast cooling results in catastrophic release of stress through microcracks in the thin film. Similarly, cracks appear for thicker GaAs layers.

The above schemes for improvement lead to layers that are within about 10% of the GaAs homoepitaxial controls, in most relevant figures of merit, and devices that exhibit similar performance. Schottky diodes, HEMTs, MESFETs, and photoconductors/photodetectors exhibit performance within a few percentage points of that for devices on homoepitaxial layers. None of the improvement schemes seems to introduce novel failure mechanisms. Indeed, comparative studies of devices on homoepitaxial and heteroepitaxial layers indicate that relability is similar, with indistinguishable MTTF, and that not only the failure mechanisms involved are identical for all temperatures considered but, in addition, each seems to be prevalent at the same temperature range for both types of epitaxy. In other words, it seems that the metal–GaAs interface is more vulnerable than the GaAs–Si interface and that no additional failure mechanisms are introduced by either the heteroepitaxy or the schemes utilized for layer improvement.

ACKNOWLEDGMENT

This work was supported in part by the European Community under ESPRIT II Project 2289 "Optical Interconnections for VLSI and Electronic Systems'' (OLIVES); P. Panayotatos wishes to also acknowledge the support of Rutgers University under the FASL program. We would also like to thank Prof. J. Stoemenos for TE microscopy and analysis and K. Tsagaraki for SE microscopy.

REFERENCES

1. L. Dekker and E. Frietman, *Proceedings of the SPIE Conference on Optical Interconnections and Networks*, 1991.
2. A. Christou, ''Optical Interconnects Trade-off Study,'' 1991 (unpublished).
3. S.F. Fang, K. Adomi, S. Iyer, H. Morkoc, H. Zabel, C. Choi, and N. Otsuka, *J. Appl. Phys.*, Vol. 68, R31 (1990).
4. L.F. Miller, *IBM J. Res. Dev.*, Vol. 13, 239 (1969).
5. C. Edge, M.J. Wale, F.A.Randle, and D.J. Pedder, *SPIE, 1177, Integrated Optics and Optoelectronics*, p. 374 (1989).
6. M.J. Wale, C. Edge, F.A. Randle, and D.J. Pedder, in *Proceedings of the 15th European Conference on Optical Communications*, Gothenburg, Sweden, p. 368 (1989).
7. H.K. Choi, G. Turner, and B-Y Tsaur, *IEEE Electron Device Lett.*, Vol. 7, 241 (1986).
8. H.K. Choi, G. Turner, T. Windhorn, and B-Y Tsaur, *IEEE Electron Device Lett.*, Vol. 7, 500 (1986).
9. H.K. Choi, J.P. Mattia, G. Turner, and B-Y Tsaur, *IEEE Electron Device Lett.*, Vol. 9, 512 (1988).
10. H. Shichijo, R. Matyi, and A. Taddiken, *IEEE Electron Device Lett.*, Vol. 9, 444 (1988).
11. H. Shichijo, R. Matyi, and A. Taddiken, and Y-C Kao, *IEEE Trans. Electron Device*, Vol. ED-37, 548 (1990).
12. J. Morse, R. Mariella, G. Anderson, and R. Dutton, *IEEE Electron Device Lett.*, Vol. 10, 7 (1989).
13. J. Morse, R. Mariella, G. Anderson, and R. Dutton, *IEEE Electron Device Lett.*, Vol. 12, 379 (1991).
14. Briones et al. ESPRIT II Project 2289 (OLIVES) deliverable report D6.32, commercial in conference.
15. Y. Gonzalez, A. Mazuelas, M. Recio, L. Gonzalez, G. Armelles, and F. Briones, *Appl. Phys. A*, Vol. 53, 260 (1991).
16. M. Tachikawa and H. Mori, *Appl. Phys. Lett.*, Vol. 56, 2225 (1990).
17. H. Kroemer, in *Heteroepitaxy on Silicon*, J.C.C. Fan and J.M. Poate, eds., Materials Research Society (MRS) Symposia Proceedings, 67, MRS, Pittsburgh, 1986, p. 3.

18. D.B. Holt, *J. Phys. Chem. Solids*, Vol. 30, 1297 (1969).
19. M. Akiyama, Y. Kawarada, and K. Kaminishi, *J. Cryst. Growth*, Vol. 68, 21 (1984).
20. R. Fischer, H. Morkoc, D.A. Newmann, H. Zabel, C. Choi, N. Otsuka, M. Longerbone, and L.P. Erickson, *J. Appl. Phys.*, Vol. 60, 1640 (1986).
21. N.A. El-Masry, J.C. Tarn, and N.H. Karam, *J. Appl. Phys.*, Vol. 64, 3672 (1988).
22. C. Choi, N. Otsuka, G. Munns, R. Houdre, H. Morkoc, S. L. Zhang, D. Levi, and M. V. Klein, *Appl. Phys. Lett.*, Vol. 50, 992 (1987).
23. N. Chand, R. People, F. A. Baiocchi, K.W. Wecht, and A.Y. Cho, *Appl. Phys. Lett.*, Vol. 49, 815 (1986).
24. N. Chand, F. Ren, A. T. Macrander, J.P. van der Ziel, A.M. Sergent, R. Hull, S.N.G. Chu, Y.K. Chen, and D.V. Lang, *J. Appl. Phys.*, Vol. 67, 2343 (1989).
25. S.J. Pearton, C.R. Abernathy, R. Caruso, S.M. Vernon, K.T. Short, J.M. Brown, S.N.G. Chu, M. Stavola, and V.E. Haven, *J. Appl. Phys.*, Vol. 63, 775 (1988).
26. G. Georgakilas, P. Panayotatos, J. Stoemenos, J.-L. Mourrain, and A. Christou, *J. Appl. Phys.*, Vol. 71, 2679 (1992).
27. K. Ishida, in *Heteroepitaxy on Silicon II*, J.C.C. Fan, J.M. Phillips, and B.Y. Tsaur, eds., MRS Symposia Proceedings, 91, MRS, Pittsburgh, 1987.
28. K. Nozawa and Y. Horkoshi, *Jpn. J. Appl. Phys.*, Vol. 30, L668 (1991).
29. T. Tao, Y. Ojada, and H. Kawanami, in *Heteroepitaxy on Silicon II*, J.C.C. Fan, J.M. Phillips, and B.Y. Tsaur, eds., MRS Symposia Proceedings, 91, MRS, Pittsburgh, 1987.
30. P. Panayotatos, A. Georgakilas, J.-L. Mourrain, and A. Christou, in *Physical Concepts of Materials for Novel Optoelectronic Device Applications I*, SPIE Vol. 1361, M. Raseghi, ed., p. 1100 (1990).
31. N. Kornilios, K. Tsagaraki, J. Stoemenos, and A. Christou, *Quality Reliab. Eng. Int.*, Vol. 7, 323 (1991).
32. S.N.G. Chu, S. Nakahara, S.J. Pearton, T. Boone, and S.M. Vernon, *J. Appl. Phys.*, Vol. 64, 2981 (1988).
33. A. Georgakilas, K. Tsagaraki, and A. Christou, *Mater. Lett.*, Vol. 10, 525 (1991).
34. A. Georgakilas, Ph.D. thesis, Physics Department, University of Crete, Heraklion, Greece, March 1990 (in Greek).
35. K. Kawabe and T. Ueda, *Jpn. J. Appl. Phys.*, Vol. 25, L285 (1986).
36. H. Takasugi, M. Kawabe, and Y. Bando, *Jpn. J. Appl. Phys.*, Vol. 26, L584 (1987).
37. D.K. Biegelsen, F.A. Ponce, A.J. Smith, and J.C. Tramontana, *J. Appl. Phys.*, Vol. 61, 1856 (1987).

38. M. Kawabe and T. Ueda, *Jpn. J. Appl. Phys.*, Vol. 26, L944 (1987).
39. P.R. Pukite and P.I. Cohen, *Appl. Phys. Lett.*, Vol. 50, 1739 (1987).
40. M. Kawabe, T. Ueda, and H. Takasugi, *Jpn. J. Appl. Phys.*, Vol. 26, L114 (1987).
41. M. Uneta, Y. Watanabe, Y. Fukuda, and Y. Ohmachi, *Jpn. J. Appl. Phys.*, Vol. 29, L17 (1990).
42. E.M. Schulson, in *Electron Microscopy and Structure of Materials*, G. Thomas, R.M. Fulrath, and P.M. Fisher, eds., University of California Press, Berkeley, 1971, p. 286.
43. D.W. Shaw, *J. Electrochem. Soc.*, Vol. 128, 877 (1981).
44. C.T. Lee and Y.C. Chou, *J. Cryst. Growth*, Vol. 91, 169 (1988).
45. A. Georgakilas, J. Stoemenos, K. Tsagaraki, Ph. Komninou, N. Flevaris, P. Panayotatos, and A. Christou, to appear in *J. of Mate. Res.*
46. A. Cornet, private communication.
47. K. Kugimiya, Y. Hirofuji, and N. Matsuo, *Jpn. J. Appl. Phys.*, Vol. 24, 564 (1985).
48. P. Pirouz, F. Ernst, and T.T. Cheng, in *Heteroepitaxy on Silicon: Fundamentals, Structure and Devices*, H.K. Choi, R. Hull, H. Ishiwara, and R.J. Nemanich, eds., MRS Symposia Proceedings, 116, MRS, Pittsburgh, 1988, p. 57.
49. A. Dimoulas, P. Tzanetakis, A. Georgakilas, O.J. Glembocki, and A. Christou, *J. Appl. Phys.*, Vol. 67, 4389 (1990).
50. J.W. Matthews, "Misfit Dislocations" in *Dislocations in Solids*, Vol. 2, F.R.N. Nabarro, ed., North-Holland, Amsterdam, 1979, Chap. 7.
51. J.C. Irvin, in *GaAs FETs: Principles and Technology*, J. Dilorenzo and D. Khandelwal, eds., Artech House, Dedham, MA, 1982.
52. A. Christou, W.T. Anderson, and A.M. Day, *Solid State Electron.*, Vol. 28, 329 (1985).
53. N.A. Papanicolaou, G.W. Anderson, J.A. Modolo, and A. Georgakilas, *Superl. Microst.*, Vol. 8, 273 (1990).

8

ELECTROMIGRATION AND STABILITY OF MULTILAYER METAL–SEMICONDUCTOR SYSTEMS ON GaAs

ARIS CHRISTOU
CALCE Electronic Packaging Research Center, University of Maryland, College Park

8.1 INTRODUCTION

The reliability of GaAs microwave devices is directly related to the integrity of the ohmic contacts [1, 2]. Industrial suppliers of GaAs devices such as field effect transistors have standardized on the use of Au/Ge or Ag/Ge based ohmic contacts mainly due to their initial low contact resistance. However, the reliability and long-term stability and electromigration effects of these contacts have not been clearly established. The major problems in GaAs ohmic contacts [3, 4] are lack of uniform wetting of the metals to the GaAs, metal segregation at the metal–semiconductor interface, the extent of an epitaxial surface layer, surface roughness, and the presence of a multitude of semiconductor–metal phases, all enhancing the occurrence of electromigration. It has also been shown [4, 5] that micro-precipitates do exist in AuGeNi contacts containing an uneven distribution of Ni and Ge and these precipitates enhance electromigration. In addition to the major problems in forming low-resistance contacts to GaAs, work elsewhere has established that

Electromigration and Electronic Device Degradation, Edited by Aris Christou.
ISBN 0-471-58489-4

AuGeNi contact resistance increases with increasing temperature, time, and bias [5, 6]. These changes are all due to the occurrence of electromigration. We report in the present investigation the identification of solid phase epitaxy regions in AuGeNi, Ag/In/Ge, and AuGeIn contacts to GaAs and on their role in electromigration. It will be shown that the extent of solid phase formation directly controls the value of the specific contact resistivity and the stability of the contacts. The minimum specific contact resistivity (r_c) occurs for optimum solid phase formation conditions. Unique to this study is the utilization of high spatial resolution Auger spectroscopy (μAES) and sputter profiling to identity microsegregation and solid phase formation in ohmic contacts. With (μAES), surface chemical changes can be analyzed with a spatial resolution of 1000 A [7, 8]. Microsegregation effects in these contacts have been studied as a function of contact alloying parameters. In addition, the crystallographic orientation of epitaxial regions was determined by selected area electron channeling in the scanning electron microscope. The stability of the contacts and the degradation are due to electromigration in the gold based metal systems.

8.2 THE OHMIC CONTACT/GaAs SYSTEM

Ohmic contact specimens were formed on (100) n/n^+ GaAs wafers having a 2.0-μm-thick epitaxial n layer ($n = 2 \times 10^{16}$ cm^{-3}). Square metal contact regions were defined by selective etching. Each wafer consisted of 10–15 contact squares of areas 10^{-5} to 2×10^{-3} cm^2 for electrical evaluation. The AuGeNi contacts consisted of 1600–1700 Å of AuGe eutectic followed by 400 Å of Ni overlay. AuGeIn samples were formed by first depositing 400 Å of In followed by 1700 Å of AuGe. Thicker AuGeIn contacts were formed by depositing AuGeIn to a thickness of 0.6 μm. A layered structure of Ag/In/Ge of total thickness of 2000 Å was deposited in order to form Ag/In/Ge contacts. The compositions of the AuGeIn contacts were 90 wt.% Au or 90 wt.% Ag, 5 wt.% Ge and 5 wt.% In, with a total thickness of 2000 Å. Sintering of each sample was accomplished either in an open tube furnace (N_2 or forming gas atmosphere) or in a vacuum tube furnace located within an ultrahigh vacuum system. A typical annealing scheme for AuGeNi is shown in Figure 8.1. All temperatures have been corrected for the temperature difference of 30°C at 450°C that exists between the contact and the substrate surface. The contact resistivity changes can also be measured as a function of temperature and time.

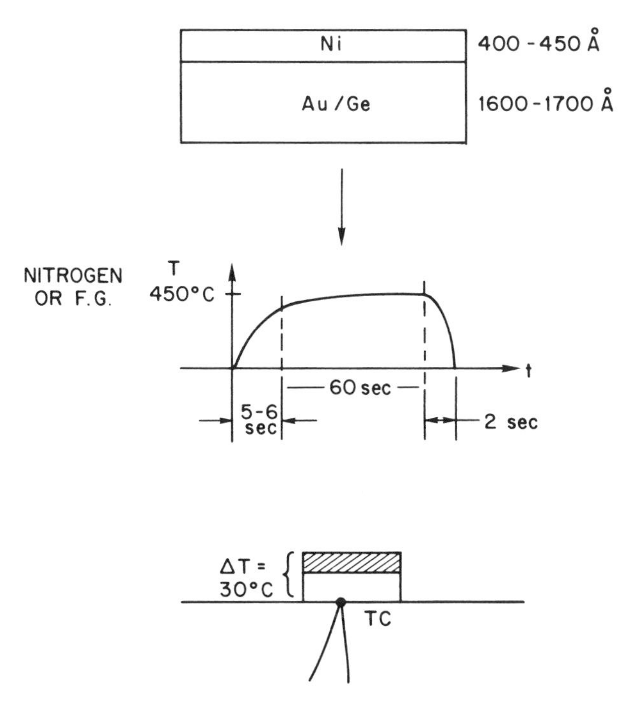

FIGURE 8.1. Schematic showing annealing scheme for the AuGeNi-GaAs contact.

The specific contact resistance, r_c, was measured using a modified form of the Cox–Stack [9] method. The total resistance of a square contact of side a in length is given by

$$R_T \simeq \frac{r_c}{a^2} + \sqrt{\frac{\pi}{4}\frac{\rho}{\pi a}} \arctan\left(4\,\frac{\pi}{4}\frac{h}{a}\right) + R_0$$

where ρ is the resistivity of the epi material of thickness h and R_0 is the thickness of the back contact, n^+ substrate, and probe resistance. The second term represents the spreading resistance term, which is only weakly dependent on the exact shape of the contact area. The spreading resistance term is, however, dependent on the reactions at the interface and can measure the degree of electromigration occurrence.

Ohmic contact surfaces were examined utilizing the μAES technique. Auger electron spectroscopy with a focused beam in a high-resolution scanning electron microscope (SEM) has provided an extremely useful diagnostic technique for GaAs surfaces. The focused electron beam in a SEM allows one to determine surface topography of the specimen,

TABLE 8.1. Critical Eutectic Temperatures and Compositions of GaAs Contacts

	AuGeNi	Ag/In/Ge	AuGeIn
AuGe eutectic composition	88/12 wt.%	95/5 wt.%	88/12 wt.%
Binary eutectic temperature	363°C (Au-Ge)	651°C (Ag-Ge)	363°C (Au-Ge)
Ternary eutectic temperature	~425°C	>651°C	~520°C
Maximum stability temperature	442°C	575°C	495°C

which can be correlated with the Auger spectra of selected area. The μAES technique has been shown to have surface resolution of 1000 Å at a beam current of 5×10^{-8} A necessary for AES analysis [7, 8].

8.2.1 Microsegregation: Metal Migration Effects

Microsegregation effects were observed on the three contacts investigated over the temperature region that resulted in ohmic behavior. For discussion purposes, we define solid phase epitaxy formation as the structures that form at a sintering temperature slightly below the eutectic temperature and that have the same orientation as the substrate [10]. Precipitation structures are those that form at sintering temperatures above the eutectic temperature. Table 8.1 summarizes the eutectic temperatures employed in the present investigation and the maximum temperature for ohmic contact stability.

The Ag/In/Ge System The Ag/In/Ge contacts exhibited solid phase facet growth up to a temperature of 650°C with a typical (100) orientation. Figure 8.2 shows facet growth and μAES analysis of Ag/In/Ge, indicating the presence of (100) oriented Ge/In regions. The (100) particles are Ga-In-Ge-As regions with surface composition as indicated by the square particle analysis. After sputtering for approximately 20 minutes, the particles become Ge rich. The agglomerated regions are shown to be Ag particles with absorbed sulfur. In both the matrix and the epitaxial particles, Ge is shown to have migrated into the GaAs substrate. By scanning the electron beam over all areas of interest, a complete map of elements as a function of depth and lateral extent has been obtained and is shown in Figure 8.3. The selected area electron channeling pattern indicates that the rectangular regions have a (100) orientation as does the GaAs substrate. The sinter temperature range for which solid

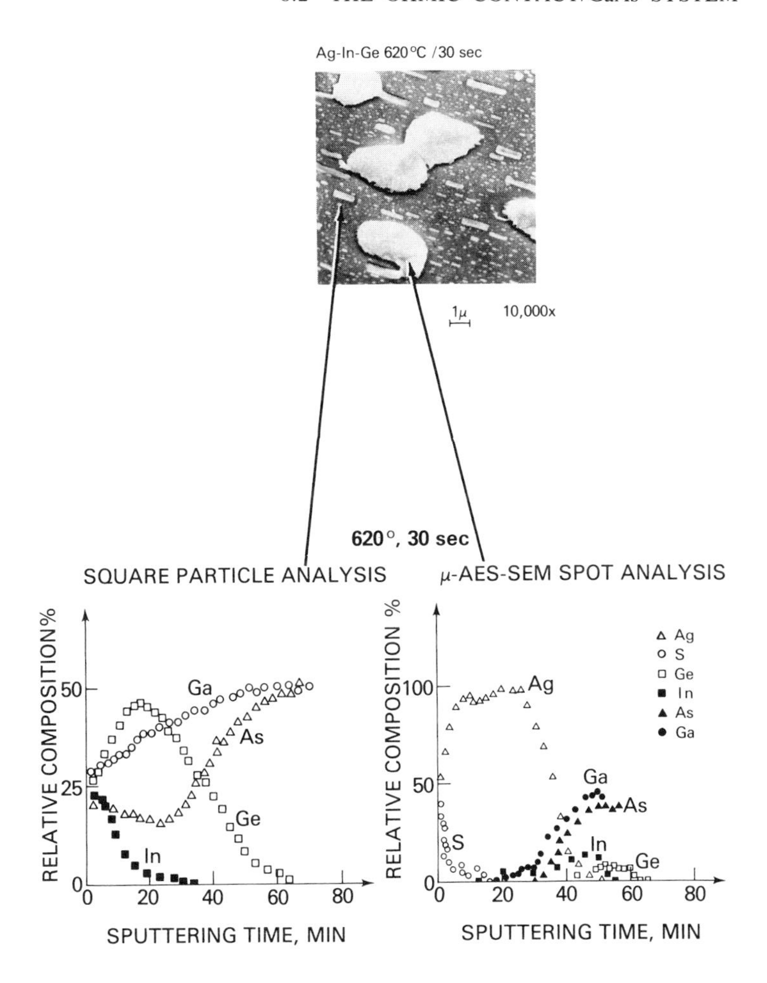

phase epitaxy was observed was determined and correlated with variation in specific contact resistivity, r_c. Figure 8.4 shows that the minimum r_c of $\sim 10^{-5}\ \Omega \cdot \text{cm}^2$ was attained within the temperature region (620°C, 1 minute) for solid phase formation, which is below the eutectic temperature of 651°C. It is suggested in Figure 8.4 that the extent of solid phase formation in Ag/In/Ge contacts also controls the contact resistance. Increases in r_c were only observed when the microstructure morphology changed from oriented to random precipitation above 640°C. The random precipitation morphology also exhibited the greatest degree of electromigration resistance.

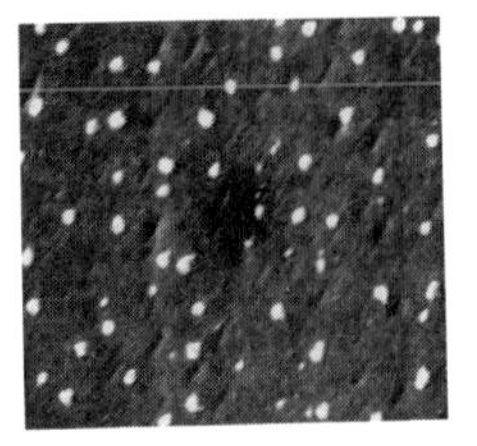

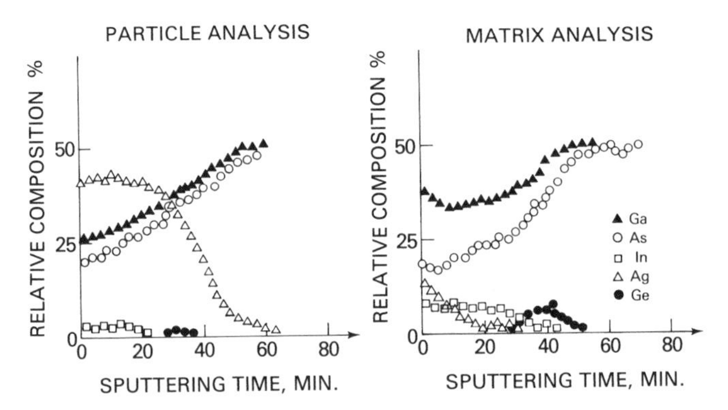

FIGURE 8.3. Cross section of Ag/In/Ge contact with selected area channeling pattern of solid phase epitaxial region.

The extent of solid phase epitaxy (as determined by dimensions of the oriented regions) was also determined as a function of sinter temperature and specific contact resistivity. A variation in average solid phase particle size from less than 0.5 to 2 μm was attained by varying sintering temperatures while maintaining the sinter duration at 60 seconds. The results summarized in Figure 8.5 indicate an excellent correlation between the 2.0-μm particle size and the low contact resistance samples. A degraded contract resistance of $\sim 10^{-1}\ \Omega \cdot \text{cm}^2$ was observed for specimens with solid phase particle size of $\sim 0.5\ \mu$m. The results for Ag/In/Ge contacts summarized in Figures 8.2–8.5 show that the extent of solid phase formation directly controls the value of the specific contact resistivity. The minimum r_c occurs for optimum solid phase formation conditions.

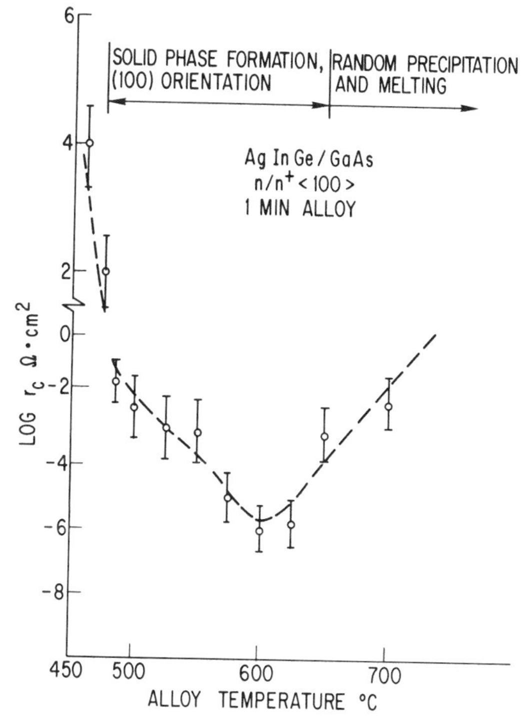

FIGURE 8.4. Variation of log r_c with alloy temperature for constant 1 minute sintering times.

The AuGeNi Contact System The AuGeNi contacts displayed typical indications of solid phase formation with facet growth of Ni-Ga and Ni-Ge in Au films having the same orientation as the (100) substrate. A typical solid phase epitaxy growth structure at the AuGeNi–GaAs interface is shown in Figure 8.6. Regions typical of solid phase formation were observed at temperatures well below the eutected temperature. Auger electron spectroscopy analysis of the (100) oriented particles and the surrounding matrix area is shown in Figure 8.7. The particle sputter profiles indicate a significant increase in nickel concentration at the contact–substrate interface (50–70 minutes of sputtering) probably in the form of Ni-Ga and Ni-Ge solid solutions. Presumably, the metal Ni is transported through the AuGe to the GaAs surface, where (100) oriented nickel–rich regions form by a previously described solid phase epitaxy process [10]. The matrix composition also shown in Figure 8.7 consists of a Au-Ga solid solution.

The variation of r_c with alloy temperature for AuGeNi was found to

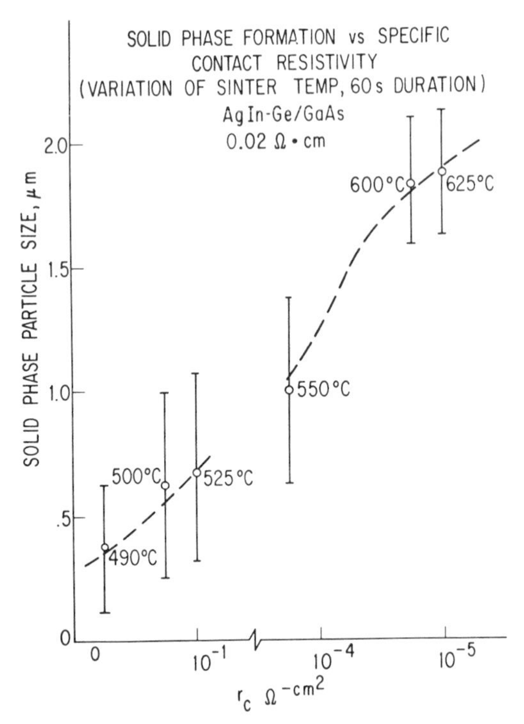

FIGURE 8.5. Variation of solid phase particle size with specific contact resistivity, r_c.

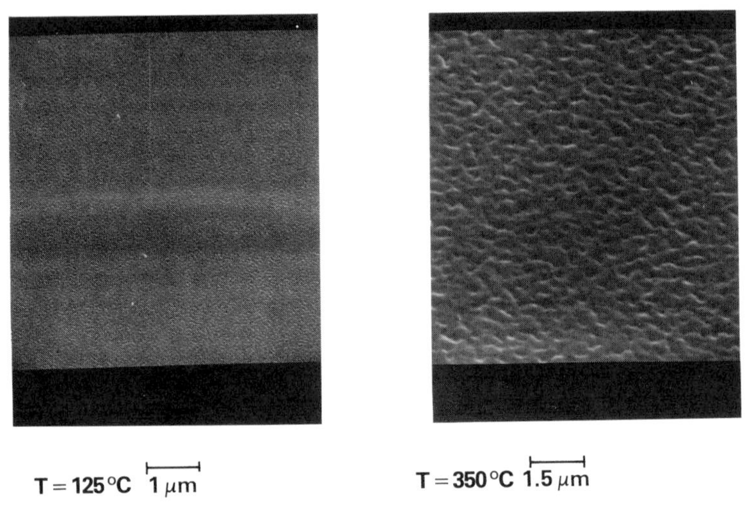

FIGURE 8.6. Solid phase epitaxy of AuGeNi contacts.

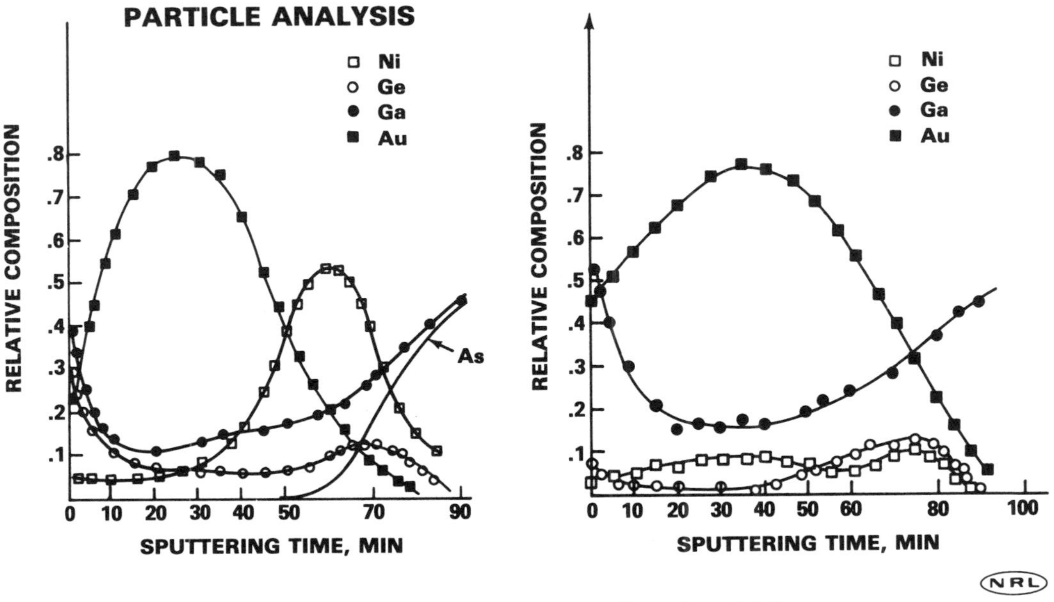

FIGURE 8.7. Microspot AES sputter profiles of AuGeNi contacts.

depend directly on the presence of solid phase regions. Figure 8.8 indicates the variation of log r_c with alloy temperature, showing three different microstructure morphologies with a degradation in r_c observed only in the regions of random facets or random precipitation (above the temperature). The size of the solid phase epitaxial regions was found to vary from less than 1 μm (350°C) for $r_c = 10^{-1}\ \Omega \cdot \text{cm}^2$ to 3.0 μm for $r_c = 10^{-6}\ \Omega \cdot \text{cm}^2$ at 425–450°C. Figure 8.9, which shows the variation of particle size with r_c indicates the presence of a particle size plateau at 3.0 μm and 420–450°C, which corresponds with the temperatures for attaining the lowest contact resistivity. This experiment provides further empirical evidence that the greater the extent of solid formation, the lower the resultant contact resistance. The lower the contact resistance, the greater is the resistance to electromigration.

The AuGeNi contact formed by annealing at 450°C for 60 seconds

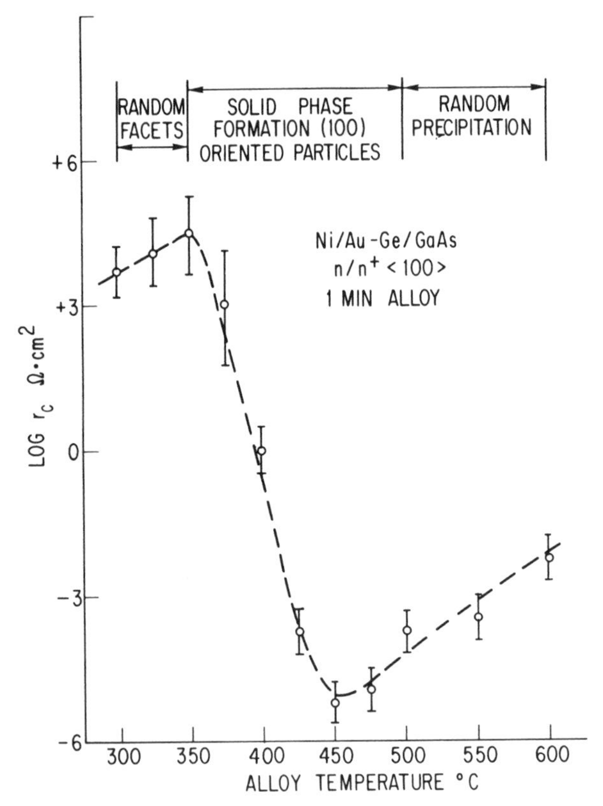

FIGURE 8.8. Variation of log r_c with alloy temperature for constant 1 minute sintering times.

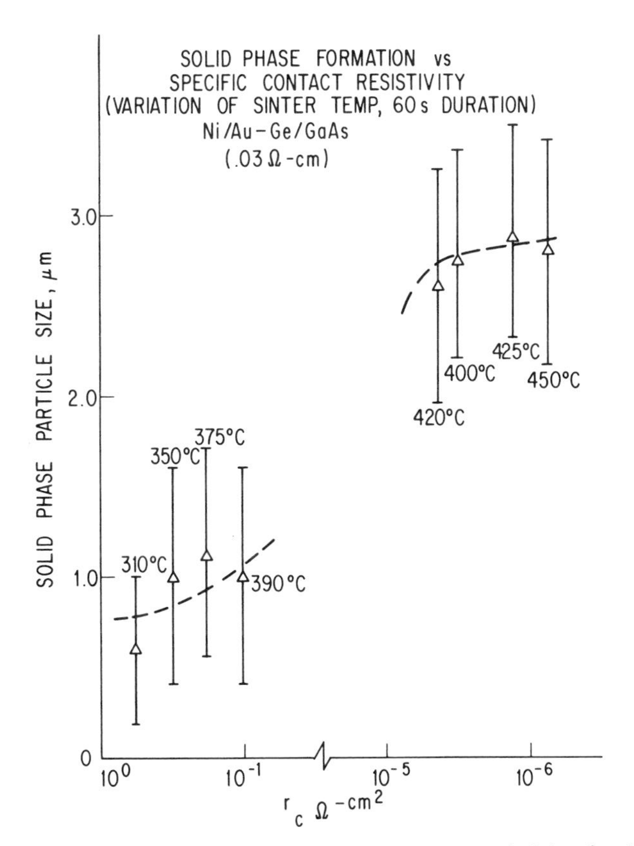

FIGURE 8.9. Solid phase formation vs specific contact resistivity for AuGeNi contacts.

was further analyzed by μAES sputter profiling in order to determine the lateral and depth distributions of Au, Ge, Ni, Ga, and As. A schematic of the results is summarized in Figure 8.10, which shows that Ge has migrated into the GaAs substrate, thereby increasing impurity concentration necessary for attaining a low contact resistivity. In addition, Ga has outdiffused to the Au overlay region and is shown to be present at the same concentration level as Au in the regions between the epitaxial particles. The AuGe, GeNi, Au, and Ge parentheses denote that these were minor constituents only with GaAs being the major component.

The AuGeIn Contact System The AuGeIn system also exhibited solid phase epitaxy formation at sintering temperatures of 400–530°C (1-min-

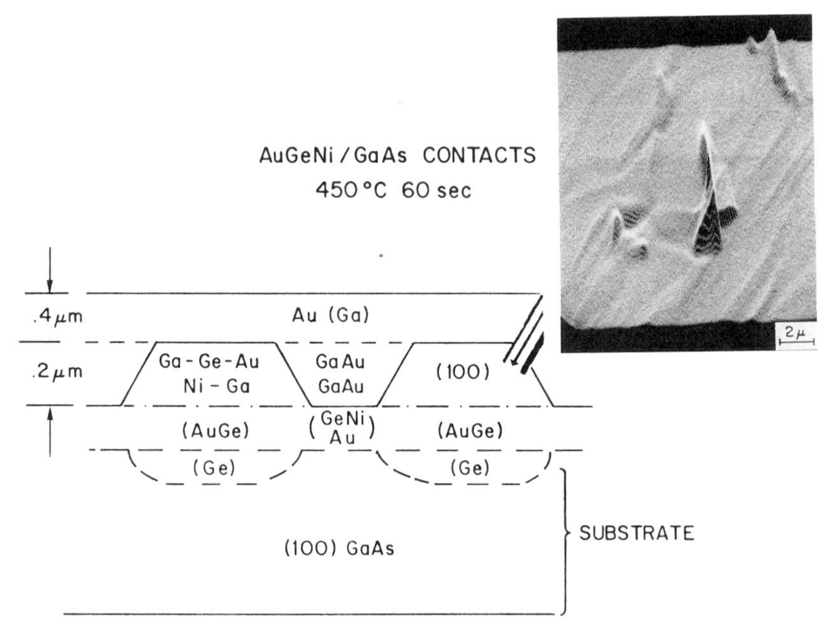

FIGURE 8.10. Cross section of AuGeNi contact with selected area channeling pattern showing a (100) orientation.

ute sinter duration). Elongated faceted regions were observed on the top surface of the 0.6-μm-thick contact and epitaxial regions were observed at the contact–GaAs interface. The epitaxial (100) regions consisted of Ge-In in addition to Ga at approximately 40% concentration. The decrease in As concentration at the interface has been accompanied by an increase in Ge, In concentration as shown in Figure 8.11. In both regions A and B, Ga outdiffusion to the contact surface has been observed. Region B also shows 7% As on the contact surface. The migration of Ge into the GaAs substrate was observed in both the A and B regions and seems to be a general phenomenon with all sintered GaAs contacts. The greater the degree of solid phase epitaxial formation at the metal–GaAs interface, the greater is the resistance to electromigration degradation. One explanation may be raised to the amount of gold present for electromigration to take place.

Lateral and depth profile mapping results of the AuGeIn contact are shown in Figure 8.12. The epitaxial (100) regions consist of equal concentrations of Ge and In. A thinner layer of Ge-In was also found in the matrix region between the epitaxial regions. The selected area electron channeling pattern taken after sputter etching to the contact–GaAs in-

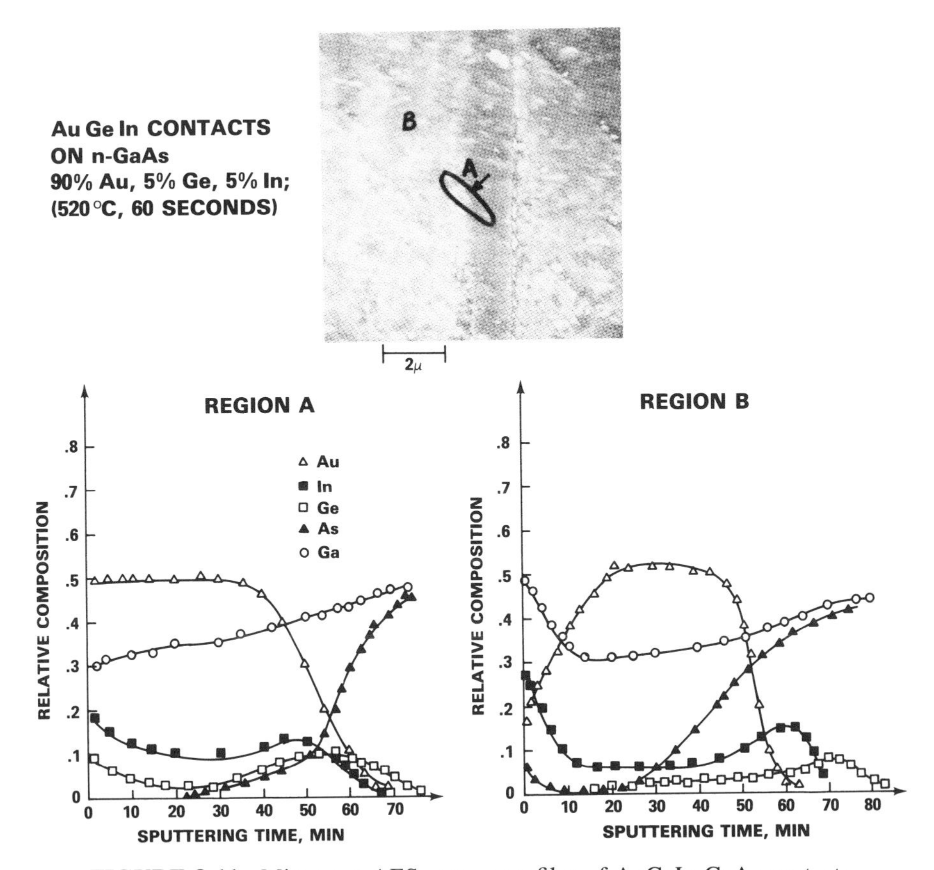

FIGURE 8.11. Microspot AES sputter profiles of AuGeIn-GaAs contacts.

terface indicates that the solid phase regions are single crystal with a (100) orientation. Channeling patterns of the matrix region indicated the presence of polycrystalline material. Outdiffusion of Ga and In is also shown to be present throughout the Au region of the contact.

8.3 ELECTROMIGRATION RESISTANCE AND SOLID PHASE FORMATION

The experimental results show that AuGeNi, Ag/InGe, and AuGeIn contacts to GaAs undergo precipitation and solid and solid phase formation over a temperature region that corresponds to ohmic behavior. The extent of solid phase formation directly controls the value of the

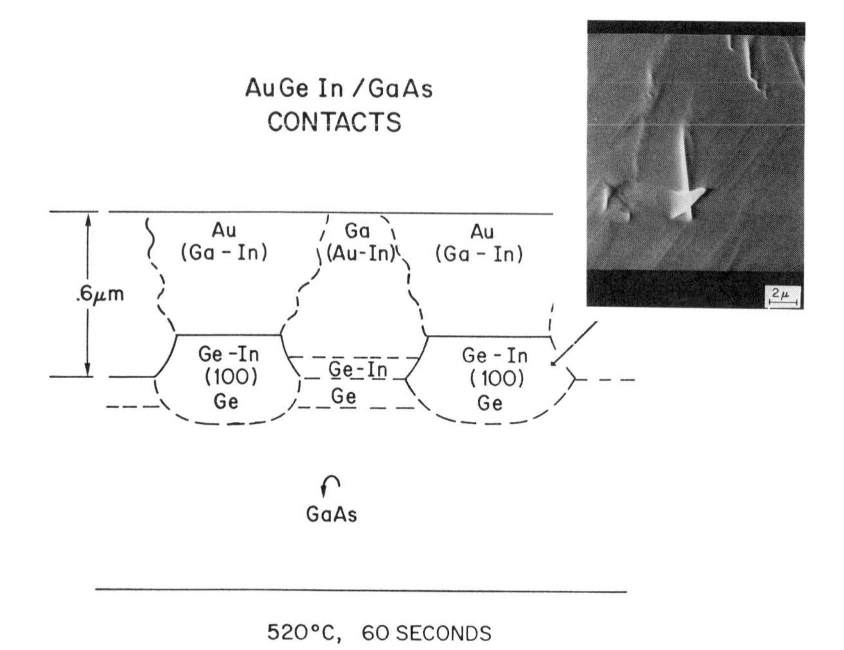

FIGURE 8.12. Cross section of AuGeIn-GaAs contact with channeling pattern insert showing (100) orientation.

specific contact resistance. Solid phase growth occurs as a result of the dissolution of GaAs in the vicinity of the eutectic temperature. As the ohmic contact composite structure is brought up to a temperature below the eutectic melting temperature, dissolution of the GaAs will take place until the solubility limit is reached. The composite structure, when cooled, undergoes precipitation and growth of the dissolved atoms onto the substrate. The epitaxial layers thus formed always incorporate either Ni or In and Ge. The cleanliness or the substrate–metal interface will determine the ultimate extent of the epitaxial layer. If the interface is not free of all contaminants, such as surface oxides, the epitaxial layer will not be laterally uniform. When electromigration occurs, the AuGa regions are the microstructural area that migrate under bias.

The driving force for solid phase growth in GaAs ohmic contacts is the reduction in chemical free energy between the amorphous or near amorphous state of the AuGe and the crystalline epitaxial state. Once AuGe and Ni have undergone interdiffusion and crystallization on the substrate, further growth occurs because of the difference in chemical potential between the regrown crystallites and the substrate. In general, the transport mechanism for epitaxial growth is basically precipitation of crystals out of a supersaturated solid solution.

In summary, the relationship between solid phase epitaxial growth, specific contact resistivity, and the occurrence of electromigration has been established. Solid phase growth results in regions free of oxide layers or contaminants and hence a lower contact resistivity. The regions of the greatest degree of solid phase formation are also the regions that exhibit the highest degree of electromigration resistance.

8.3.1 The Ge-GaAs System and Migration

Gallium arsenide Schottky barrier mixer diodes have demonstrated low noise figure; however, reported barrier heights have been substantially greater than those of most silicon Schottky diodes. The barrier height determines the RF local oscillator power required to obtain optimum performance. Consequently, in microwave systems having limited local oscillator power, a low barrier height device is required [11]. This chapter reports on the application of thin epitaxial Ge films to obtain low barrier height GaAs mixer diodes and on their stability and electromigration effects. The thin Ge films have been used in the present investigation to control the field at the surface of GaAs and, consequently, reduce the effective barrier height of a Schottky barrier on *n*-GaAs [12]. Previously, low barrier Schottky diodes have been developed by implantation of Ge into epi-GaAs [13]. However, these diodes required encapsulation and high-temperature annealing in order to eliminate ion implantation damage. When subjected to microwave pulses these structures tend to burn out by metal migration enhanced by local fields.

8.3.2 Epitaxial Ge-GaAs Diodes

The epitaxial Ge-GaAs diodes have been developed by depositing Ge films 150 Å thick by electron beam evaporation in an oil-free vacuum system onto heated GaAs surfaces [14]. The (100) GaAs substrate used was Te doped to 1×10^{18} cm^{-3} and the epitaxial GaAs was 0.27 μm thick and doped to 1.1×10^{17} cm^{-3}. The epitaxial layer was grown by conventional $AsCl_3$ process. Substrate temperatures for Ge deposition were in the range of 325–425°C. At 325°C substrate temperature, near ideal electrical characteristics with a barrier height of 0.5 eV and ideality factor of 1.1 were obtained. Figure 8.13 shows typical I–V characteristics as a function of deposition temperature. Electrical properties of the films indicated that, at $T = 325$°C, the highest carrier concentra-

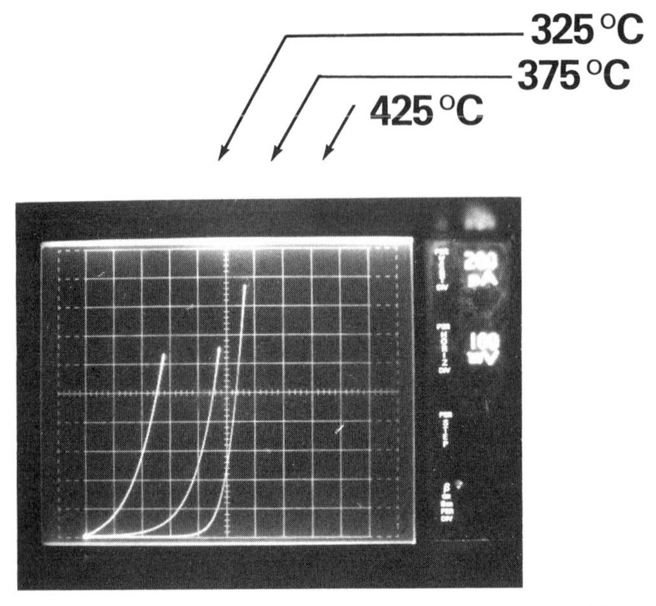

FIGURE 8.13. Typical IV characteristics of Ge-GaAs diodes with Pt-Ti-Mo-Au metallization.

TABLE 8.2. Electrical Properties of Thin epi-Ge on GaAs Substrates

Substrate	Deposition Temperature	Carrier Concentration	Mobility ($cm^2/V \cdot s$)
GaAs (100)	325°C	3×10^{18}	80
GaAs (100)	370°C	1×10^{18}	80
GaAs (100)	425°C	6×10^{17}	130

tion (Table 8.2) of 3×10^{18} cm^{-3} was attained. The barrier height of 0.5 eV represents approximately a 0.2–0.25 eV decrease with respect to standard Au–GaAs diodes. All mixer diodes were processed with Ge deposited at a substrate temperature of 325°C. The lower barrier height is directly responsible for the occurrence of Ge-GaAs local interaction.

8.3.3 Mixer Diode Fabrication and Measurement

Prior to Ge deposition, the wafer was dip etched in hydrochloric acid and rinsed in alcohol to inhibit oxide growth. The technique for epitaxial Ge deposition at 325°C substrate temperature has been published previously [14]. Following Ge deposition, a layer of SiO_2 was grown at

325°C until a thickness of 8000 Å was achieved. Circular windows 7.5 μm in diameter were etched in the oxide. Platinum, titanium, molybdenum, and gold layers were sputtered sequentially on the wafer. Using a photoresist process, a second mask defined an area of 1.25 μm diameter centered on the original areas, and excess Pt, Ti, Mo, and Au were then etched. Ohmic contact to the n^+ side was obtained by evaporating Au-Ge at 200°C.

Noise figure, noise temperature ratio, RF impedance (VSWR), and IF impedance are the essential parameters of a microwave mixer diode. The stability of these parameters degrades by localized electromigration. Once a diode mount is specified, these parameters depend on (a) RF frequency of the local oscillator, (b) local oscillator power level, and (c) temperature. Epi-Ge-GaAs mixer diodes were reexamined for noise figure, VSWR, IF impedance, and rectified current in a waveguide test mount. These devices were tested at 9.375 GHz and the local oscillator power was varied from −8 to +8 dBm. For comparison, the standard Pt-Ti-Mo-Au/GaAs and the ion implanted GaAs diodes were also tested. All noise figure measurements were made by the standard technique using a gas discharge tube and a 30 MHz IF amplifier. The noise figure (NF) of the NF amplifier varied from 1.2 to 1.5 dB depending on the source resistance. Results for NF are given in Figure 8.14 and show that epitaxial Ge-GaAs diodes exhibit a noise figure of

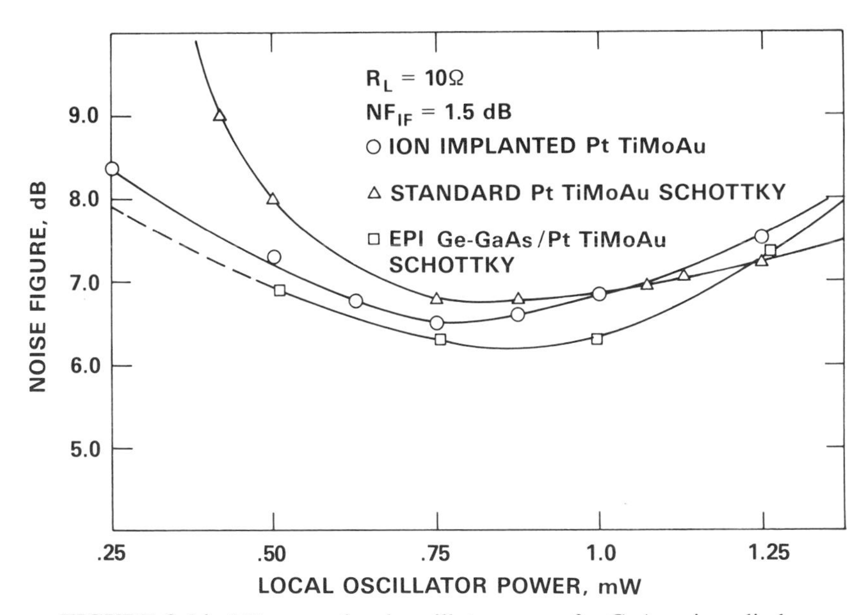

FIGURE 8.14. NF versus local oscillator power for GaAs mixer diodes.

6.0 dB (SSB) at 0.75 mW while the implanted diodes exhibit a NF of 6.5 dB. The standard diodes show very high noise figure values at 0.4 mW power level, whereas the ion implanted diodes exhibit a NF of 7.5 dB and the epitaxial Ge-GaAs diodes exhibit a NF less than 7.0 dB. The Ti-Mo-Au/Ge-GaAs diodes tended to withstand less than 5 W of CW power at X-band. Degradation occurred by electromigration at localized mesa edges.

IF impedance was measured by the 1 kHz bridge method and local oscillator power was varied from −6 to +6 dBm. A load resistance of 100 Ω was used for the measurements. The measurements show that the epitaxial Ge, implanted, and standard diodes have similar IF impedance at high local oscillator power levels. However, epitaxial Ge-GaAs mixer diodes exhibit lower IF impedance than both the implanted and the standard GaAs diodes. The VSWR versus local oscillator power is shown in Fig. 8.15. The results again show that low barrier height Ge-GaAs Schottky diodes exhibit lower VSWR compared to standard and implanted diodes at low local oscillator power levels.

The above results show that epitaxial Ge-GaAs Schottky diodes with Pt-Ti-Mo-Au metallization exhibit superior mixer performance in comparison with implanted and standard GaAs Schottky diodes at low local oscillator power levels. These diodes also exhibited the greatest degree of stability. This is due to the lower barrier of epi-Ge-GaAs Schottky diodes (ϕ = 0.50 eV) compared to 0.60 eV for implanted diodes and

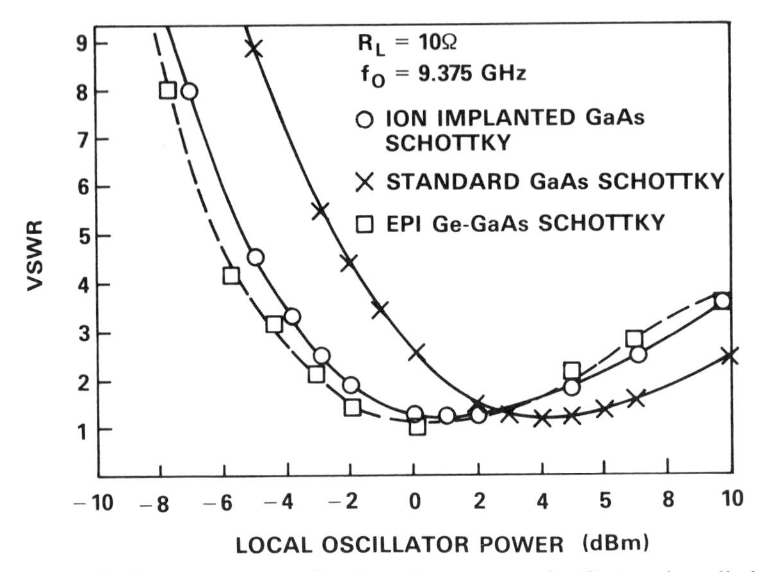

FIGURE 8.15. VSWR versus local oscillator power for GaAs mixer diodes.

0.77 eV for standard Pt-T-Mo-Au diodes. The lowering of the barrier due to image forces at the metal–semiconductor interface is given as [12].

$$\Delta\phi = \left\{\frac{q}{4\pi\epsilon_s}\right\}^{1/2} \left\{\frac{2qN_1(\phi_B - V)}{\epsilon_s}\right\}^{1/4}$$

where N_1 is the carrier concentration of the thin Ge layer, V is the applied voltage, and ϵ_s is the dielectric constant. The epitaxial Ge layer must be thin, smaller than the depletion width. The barrier lowering $\Delta\phi$ is optimized by varying the carrier concentration N_1 and the thickness of the epitaxial Ge layer. Therefore, the lower barrier height mixer requires less local oscillator power in order to operate in the nonlinear region. Further lowering of the diode noise figure can be accomplished by decreasing the thickness of the Ge film while maintaining film continuity. In addition, new deposition and anneal parameters are being investigated in order to further reduce the barrier height.

8.4 THE Al/GaAlAs METAL–SEMICONDUCTOR SYSTEM

We report in the present chapter the formation of GaAlAs epitaxy on (100) n-GaAs ($1 \times 10^{16}/\text{cm}^{-3}$) heated substrates by molecular beam epitaxial technique in an ultrahigh vacuum system. It is shown that such a system degrades by electromigration only when Al is available for electromigration to take place. The structural and electrical properties of Al films on GaAs surfaces are of current interest in the fabrication of transferred electron devices and field effect transistors [15, 16]. In addition, the epitaxial growth of metals on semiconducting substrates by the solid phase epitaxy process is presently under investigation at a number of laboratories [17–20]. Up to the present time, no work has been reported on the growth of aluminum single crystal films by vacuum deposition on heated (100) GaAs substrates under typical semiconductor processing conditions. Such a study is essential for the determination of the role of interfacial impurities in the formation of Al-GaAs Schottky barriers, and on the occurrence of electromigration in these Schottky gates.

The possibility of Al epitaxy on GaAs is practical since their lattice constants differ by a factor of approximately 1.4. From the point of view of structure, the lattice registry of the (100) Al lattice can be accommodated on (100) GaAs by a 45° rotation. Such a rotation must also accommodate the fcc Al lattice at the interface. The observed growth, however, of the layered GaAs-Al-GaAs ··· structure by molecular beam epitaxy is (110) Al on (100) GaAs with [100] Al parallel to [110] GaAs [21, 22]. In addition, the possibility of forming GaAlAs by hot deposition is also highly probable due to the lattice match that exists. The presence of metal epitaxy should prevent the occurrence of electromigration.

In the present investigation, our approach is as follows: (a) The GaAs substrate is desorbed at vacuum levels of 2×10^{-10} torr by annealing at 353°C for 15 minutes. (b) The aluminum film is vacuum deposited on the substrate within a temperature range of 125–450°C. (c) Immediately after deposition and without breaking vacuum, the Al films were evaluated in situ structurally by LEED and by Auger electron spectroscopy. Complete analysis of the films was accomplished by reflection electron diffraction and by micro-spot Auger electron spectroscopy [23]. Scanning electron microscopy (SEM) was used to characterize major topographical features. The substrates were (100) GaAs and were chemically polished in NaClO and etched in H_2SO_4 and H_2O_2.

8.4.1 Deposition and Electromigration Effects

Al depositions (2000 Å) on GaAs yielded a strong [100] texture at deposition temperatures of 200°C. Depositions at temperatures above 200°C (200, 350, and 400°C) exhibited a whitish surface appearance characteristic of terminal growth structures, which cause the optical diffraction phenomena to be observable. Films deposited at 350°C gave discernible diffraction patterns characteristic of single crystal layers. These films were resistant to Al electromigration. Figure 8.16 shows the reflection electron diffraction pattern for aluminum films deposited at 350°C, indicating a (100) structure with relaxation effects characteristic of high-quality films. The square-arrayed pattern corresponding to (100) GaAlAs was viewed from the [100] GaAs azimuth. The (100) is the dominant orientation with a twin reflection at (110). The LEED pattern in addition to GaAlAs showed a 2×8 Al surface structure at 350 and 400°C. The GaAs system does not have a 2×8 structure [24] but does have an 8×2 structure with the two structures differing in symmetry by a structural rotation of 90°. To examine for surface and bulk film related effects, Auger, sputter-Auger, and micro-spot Auger spectroscopy (MAS) were carried out. Furthermore, through electromigration testing, the microstructure with the greatest degree of Al epitaxial formation showed the greatest degree of resistance to electromigration.

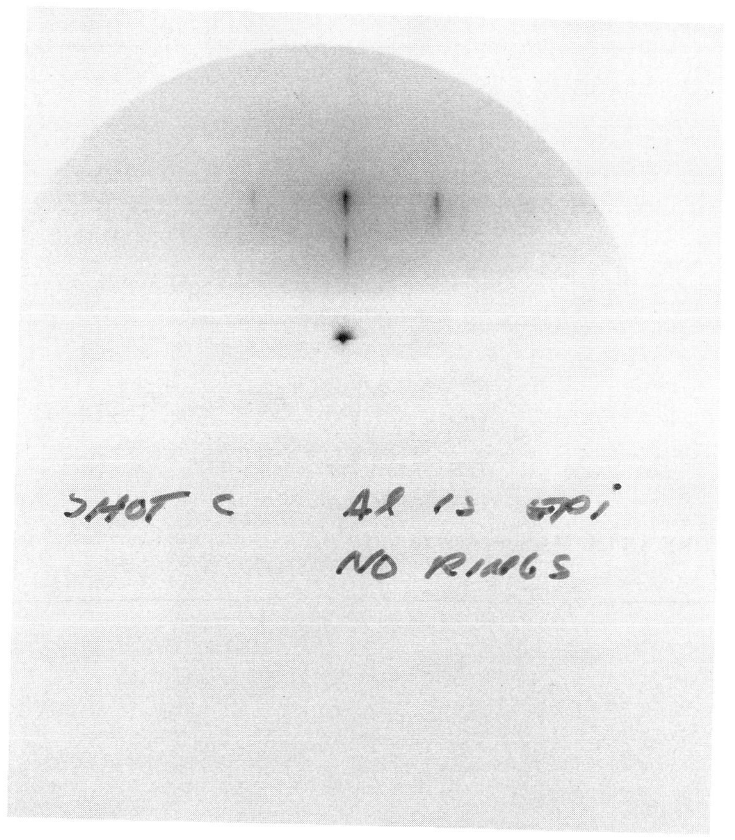

FIGURE 8.16. Reflection electron diffraction pattern for Al deposited at 350°C indicating a (100) structure.

MAS analysis of the Al-GaAs structure deposited at 125 and 250°C is shown in Figures 8.17a and 8.17b. At 125°C, a shallow surface oxide was observed, and also shown is a sharp Al/GaAs interface without the presence of interfacial impurities. Deposition at 250°C resulted in the probable formation of a GaAlAs compound at the metal–semiconductor interface. The extent of the GaAsAs region (as indicated by equal concentrations of Ga, Al, and As) was 150 Å at 250°C. An outdiffusion of Ga and As was observed with a 20% and 10% accumulation of Ga and As, respectively, on the free surface aluminum.

Figure 8.18 shows MAS and SEM analysis of films deposited at 400°C The surface morphology of the films deposited at 350°C showed a similar structure. The surface roughness of the epitaxial films necessitated the microanalysis of two different regions in order to completely characterize the films. Region B, the valley as shown in the SEM mi-

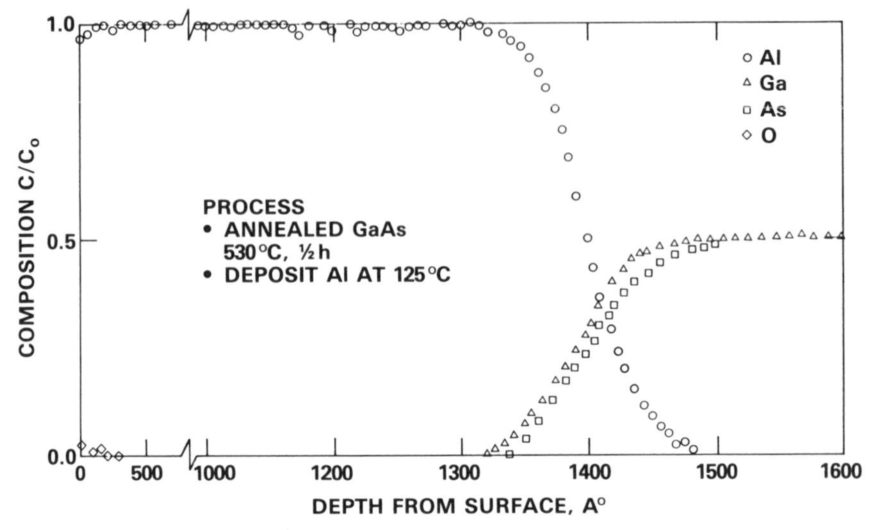

FIGURE 8.17*a*. MAS profiles of the Al-GaAs structure deposited at 125°C.

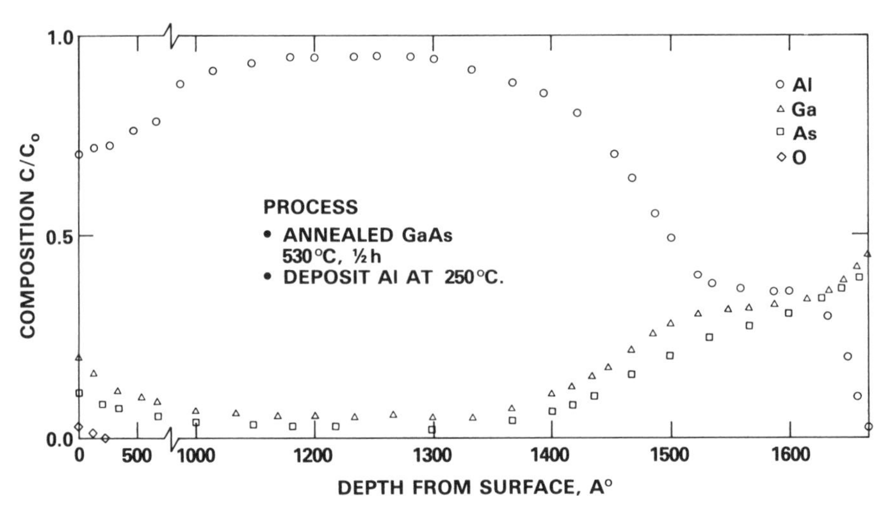

FIGURE 8.17*b*. MAS profiles of the Al-GaAs structure deposited at 250°C.

crograph, consists essentially of equal concentrations of Ga, Al, and As up to a depth of 1800 Å, and aluminum on the front surface. The mound regions (A) consist of a GaAlAs interface at 1400 A depth and a similar GaAlAs at the surface (300 Å). The bulk of the film consisted of Al up to 72% composition but with significant amounts of outdiffused Ga and As. Oxygen was not present at any of the metal–semiconductor inter-

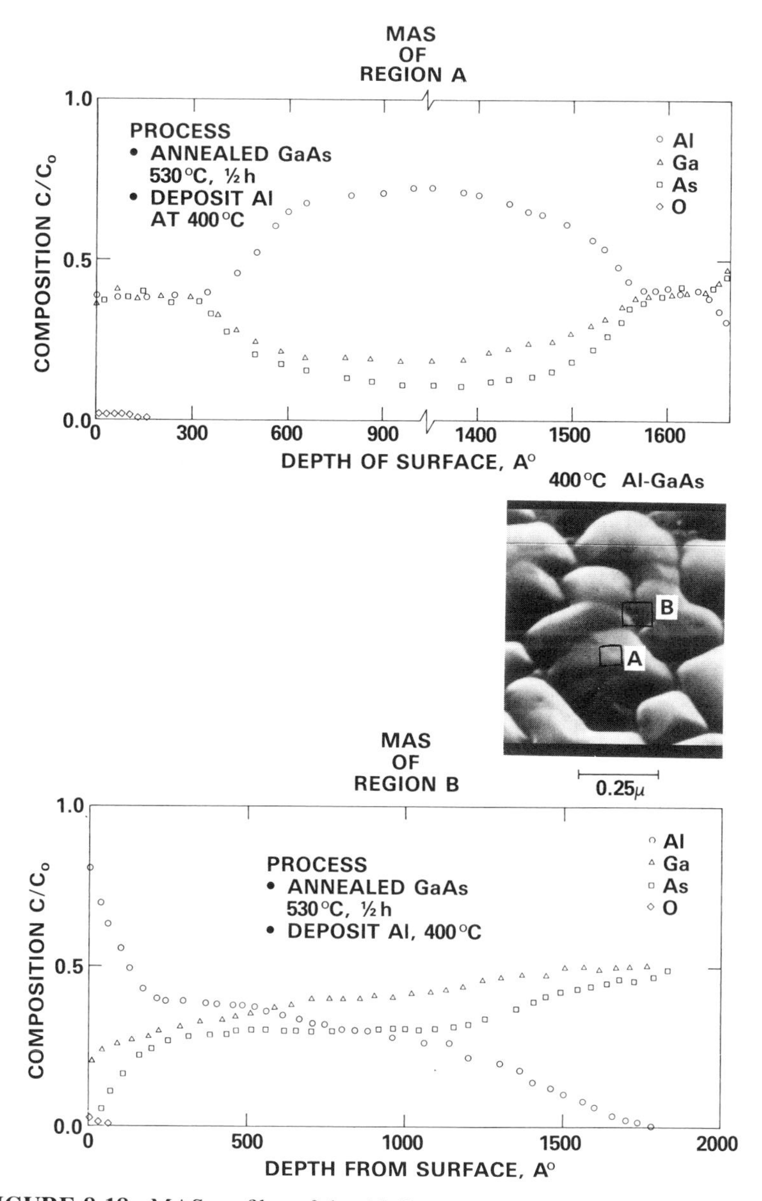

FIGURE 8.18. MAS profiles of the Al-GaAs structure deposited at 400°C.

faces analyzed nor within the bulk of the aluminum film. This observation is in contrast with previous Al-GaAs depositions at room temperature followed by anneal treatments up to 350°C [25].

Electrical measurements were made on samples deposited at 356°C and exhibited single crystal reflection electron diffraction patterns. Forward *I–V* characteristic for Schottky barriers with epi-Al-GaAlAs-GaAs structures is shown in Figure 8.19. A barrier height of 0.76 eV and ideality factor of 1.03 were measured. The barrier height for Al/GaAs Schottky structures [25] deposited without substrate heating was measured to be 0.7 eV and increased to 0.8 eV after annealing at 250°C for 1 hour. Prolonged annealing at 250°C resulted in a further increase in barrier height to 0.9 eV after 24 hours. However, after annealing at 450°C for 1 hour, the barrier height decreased to 0.8 eV. In contrast, the epi-Al-GaAs structures maintained a barrier height of approximately 0.75 eV even after subsequent anneal treatment at temperatures of 125–

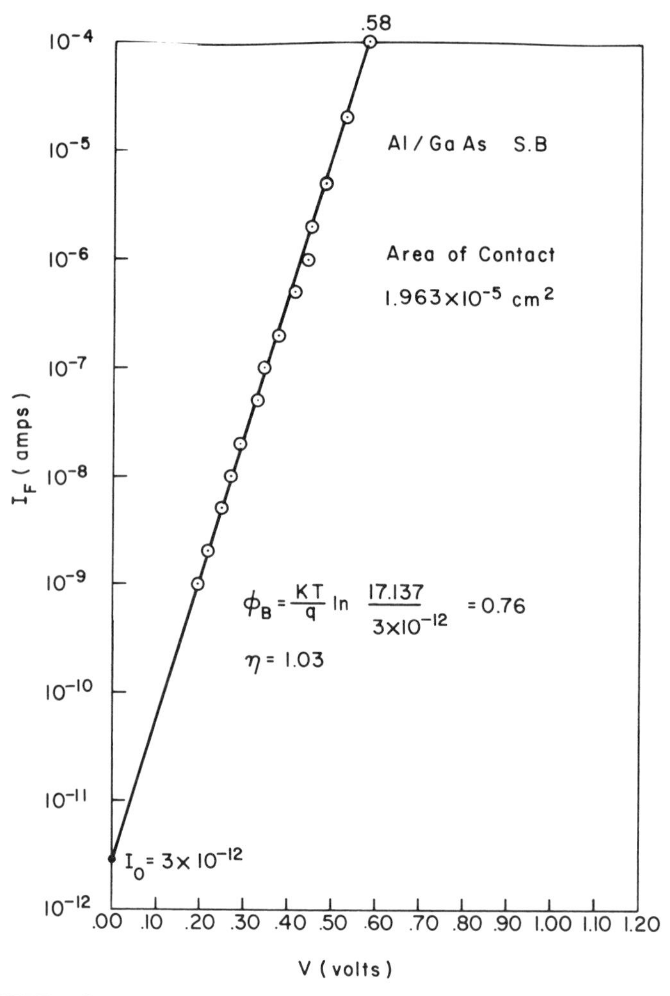

FIGURE 8.19. Forward *I-V* curve for Al-GaAs deposited at 350°C.

450°C. In addition, these films when tested at a current density of 1×10^{-6} A/cm^2 exhibited an MTF of 1000 hours at a temperature of 200°C.

Al-gate reliability and electromigration are therefore serious problems, especially when current densities exceed 10^5 A/cm^2 [26]. The problems are enlarged in the presence of metal inhomogeneities such as those present in nonepitaxial layers. Electromigration in Al/GaAs or Al/AlGaAs structures induces voids and openings in gate fingers, resulting in drain current variations. The current densities at which electromigration is observed in epitaxial Al gates are much higher and approach the 10^6 A/cm^2 observed for Au gates [27, 28]. The basic failure observed has been gate interruptions, resulting in failed devices after 200 hours of forward gate testing. Accompanying Al electromagnetism is the Au-Al interaction, observed in devices which had Au bond pads. In addition, multilayer structures such as Ti/Al/Ti, Ti/Al, and Al/Cr [29] have also been applied in order to improve resistance of Al gates to electromigration. However, Ti is prone both to oxidation and to interaction with Al, and Cr also cannot prevent the outdiffusion of Ga and As from the substrate.

8.5 CONTACT DEGRADATION

The contact resistance of the source and the drain directly affects all device parameters, especially transconductance and noise figure. The AuGeNi system degradation as a result of elevated temperature annealing (up to 350°C) results in source and drain resistance increases. The detected increase in contact resistance has been explained by the following effects [30–33].

1. The outdiffusion of Ga after an initial reaction with Au.
2. The overlayer metallization can also diffuse through the AuGeNi and react with the GaAs [34, 35].
3. The sloping effect results in a change in contact resistivity due to Au and Ni diffusion [36, 37].
4. The total contact area has been reduced due to the accompanying diffusion and reactions.

Similar effects can also occur as a result of storage tests. However, the time necessary for the reactions increases due to the non-bias and lower temperature conditions.

8.6 CONCLUSIONS

Three different metal–semiconductor systems deposited on GaAs have been assessed in terms of resistance to electromigration degradation. In the case of ternary alloys with Ge, the degree of electromigration resistance is directly proportional to the degree of AuGa formation. The presence of the AuGa microstructure induces electromigration to occur at lower temperatures and current densities.

In the case of Ge-GaAs with Ti-Mo-Au top surface metallization, local metal migration effects occur in regions with the lowest barrier height between the metallization and GaAs. The formation of an epitaxial Ge and GaAs layer results in barrier height lowering and metallization burn-out may occur through avalanche multiplication.

Finally, the epitaxial Al-GaAs system also presents electromigration resistance characteristics proportional to the degree of epitaxial aluminum formation. Epitaxially grown layers are more resistant to electromigration than polycrystalline aluminum films on GaAs.

REFERENCES

1. A.M. Andrews and H. Holonayak, *Solid State Electron.*, Vol. 15, 601 (1972).
2. G.Y. Robinson, *Solid State Electron.*, Vol. 18, 331 (1975).
3. C.R. Paola, *Solid State Electron.*, Vol. 18, 1189 (1970).
4. P.C. Moutou, J.J. Godard, J.M. Montel, and B. Dixneuf, *1975 Cornell Conference Proceedings*, p. 225 (1975).
5. H.M. Macksey, *Gallium Arsenide and Related Compounds*, Vol. 33b, 254 (1976), Institute of Physics Conference Series.
6. A. Christou and K. Sleger, 1977 Cornell Conference on Active Microwave Devices, Aug. 17–19, 1977, Ithaca, NY.
7. A. Christou, in *Scanning Electron Microscopy*, O. Johari and I. Corvin, eds., ITT Research Institute, Chicago, pp. 149–156.
8. A. Christou, *J. Appl. Phys.*, Vol. 47, No. 12, 5464 (1976).
9. R.H. Cox and H. Stack, *Solid State Electron.*, Vol. 10, 1213 (1967).
10. D. Sigurd, G. Ottaviani, V. Marrello, J.W. Mayer, and J.O. McCaldin, *J. Non-Cryst. Solids*, Vol. 12, 135 (1972).
11. W.J. Moroney and Y. Anand, "Low Barrier Height GaAs Microwave Schottky Diodes using Gold-Germanium Alloy," 1970, *Proceedings of the International Symposium on Gallium Arsenide*, Institute of Physics Society Conference, pp. 259–267 (1970).

12. J.M. Shannon, "Decreasing the Effective Height of a Schottky Barrier Using Low Energy Ion Implantation," *Appl. Phys. Lett.*, Vol. 24, 488–492 (1974).
13. Y. Anand, A. Christou, and H. Dietrich, "Low Barrier Height Ion Implanted GaAs Mixer Diode," in *Proceedings of the Conference on Active Semiconductor Devices and Circuits*, Cornell University, Ithaca, NY, (1977).
14. A.T. Anderson, Jr., A. Christou, and J.E. Davey, "Development of Ohmic Contacts for GaAs Devices Using Epitaxial Ge Films," *IEEE J. Solid State Circuits*, Vol. SC-13, 430–435 (1978).
15. H. Kim, G. Sweeney, and T.M.S. Heng, *Gallium Arsenide and Related Compounds*, Institute of Physics Conference Series 24, p. 307, (1975).
16. D.A. Abbott and J.A. Turner, *IEEE Trans.*, Vol. MTT-24, 317 (1976).
17. J.E. Davey, A. Christou, and H.M. Day, *Appl. Phys. Lett.*, Vol. 28, 365 (1976).
18. D. Sigurd, G. Ottaviani, V. Marrello, J.W. Mayer, and J.O. McCaldin, *J. Non-Cryst. Solids*, Vol. 12, 135 (1972).
19. D. Sigurd, G. Ottaviani, H.J. Arnal, and J.W. Mayer, *J. Appl. Phys.*, Vol. 45, 1740 (1974).
20. H. Sankur and J.O. McCaldin, *Appl. Phys. Lett.*, Vol. 22, 64 (1973).
21. L.L. Chang, L. Esaki, W.E. Howard, and R. Ludeke, *J. Vac. Sci. Technol.*, Vol. 10, 11 (1973).
22. R. Ludeke, L.L. Chang, and L. Esaki, *Appl. Phys. Lett.*, Vol. 23, No. 4, 201 (1973).
23. A. Christou, *J. Appl. Phys.*, Vol. 47, 5464 (1976).
24. A. Christou and H.M. Day, *J. Appl. Phys.*, Vol. 47, 4217 (1976).
25. A.Y. Cho, *J. Appl. Phys.*, Vol. 42, 2074 (1971).
26. P.M. White, G.C. Roger, and B.S. Hewitt, "Reliability of Ku-band GaAs power FETs," in *Proceedings of the IRPS*, pp. 297–301 (1983).
27. E.D. Cohen and A.C. Macpherson, "Reliability of Gold Metallized Commercially Available Power GaAs FETs," in *Proceedings of the IRPS*, pp. 156–160 (1979).
28. M. Otsubo, Y. Mitsui, M. Nakatani, and H. Wataze, *Proc. IEDM*, pp. 114–117 (1980).
29. V. Singh and P. Swarup, *Thin Solid Films*, Vol. 97, 277–286 (1982).
30. B.R. Sethy and H.L. Hartnagel, *J. Phys. D. Appl. Phys.*, Vol. 18, L9–L13 (1985).
31. M. Ogawa, *J. Appl. Phys.*, Vol. 51, 406–412 (1980).
32. A. Callegari, E.T.S. Pan, and M. Murakami, *Appl. Phys. Lett.*, Vol. 46, 1141–1143 (1985).
33. Y.C. Shih, M. Murakami, E.L. Wilkie, and A.C. Callegari, *J. Appl. Phys.*, Vol. 62, 582–590 (1987).

34. E.D. Marshall, B. Zhang, L.C. Wang, P.F. Jiao, W.X. Chen, T. Sawada, and S.S. Lau, *J. Appl. Phys.*, Vol. 62, 942–947 (1987).
35. Y.C. Shih, M. Murakami, and W.H. Price, *J. Appl. Phys.*, Vol. 65, 3539–3545 (1989).
36. P.E. Hallali, M. Murakami, W.H. Price, and M.H. Norcott, *J. Appl. Phys.*, Vol. 70, 7443–7448 (1991).
37. M. Murakami, W.H. Price, J.H. Greiner, and J.D. Feder, *J. Appl. Phys.*, Vol. 65, 3546–3551 (1989).

9

ELECTROTHERMOMIGRATION THEORY AND EXPERIMENTS IN ALUMINUM THIN FILM METALLIZATIONS

Aris Christou

CALCE Electronic Packaging Research Center, University of Maryland, College Park

9.1 INTRODUCTION

Metal migration in transistors and integrated circuits as a failure mode was first reported by Huntington and co-workers [1–3]. The investigations by Penny [4] related Huntington's work to semiconductor products. Previous investigations have indicated that a surface bridge short can be formed in transistors during RF testing [5–10]. The phenomenon is a field assisted migration of aluminum between base and emitter contacts along the oxide–silicon interface. This chapter summarizes the simulations where localized melting is initially induced by high electric fields and channeling between the base and emitter fingers proceeds by a combination of electromigration and thermomigration. It will be shown that this combination of effects is both a necessary and sufficient condition for the phenomenon known as metal migration to occur during high-power testing of test structures and power transistors. This phenomenon can occur in MESFETs and power MOSFETs as well as in bipolar transistors [11–15].

Electromigration and Electronic Device Degradation, Edited by Aris Christou.
ISBN 0-471-58489-4

9.2 BACKGROUND AND OBJECTIVES

The objective of the present chapter is to investigate the theory of electromigration and thermomigration in thin film aluminum and to apply the theory in deriving equations from which failure times can be obtained for Al/Si interdigital semiconductor systems exposed to high power densities. The same concepts may also be extended to simulate subsurface shorts.

The theory describing the mechanism of mass transport relates the basic forces on an aluminum ion in a conductor. Localized breakdown in the transistor develops a localized region of high temperature within the conducting junction. The hot spot is caused by the nonuniform current conduction in the semiconductor junction. Localized melting at the hot spot will establish a high temperature gradient between the base and emitter. The temperature gradient will affect the rate of migration of aluminum into silicon across a planar interface. This penetration or thermal mass transport will enhance the mass transport due to an electric field gradient. In the case of power MESFETs, the high field region will be present between the drain and gate.

The test samples used during high-power microwave testing were the interdigitized transistors. They consisted of small-grain aluminum conductor deposited on silicon windows. Silicon dioxide was thermally grown on the silicon wafer (Figure 9.1). Experimental microwave test results led to the formulation of an equation relating power dissipation

FIGURE 9.1. Interdigitated structure for the investigation of electrothermomigration of Aluminum on silicon. Metallization widths are 3 μm and 5 μm.

(P) to failure time (MTF):

$$\frac{1}{\text{MTF}} = AP^n$$

where n was a function of the test configuration. The mass transport causing the failure was the base–emitter short across the Si–SiO_2 interface (see Figures 9.2 and 9.3). Migration across the Si–SiO_2 interface

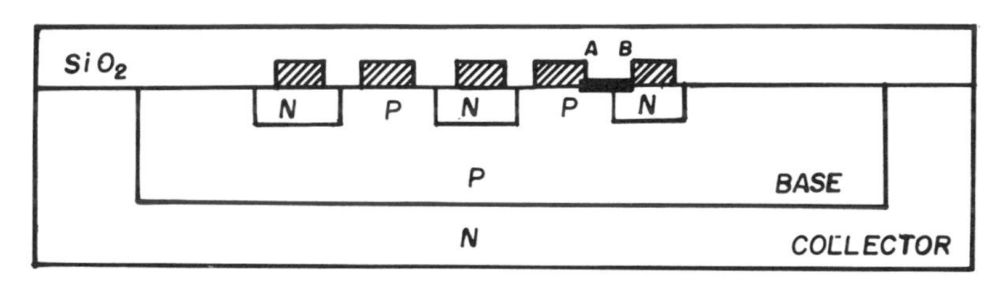

(a)

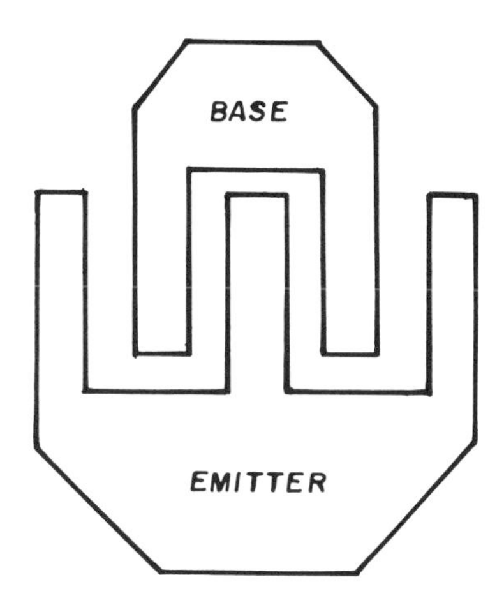

(b)

FIGURE 9.2. Base-emitter metallization system showing the location of the base-emitter short.

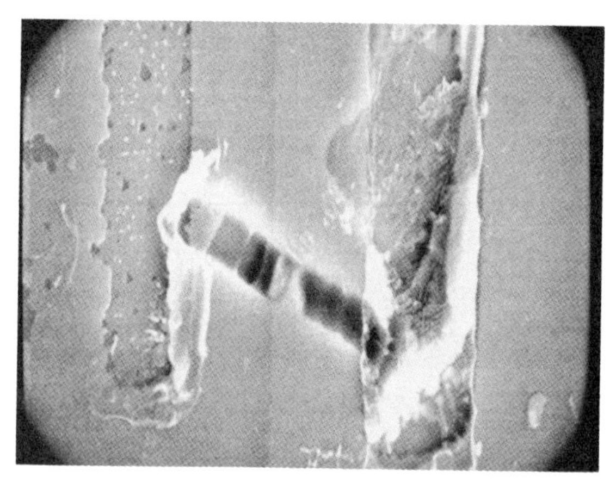

FIGURE 9.3. SEM images showing the base-emitter short. The bottom figure shows the shorted region after etching of the metallization. The base-emitter separation is 4 μm.

is the favored process as confirmed by the thin film diffusion experiments discussed in this chapter (see Section 9.6.1).

9.3 ELECTROMIGRATION IN TRANSISTOR METALLIZATIONS

9.3.1 Total Force on an Ion

When an electric field is applied to a thin film conductor, the total force on an activated ion, that is, an ion free from the crystal-line lattice, is made up of two forces: an electric field force and a momentum exchange force. The electric field force is due to the field interaction with the effective ionic charge. Since the effective ionic charge is diminished by the screening effect of the ion's bound electrons, the magnitude of the electric field force is small. The momentum exchange force arises from collisions of the current carriers with the ions and is of significant magnitude. These two forces act in opposite directions, the electric field force in the direction of the electric field, and the momentum exchange force in the opposite direction since it is brought about by a current flowing in the direction opposite to the applied field.

Several workers [1–3] in the area have derived expressions for the total force due to an applied electric field. As treated by Ho and Huntington [1], the total momentum exchange between the current carriers and the moving ions is related to the effective contribution to the resistivity associated with the ions in the activated configuration for diffusion. Thus the total force may be written as

$$F = eEZ\left(1 - \frac{1}{2}\frac{N}{P}\frac{\rho d}{N_d}\frac{m^*}{m}\right) \tag{9.1}$$

where d is the defect resistivity (i.e., the partial resistivity due to scattering by activated defects), N_d is the density of defects, ρ is the partial resistivity due to scattering from lattice ions, N is the density of ions, m is the free-electron mass, m^* is the effective electron mass, Z is the electron–atom ratio, E is the applied electric field, and e is the electronic charge. The first term in brackets is the electrostatic field force and the second term is the momentum exchange force.

In Fiks work [2] two assumptions are made:

1. The electrons are assumed to be free.
2. In each collision they lose all their field-acquired momentum to the activated ion. The force is then given by

$$F = eE(Z - n_e l_e A_e) \tag{9.2}$$

where n_e is the number of electrons per unit volume, l_e is the electronic mean free path, and A_e is the electronic scattering cross section.

Fiks indicates that an extension of this model to include metals with both holes and electrons would yield an expression for the force of the form

$$F = eE(Z - n_e l_e A_e - n_h l_h A_h) \tag{9.3}$$

where the h subscript denotes holes.

Bosvieux and Friedel [3] approached the problem quantum mechanically, using the Born approximation to calculate the perturbation of the electronic wave function in the vicinity of a moving ion. The net electrostatic force was then evaluated from the corresponding change in charge density. Their resultant force expression agrees very closely with those of Huntington and Fiks. In fact, their F is identical to Equation (2) except for the factor $\frac{1}{2}$ provided the diffusion is by means of the vacancy mechanism.

Each of these expressions for the total force F differs from each other in detail but all have the same basic form:

$$F = eEZ^*$$

where Z^*e is the effective ionic charge, and Z' represents the net effect of the electrostatic field force and the momentum exchange force.

9.3.2 Ionic Drift Velocity

The relationship between the ionic drift velocity v_d under the influence of the applied electric field force Z^*eE is of the form

$$v_d = \mu Z^* eE \tag{9.4}$$

where μ is the ionic mobility. But from the Nernst–Einstein relation,

$$\mu = D/kt \tag{9.5}$$

where T is the absolute temperature, k is Boltzmann's constant, and D is the diffusion coefficient. Note that the temperature dependence of D is of the form

$$D = D_0 \exp(-\Delta H/kT) \tag{9.6}$$

where ΔH is the activation energy for a given diffusion process. Substituting Equation (9.5) into (9.4) gives

$$v_d = \frac{DZ^* eE}{kT} \tag{9.7}$$

9.3.3 Total Ion Flux

The total ion flux U through a conductor (i.e., the total number of ions crossing a unit area per unit time) is

$$U = Nv_d \tag{9.8}$$

where N is the number of activated ions per unit volume. Using Equations (9.6) and (9.7) and writing the applied field as $E = \rho J$, where ρ is the electrical resistivity and J is the current density, we have for the flux

$$U = ND_0 e \frac{\rho Z^*}{kT} j \exp(-\Delta H/kT) \tag{9.9}$$

After the ions have acquired sufficient thermal energy to be in the activated state (i.e., free of the lattice), this expression describes their motion under the influence of the electric field and momentum exchange forces. If the majority charge carriers are electrons, then the activated ions will move in the direction opposite to the applied electric field.

9.3.4 Ion Flux Divergence

The existence of an ion flux in a conductor will not cause the conductor to fail. In order for a failure to occur there must exist a nonvanishing ion flux divergence, that is,

$$\nabla \cdot U = 0 \tag{9.10}$$

If $\nabla \cdot U > 0$, then more ions will be flowing out of a given volume of the conductor than are flowing into it. This will result in the formation of voids in the volume, with aggregation of metal in the regions outside it. Under the proper conditions, void formation eventually leads to open circuit failure of the conductor.

Now let us examine the possible causes of a nonzero flux divergence. According to the continuity equation

$$\frac{dN}{dt} = -\nabla \cdot U \tag{9.11}$$

the divergence of the ion flux for a given volume is just equal to the time rate of change of the ion density within the volume. In one dimension Equation (9.11) has the form

$$\frac{dN}{dt} = -\frac{dU}{dx} \tag{9.12}$$

where electron flow and hence ion flux is in the x direction. If $U = U(C, T)$, where C is the local composition, then Equation (9.12) becomes

$$\frac{dN}{dt} = -\left[\frac{dU}{dC}\frac{dC}{dx} + \frac{dU}{dT}\frac{dT}{dx}\right] \tag{9.13}$$

Thus we see that any gradients in composition of temperatures along the x direction (length) of the conductor will result in a nonzero ion flux divergence.

9.3.5 Failure Mechanism

Since atoms can be more easily removed from the structurally inhomogeneous grain boundaries than from within grains, there exists at these boundaries a nonzero ion flux divergence. This flux divergence causes ionic migration away from the boundaries, electrical isolation (open circuit failure) occurring when such migration has removed one atomic layer from the surface of every grain in a given cross section. This removal process reduces the original cross-sectional area of the conductor, A_0, to a final effective area, KA_0. For spherical grains of radius R and average interatomic distance a, K is given by the product function ratio of the effective radii,

$$K = \frac{\prod (R - a)^2}{\prod R^2} \tag{9.14}$$

or

$$K = \left(1 - \frac{a}{R}\right)^2 \tag{9.15}$$

By expanding Equation (9.15) and discarding higher order terms, we obtain

$$K = 1 - \frac{2a}{R} \tag{9.16}$$

For a conductor of cross-sectional area A, length l, and containing a total of n ions, the ionic density is

$$N = \frac{n}{AT}\frac{dN}{dt} = \frac{dN}{dA}\frac{dA}{dt} \tag{9.17}$$

If only the cross-sectional area is a function of time, then

$$\frac{dN}{dt} = -\frac{n}{A^2 l}\frac{dA}{dt} \tag{9.18}$$

or, using Equation (9.17),

$$\frac{dN}{dt} = -\frac{N}{A}\frac{dA}{dt} \tag{9.19}$$

This result may be inserted into the continuity equation as given by (9.12) to yield

$$\frac{dA}{dt} = +\frac{A}{N}\frac{dU}{dx}, \quad dt = \frac{N}{A\dfrac{dU}{dx}}\,dA \tag{9.20}$$

an equation relating the time rate of change of the cross-sectional area to the flux gradient

$$\frac{dU}{dx} = \frac{N}{A}\frac{dA}{dt} \tag{9.21}$$

Thus from Equation (9.20) the failure time is given by

$$t_f = \int dt = \int_{A_0}^{KA_0} \frac{dA}{\dfrac{A}{N}\dfrac{dU}{dx}} \tag{9.22}$$

The linearity of the heat diffusion equation indicates that the temperature gradient will be a linear function of the power density. Now for a conductor of length l the power density is $J^2\rho l$, so that the temperature gradient is given by [6]

$$\frac{dT}{dx} = GJ^2\rho l \tag{9.23}$$

where ρ is the resistivity and G is a constant depending only on the thermal properties of the conductor material and geometry.

Assuming that ρ and Z^* are linear functions of temperature [6], that is,

$$\rho = \rho_0 = \rho_1 T \tag{9.24}$$

$$A^* = Z_0 - Z_1 T \tag{9.25}$$

Taking the derivative of Equation (9.9) with respect to T yields

$$\frac{dU}{dt} = \frac{ND_0 eJ}{k} \exp\left(\frac{\Delta H}{kT}\right)\left\{\frac{\rho_1 Z^*}{T} - \frac{Z_1}{T} - \frac{\rho Z^*}{T^2} + \frac{\rho Z^* \, \Delta H}{KT^3}\right\} \tag{9.26}$$

Substituting this result along with Equation (9.23) into Equation (9.21) gives

$$\frac{dU}{dx} = NBJ^3 \tag{9.27}$$

where

$$B = D_0 eGl \exp\left(\frac{-H}{kT}\right)\left\{\frac{Z^* \rho H}{K^2 T^3} + \frac{\rho_1 Z^*}{kT} - \frac{\rho Z_1}{kT} - \frac{\rho Z^*}{kT^2}\right\} \tag{9.28}$$

Using Equation (9.27) and the fact that $J = I/A$, the failure time given by Equation (9.22) becomes

$$t_f = \frac{-1}{BI^3} \int_{A_0}^{KA_0} A^2 \, dA = \frac{1 - K^3}{3BJ_0^3} \tag{9.29}$$

where J_0 is the initial current density. But from Equation (9.26) we have

$$K^3 = \left(1 - \frac{a}{R}\right)^3 \tag{9.30}$$

or

$$K^3 \simeq 1 - \frac{6a}{R} \tag{9.31}$$

Thus Equation (9.29) has the form

$$t_f = \frac{2a}{BRJ_0^3} \tag{9.32}$$

A cursory examination of this equation might lead one to conclude that the time-to-failure varies inversely with grain size, thus contradicting much experimental evidence in the area (e.g., 6). However, the denominator of the expression contains the initial current density, which is itself a function of grain size. Therefore the explicit terms in the equation do not necessarily reflect the actual failure time–grain size relationship. The true nature of the relationship is contained in the implicit variation of the grain size–current density product.

9.4 THERMOMIGRATION IN ALUMINUM

9.4.1 Drift Velocity

It is the purpose here to derive an approximate quantitative expression for the rate of migration of aluminum into silicon across a planar interface between the two materials under the influence of a thermal gradient. This penetration is usually referred to in the literature as thermomigration or thermal mass transport.

The force acting on the Al ions due to the thermal gradient has the form

$$F = \frac{-Q^* \nabla T}{T} \tag{9.33}$$

where Q^* is the heat of transport, defined to be the quantity of heat flow needed to maintain steady-state conditions with unit mass transport, and T is the absolute temperature. The ionic drift velocity (and hence the rate of penetration) under the influence of this force is

$$v_d = \mu F = -\frac{\mu Q^* \nabla T}{T} \tag{9.34}$$

In this equation D is the diffusivity of Al into Si and is expressed in the form $D = D_0 \exp(\Delta H / Kt)$, where ΔH is the activation energy for the process, K is Boltzmann's constant, and D_0 is the apparent value of D at infinite temperature.

9.4.2 Temperature Dependence and Failure Time

The temperature dependence of failure may be calculated from the equation for ionic drift resulting in Al migration into the silicon regions.

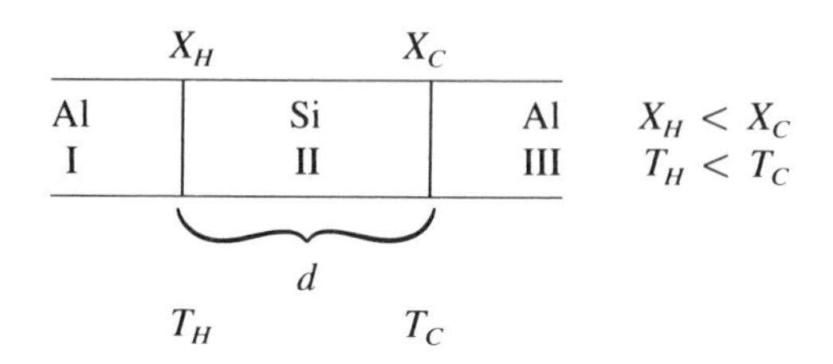

$$v_d = (D_{\rm I} - D_{\rm II}) \cdot \frac{\partial N}{\partial x} = -\frac{\mu Q^* \nabla T}{T} \tag{9.35}$$

For a one-dimensional system arranged as shown above, failure brought about by the thermal gradient occurs when Al ions transit the distance from the boundary at coordinate X_H and high temperature T_H to the boundary at coordinate X_C and low temperature T_C. The migration rate of the Al from region I is given by the drift velocity v_d in Equation (9.35).

The temperature in Equation (9.35) may be expressed as

$$T(x) = T_H - (\nabla T)x \tag{9.36}$$

where it is assumed that

$$\nabla T = \frac{T_H - T_C}{X_H - X_C} = \text{constant} \tag{9.37}$$

The dependence of v_d on the linear coordinate X shows up in the diffusivity D as an exponential and in T_2 as a quadratic term. For sufficiently small intervals in the X coordinate it will be assumed that both these dependencies combine to give a linear variation of v_d with X. Thus to a good approximation

$$v_d = \frac{1}{2}(v_{dH} + v_{dC}) = \frac{Q^* D \, \nabla T}{2k}\left[\frac{1}{T_H^2} + \frac{1}{T_C^2}\right] \tag{9.38}$$

The time (failure time) required for the transit of Al from X_H to X_C is then

$$t_f = \frac{X_H - X_C}{v_d} = \frac{2k(X_H - X_C)}{Q^* D \nabla T} \left[\frac{T_H^2 T_C^2}{T_H^2 + T_C^2} \right] \tag{9.39}$$

9.4.3 Discussion

This expression for the failure time due to thermomigration has a number of drawbacks. First, it should only be regarded as a quantitative guess, in as much as it only approximately describes the thermal mass transport failure mechanism. Second, in order to completely describe the thermomigration process one must solve the continuity equation

$$\frac{dc(x, t)}{dt} = -\nabla \cdot J \tag{9.40}$$

where $c(x, t)$ is the instantaneous concentration of Al ions at position x and time t, and J is the Al ion flux. For thermomigration the flux may be written as [8]

$$J = D \nabla C + \frac{Q^* DC \nabla T}{kT^2} \tag{9.41}$$

Thus Equation (9.40) has the final form

$$\frac{dC}{dt} = D \frac{d^2 C}{dx^2} + \frac{dD}{dx}\frac{dc}{dx} - \frac{Q^* D \nabla T}{kT^2}\frac{dc}{dx} - \frac{Q^* c}{K}\frac{d}{dx}\left(\frac{D \nabla T}{T^2}\right) \tag{9.42}$$

Evaluating the solution to Equation (9.42) in the region of the interface at X_c, failure occurs when the concentration of Al ions $C(X_c, t)$ reaches such a value that the semiconducting characteristics of the Si have been degraded.

9.5 FAILURE MECHANISM

9.5.1 Combined Effect

Having described the theories governing electromigration and thermomigration, it would be interesting to speculate as to their relative importance as failure mechanisms in high-power bipolar transistors or even insulated gate power field effect transistors (IG FETs).

In Section 9.4 it has been indicated how thermomigration would bring about the failure of a system that very much resembles the interdigital metallization in Al/Si transistors. In describing this thermomigration failure, it was assumed that Al ions migrated through the body of the silicon region separating the two Al metallizations. It has been shown that the Al ion migration path in transistors, which had failed due to exposure to microwave radiation, is along the Si–SiO_2 interface. If this failure mode is due to the existence of a thermal gradient, then the path differs from the assumed path in the derivation of Section 9.4. However, the expression for the thermomigrative failure time has fairly general applicability. For paths along interfaces the only parameter in this equation that would change appreciably would be the diffusivity D. This change is brought about by the dependence of D on the activation energy ΔH.

It is unlikely that either electromigration or thermomigration is the sole cause of interdigital transistor failure. More probably the failure is due to a combination of the two, initial electromigration being due to E–B electrical potential gradients established by Joule heating at compositional inhomogeneities and localized hot spots. In bipolar transistors localized breakdown would probably be initiated at the base side of the emitter–base junction.

Since the primary failure mechanism of these two types of migration in Al/Si interdigital structures is an electrical short produced by metal migration, any general description of their combined effect would have to determine the way in which the Al ion density varies in space and time in the region between metallizations. Such a description is given by the equation

$$\frac{dn(x, t)}{dt} = -\nabla \cdot J \tag{9.43}$$

where $n(x, t)$ is the instantaneous Al ion density at position x and time t, and J is the ionic flux. The ionic flux, written in terms of the electromigration and thermomigration forces, is given by

$$J = -D\,\nabla n + \frac{DNZ^{*}eE}{kT} + \frac{Q^{*}Dn\,\nabla T}{kT^2} \tag{9.44}$$

where the second term is the flux due to the applied electric field E, the third term is the flux due to the temperature gradient ∇T, $Z^{*}e$ is the effective charge of the diffusion species, Q^{*} is the heat of transport, and D is the diffusivity ($D = D_0 \exp(-\Delta H/kT)$).

9.5.2 Computer Simulations

By substituting Equation (9.44) into (9.43) a differential equation describing mass transport can be obtained [Equation (45)]. Solution of the equation

$$\frac{dn}{dt} = D\frac{d^2n}{dx^2} + \frac{dD}{dx}\frac{dN}{dx} - \left(\frac{Z^*eE}{kT} + \frac{Q^*\nabla T}{kT^2}\right)\left[n\frac{dD}{dx} + D\frac{dn}{dx}\right] \tag{9.45}$$

was carried out for the cases of $dD/dx = f(x, n, p)$, $dD/dx = 0$. If one assumes that $dD/dx = 0$ (which is reasonable for short diffusion distances), then Equation (9.45) reduces to

$$\frac{dn}{dt} = D\frac{d^2n}{dx^2} - \left(\frac{Z^*eE}{kT} + \frac{Q^*\nabla T}{kT^2}\right)\left[D\frac{dn}{dx}\right] \tag{9.46}$$

where $n(x, t)$ is the instantaneous Al ion density at position x.

The constants for Equation (9.46) are

$$D = 8.0 \exp\left[\frac{-3.47 \text{ eV}}{kT}\right] \text{cm}^2/\text{s} \qquad 1000°\text{C} < T < 1400°\text{C}$$

$$D = 2.8 \exp\left[\frac{-3.8 \text{ eV}}{kT}\right] \text{cm}^2/\text{s} \qquad 600°\text{C} \leq T \leq 1000°\text{C}$$

$$D = 4.8 \text{ cm}^2/\text{s} \qquad T \leq 600°\text{C}$$

and $Q^* = 0.8$ eV, $k = 8.6139 \times 10^{-5}$ eV/K, $\nabla T = 78.4$ K/cm, $Z^*e = 14.4 \times 10^{-19}$ C, and $E = 2.74$ V/cm. These constants were determined from molten Al droplet experiments migrating within a silicon matrix.

The boundary conditions for the problem are

$$\left.\frac{\partial n(x, t > 0)}{\partial x}\right|_{x=0} = 0$$

$$n(x, 0) = \begin{cases} N_0 & 0 \geq x > h \\ 0 & x > h \qquad (h = 1) \end{cases}$$

The failure criterion was that one grain surface would lose one atom layer. This implies that failure occurs when concentration of Al ions n reaches a value where the semiconducting characteristics of the silicon have been degraded. Figures 9.4 and 9.5 show the results of the calculations.

The results show that for the electrothermomigration model, the failure time is a sensitive function of hot spot temperature. The failure sequence is therefore:

1. Establishment of electromigration due to base–emitter electrical potential gradients.
2. Thermomigration domination as temperature gradients are established by Joule heating and localized melting.

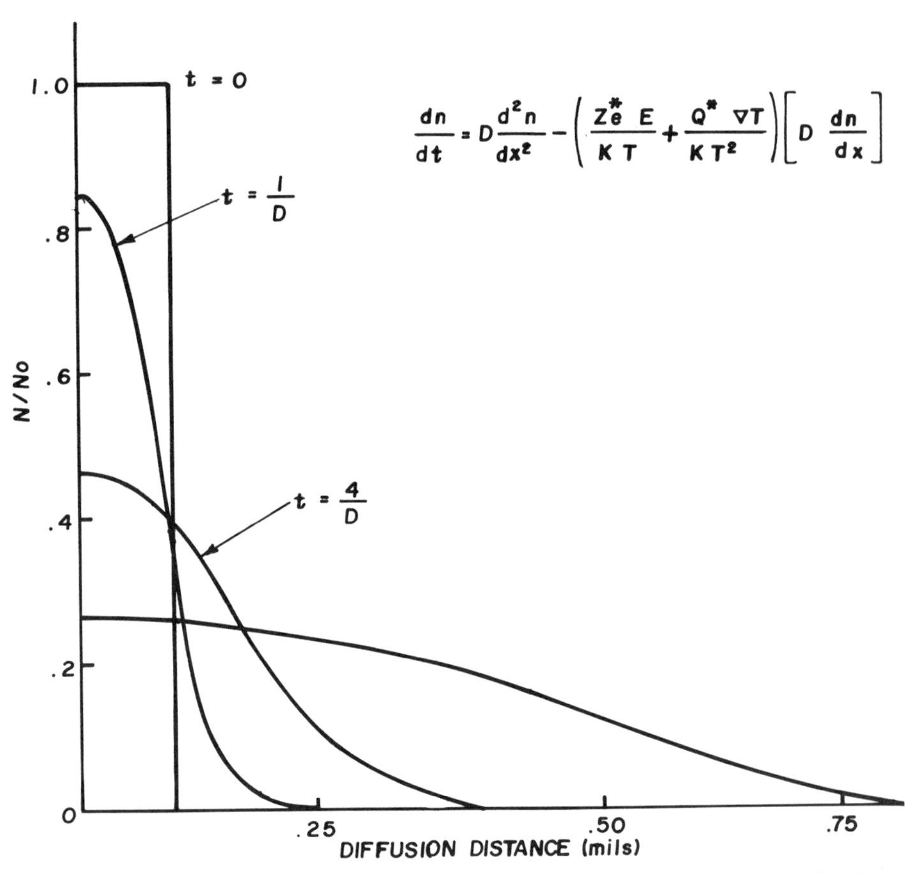

FIGURE 9.4. Diffusion of aluminum into silicon. The aluminum is assumed to be a limited source.

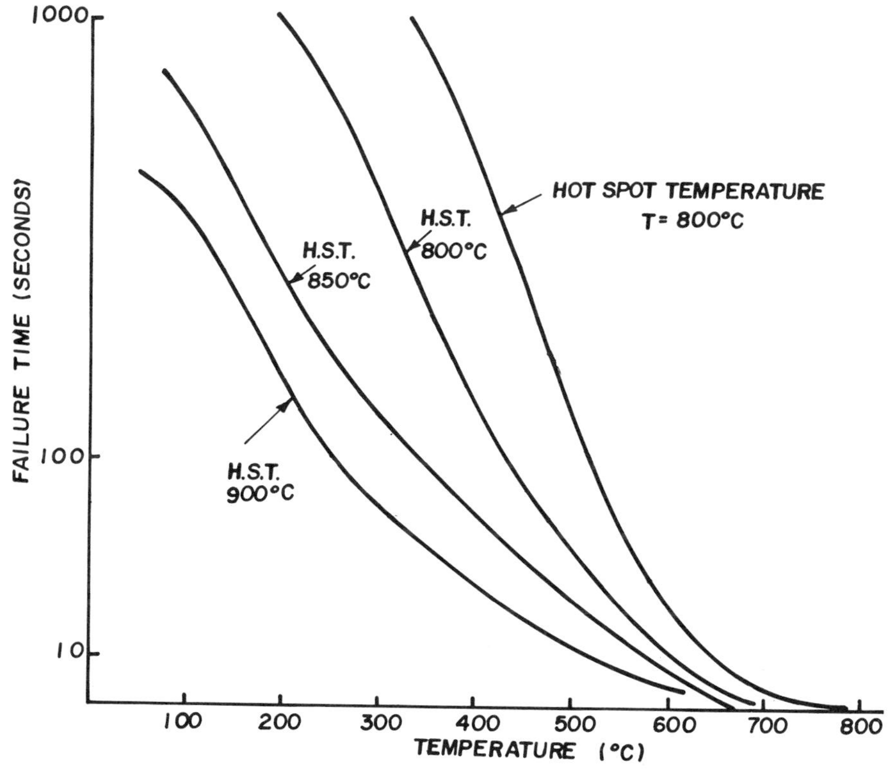

FIGURE 9.5. Failure time dependence on hot spot temperature as a variable.

The solution of Equation (9.45) for the case $dD/dx \neq 0$ was obtained by a Runga numerical integration scheme. The results summarized in Table 9.1 are in close agreement with results obtained with Equation (9.46). This shows that the spatial variation of the diffusion coefficient is a second-order effect. Figures 9.6 and 9.7 show the result of localized melting of the metallization in the case where the magnitude of the electric field is not large enough to result in a bridging short.

9.6 ASSOCIATED DIFFUSION EXPERIMENTS

9.6.1 Diffusion Experiments

A preliminary investigation of the thin film diffusion rates of aluminum on various silicon substrates has also been undertaken. Aluminum films approximately 2000 Å in thickness were vapor deposited on 200 × 200 mil (100) silicon chips, which were heated to 100°C under a dynamic

TABLE 9.1 Solution of Equation (45) for $dD/dx \neq 0$, $E = 2.7$ V/cm

Failure Time (seconds)	Hot Spot	*B-E* Temperature Gradient[a]
>10,000	100°C	550°C/cm
	220°C	
5,900	300°C	
730	500°C	
36	600°C	
0.01	700°C	
0.004		
1,000	100°C	700°C/cm
800	200°C	
638	300°C	
229	400°C	
150	500°C	
3	600°C	
0.003	700°C	

[a] Measured over a metallization separation distance of 5 μm.

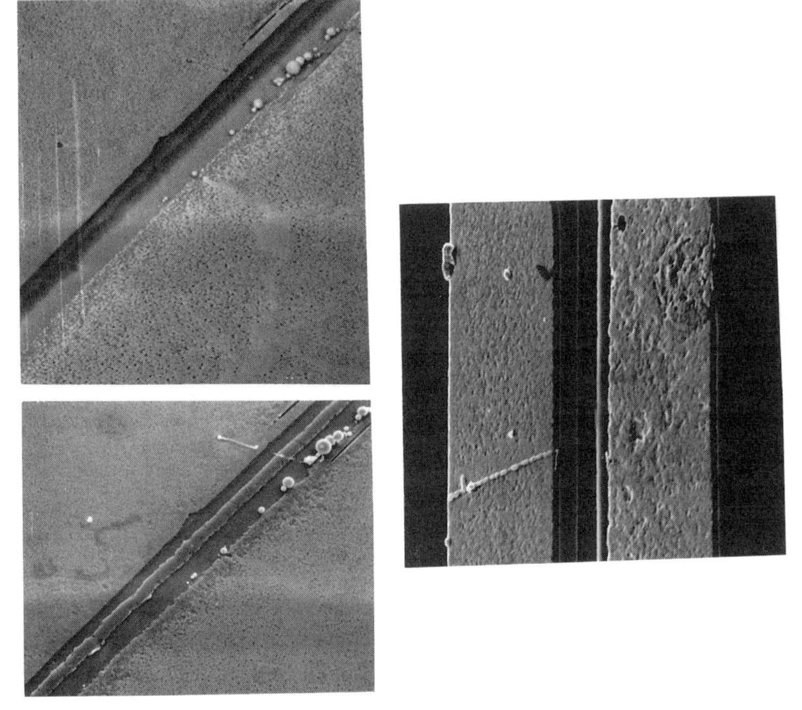

FIGURE 9.6. SEM micrograph showing localized melting prior to bridging the interconnects.

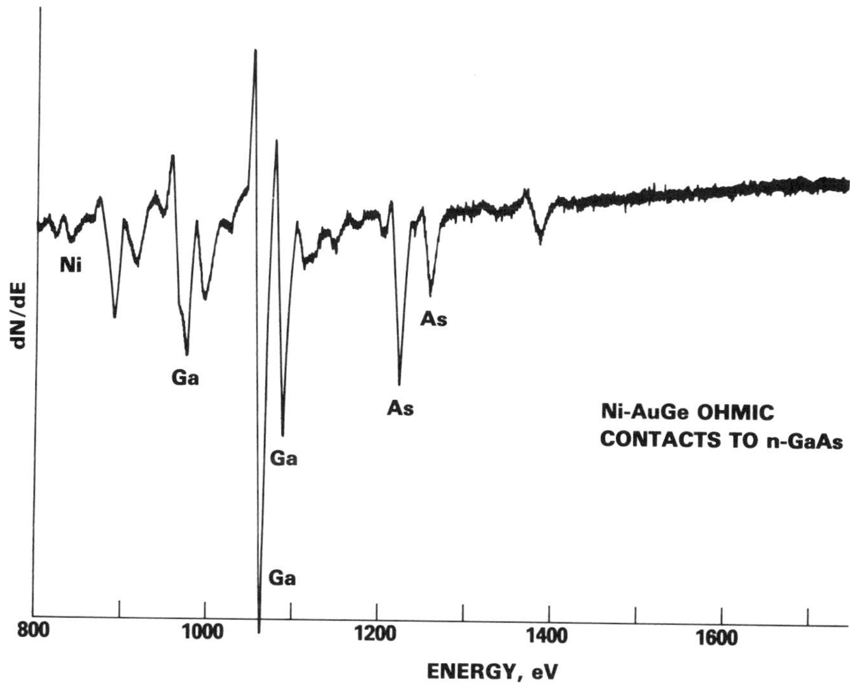

FIGURE 9.7. Auger electron spectroscopy of the gold droplets showing interaction with the semiconductor substrate.

pressure less than 10^{-7} torr. These metallized chips were then given a sequence of diffusion anneals, mounted in a cold-setting plastic, and lapped at a 10° angle. The displacement of the silicon–aluminum interface was measured with the scanning electron microscope with distance calibration based on latex spheres. In these experiments, the four substrates—silicon, silicon dioxide, *n*-silicon/*p*-silicon interface, and *n*-silicon/silicon dioxide interface—were compared. The data, summarized in Fig. 9.8, indicate chemical diffusion to proceed most rapidly at the solid state interfaces in which different materials are joined. Of particular interest are the data for the epitaxial $Si/Si/SiO_2$ interface for which aluminum diffusion proceeds at least three times faster than in any other substrate throughout the temperature range of 575°C (the Si-Al eutectic temperature). Note, however, that the reaction zone for the Si/SiO_2 system at 575°C spans only 0.04 μm/min so that either eutectic transformation or metallization fusion is required to yield the observed channel formation rates. Figures 9.9 and 9.10 show the penetration versus diffusion time data for the SiO_2/Si system and for the *n*-Si/*p*-Si system.

DEPOSITED METAL	SUBSTRATE	DISPLACEMENT	ACTIVATION ENERGY
Al	Si	6.0 Å/MIN	9.6 K CAL
Al	SiO_2	7.4 Å/MIN	40 K CAL
Al	Si/Si (p-n)	112 Å/MIN	2.3 K CAL
Al	Si/SiO_2	250 Å/MIN	10 K CAL

FIGURE 9.8. Diffusion for thin film aluminum into various types of silicon substrates.

The penetration distance was also measured by Auger sputter profiling of 2000-Å aluminum films on (100) silicon. The diffusion lengths measured as a function of time varied as the square root of the diffusivity and time for temperatures up to 500°C. These experiments indicated that diffusion was typically bulk diffusion up to 500°C as expected from the experimental conditions.

9.6.2 Failure Time and Linear Dimensions

Equation (9.32) has the shortcoming that it does not give a completely explicit relationship between failure time t_f and the linear dimensions of the conducting stripe. The following discussion will show how this relationship might be incorporated into the theory.

Consider a conducting stripe to be made up of a series of unit volume elements each having dimensions that are small with respect to the overall dimensions of the stripe. These unit elements are so constructed that each contains one, and only one, defect, that is, any macroscopic or microscopic structural inhomogeneity producing a flux divergence. Assume that the density of defects (and hence the number of defects along any arbitrarily chosen linear dimension) in the conductor is constant. The degree of severity of each defect lies between some minimum and maximum determined by the magnitude of the flux divergence the defect produces. The more severe the defect, the shorter the time interval required for the flux divergence to completely evacuate all conducting material from the unit volume element, thus producing a unit void. Nucleation of these voids along the stripe's width dimension eventually produces an open circuit.

Now at some arbitrary point along the length of the conducting stripe consider the unit volume elements (each containing a defect of some

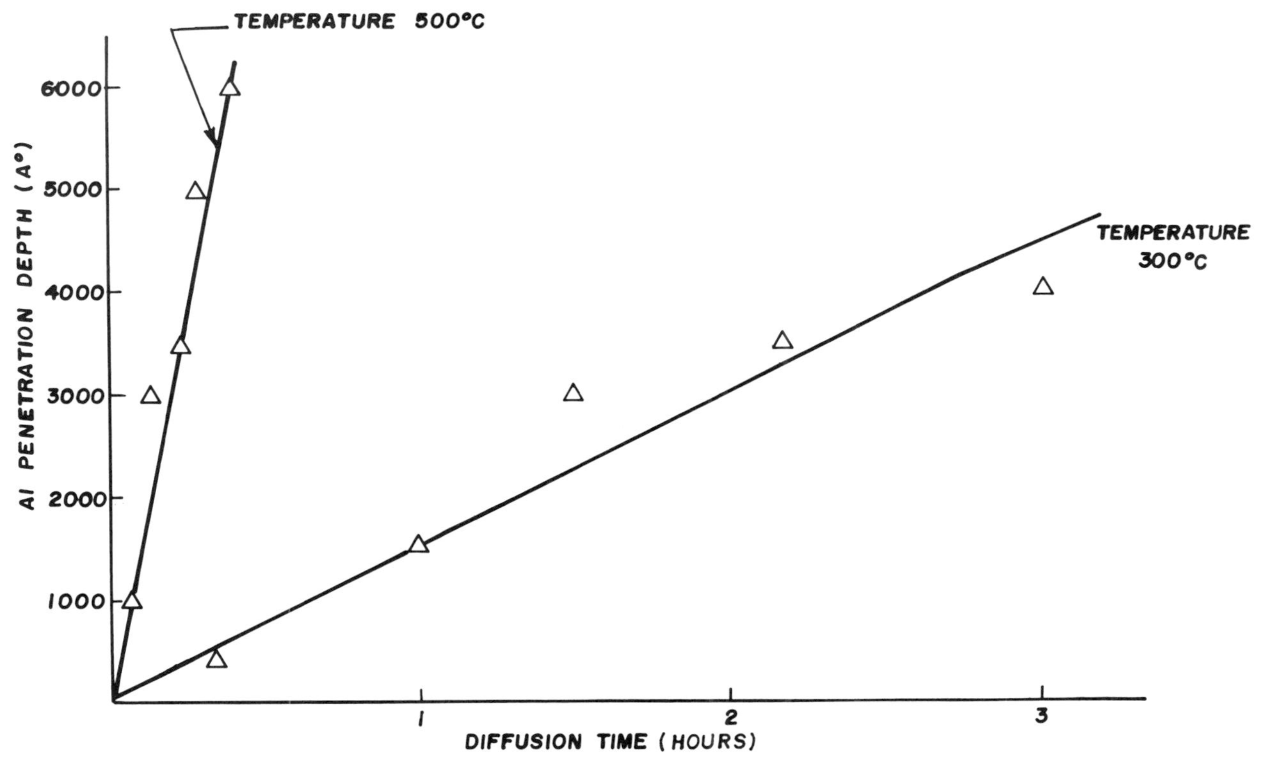

FIGURE 9.9. Diffusion of aluminum into SiO_2/Si(100) substrates.

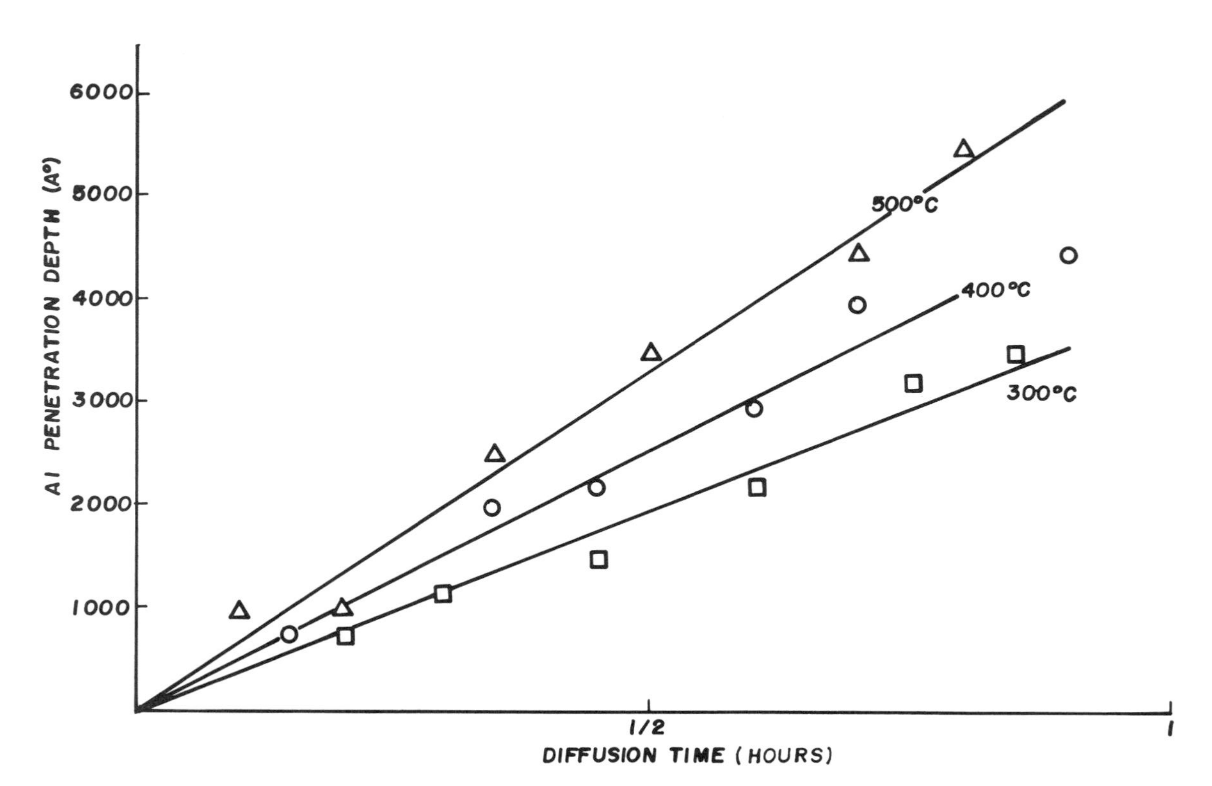

FIGURE 9.10. Diffusion of Al thin films in Si(100).

degree of severity) to be aligned across the width of the conductor. Since an electrical open in the conductor must occur along its width, the time to failure of the stripe is governed by that unit cell along the width which has the least severe defect. If one assumes that all degrees of defect severity are equally likely probabilistically, then, as the width increases, the total number of defects becomes large and hence the probability of aligning less severe defects across the width increases. Therefore it would be expected that the failure time of the stripe would increase with an increase in width. The same argument and conclusion are also applicable when considering the relationship between failure time and the thickness of the stripe.

A similar argument may be applied to the dependence of the failure time on the length of the stripe. However, considering unit elements aligned along the length, the failure time depends on the element having the *most* severe defect. Hence, assuming that the number of defects per unit length is constant, a longer stripe has a higher probability of containing more severe defects. So the failure time of the stripe would decrease with an increase in length.

On the basis of the above arguments it seems likely that the dependence of the failure time on the linear stripe dimensions would be of the form

$$t_f = A \frac{W\Theta}{l} \tag{9.47}$$

where W is the stripe width, Θ is the thickness, l is the length, and A is a constant. However, empirical evidence has shown [7] that as the length of the conductor is increased, the failure time initially decreases and then attains saturation at longer lengths. This saturation level is a function of the width and thickness parameters. The nature of this failure time–length dependence has been found to have the empirical form

$$t_f = A' \exp\left(\frac{\alpha}{l}\right) \tag{9.48}$$

where α is a constant dependent on width and thickness. Thus the relationship between the failure time and the linear dimensions is probably of the form

$$t_f = AW\Theta \exp\left(\frac{\alpha}{l}\right) \tag{9.49}$$

Equations (9.32) and (9.49) may be integrated numerically in order to give the final coherent expressions for failure time. Equation (9.49) also indicates that for the short widths found in present bipolar power transistors (2–3 μm), the t_f is short enough so that the failure may be characterized as a catastrophic infant mortality type.

9.7 CONCLUSIONS

Using a failure criterion of one monolayer of aluminum bridging a channel, calculations of failure times show that failure time is a sensitive function of hot spot temperature.

The results show that for the electrothermomigration model, the failure sequence is:

1. Establishment of electromigration due to base–emitter potential gradients.
2. Thermomigration domination as temperature gradients are established by Joule heating and localized melting.

The failure times typically are in the tens of minutes range once the localized hot spot temperature exceeds 500°C. Such failures may also be categorized as catastrophic infant mortalities. However, such catastrophic failure mechanisms may also appear during the period of constant failure rate since it is probabilistic in nature.

REFERENCES

1. P.S. Ho and H.B. Huntington, *J. Phys. Chem. Solids*, Vol. 27, 1319–1322 (1966).
2. V.B. Fiks, *Sov. Phys. Solid State*, Vol. 1, 14 (1959).
3. C. Bosvieux and J. Friedel, *J. Phys. Chem. Solids*, Vol. 23, 123–136 (1962).
4. R.V. Penney, *J. Phys. Chem. Solids*, Vol. 25, 335–345 (1964).
5. R. Amadori, W. Peters, and V. Puglielli, U.S. Naval Weapons Laboratory, TN-F/97-71, May 1971.
6. H.S. Carslaw and J.C. Jaeger, *Conduction of Heat in Solids*, Oxford University Press, London, 1959.
7. H.B. Huntington, *J. Phys. Chem. Solids*, Vol. 29, 1641–1651 (1968).
8. R. Rosenberg and M. Ohring, *J. Appl. Phys.*, Vol. 42, No. 13, 5671–5679 (Nov. 1971).

9. J. Black, *Metallization Failures in Integrated Circuits*, Tech. Report No. RADC-TR 68-243, Dec. 28, 1966.
10. J.D. Venables and R.G. Lye, in *Proceedings of the 10th IEEE IRPS*, pp. 159–164 (1972).
11. P.F. Tang, *Modeling of Electromigration Phenomena with Application to GaAs on Au*, Ph.D. thesis, Carnegie–Mellon University, 1990.
12. J. Cho and C.V. Thomson, *Appl. Phys. Lett.*, Vol. 54, No. 25, 2577 (1989).
13. C.A. Ross and J.E. Evetts, *Scripta Metall.*, Vol. 21, 1077–1082 (1987).
14. B.N. Agarwala, M.J. Attardo, and A.P. Ingraham, *J. Appl. Phys.*, Vol. 41, No. 10, 3954 (1970).
15. K. Hinode and Y. Homma, in *Proceedings of the IEEE IRPS*, p. 25 (1990).

10

RELIABLE METALLIZATION FOR VLSI

M.C. PECKERAR
Surface and Interface Science Branch,
Electronics Science and Technology Division,
Naval Research Laboratory, Washington, DC

10.1 INTRODUCTION

For over two decades there has been a log-linear reduction in the minimum feature size appearing on an integrated circuits. To date, the evolutionary trend has been so reliable that it has been referred to as "Moore's law." The history of minimum feature size reduction is shown in Table 10.1. Minimum dimension features are required for MOS transistor gates and for bipolar transistor emitter contacts. Interconnect metallization rarely needs to approach these dimensions. In fact, metallization feature sizes are typically two to three times the minimum feature size. But still, as devices are scaled to fit into ever smaller areas, metal interconnects must shrink proportionately to accommodate the new design geometry.

As the cross-sectional area of a current-carrying wire reduces, the current density in the wire also increases. There are limits to how much absolute current flow can be reduced. Microcircuits on the chip must communicate without noise corruption of the signals they generate and the microcircuits themselves must, at some point, charge contact pads and drive external devices. High current densities will cause Joule heat-

Electromigration and Electronic Device Degradation, Edited by Aris Christou.
ISBN 0-471-58489-4

TABLE 10.1. Projected Memory Chip Parameters for Prototype Production

Year	Megabits	Area (mm)	L (mm)	W (mm)	Design Rules (μm)	Pixels/Field $\times 10^{11}$
1985	1	50	10	5	1.0	0.5
1988	4	100	14	7	.7	2
1990	16	200	20	10	.5	8
1993	64	400	28	14	.35	32
1996	256	800	40	30	.25	128
1999	1000	1600	56	28	.18	512

Remember: There are three levels of chip development:

- Proof-of-principles demonstration
- Prototype (limited volume) production
- Volume manufacture

Each is separated by a 2–3 year time interval.

ing of the interconnects. Also, when current densities exceed 10^5 A/cm^2, ''friction'' between the transported electrons and the metal lattice can cause migration of mass (called electromigration) along the interconnect. Both of these problems lead to time-dependent failure. Heating accelerates most bulk failure mechanisms occurring in the device substrate. Electromigration causes metal voids, or ''open circuits,'' in the interconnects.

In this chapter, we review the major metallization schemes currently in use in integrated circuit technology. This serves to orient the reader to the materials systems employed. Next, we present some of the major reliability problems associated with these schemes. A number of time-dependent failure mechanisms is possible. Simple corrosion and stress-induced corrosion can be major problems. This chapter, however, focuses on electromigration-induced failure, as it is the most significant of these problems in commercial practice.

There are interesting tie-ins between failure statistics and the distribution of variables pertaining to physical characteristics of the metal film. Specifically, it is possible to relate the distribution of metal grain sizes to the shape of the failure rate versus time curve. This relationship is elaborated on later in the discussion of modeling reliability. The chapter concludes with a discussion of current trends and future plans for advanced metallization systems.

For an introduction and review of the basic processes used to fabricate metallized structures, the reader is referred to key references in the field [1–4].

10.2 MAJOR TRENDS IN THE DESIGN OF MULTILEVEL METALLIZATION SYSTEMS

In the past, a single layer of metal may have sufficed for all interconnect functions in an integrated circuit. We can summarize these functions as follows:

- Emitter/collector/base contacts (in bipolar technology)
- Source/drain/gate contacts (in MOS technology)
- Powerbus and ground
- Internal clock, control, and data lines
- Input/output lines

In the past, the metallization may have been used as an active gate in MOS transistors.

Currently, the gate-containing layer of interconnects in MOS integrated circuits is made of polycrystalline silicon. This material can withstand relatively high temperatures (about 1000°C). ''Self-aligned'' processes are possible with polysilicon, in which the material is defined and then doped by ion implantation. As the material will stand up to high temperatures, the implanted doping can be heated to make it electrically active. The implant boundary is perfectly aligned with the edge of the polysilicon and substrate scattering and diffusion occurring during activation define the degree to which the doping undercuts the feature. This process is illustrated in Figure 10.1. In more advanced processes,

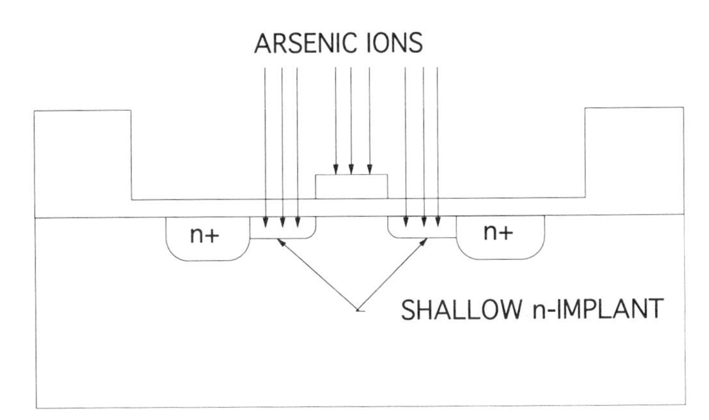

FIGURE 10.1. An ion-implant operation self-aligning the source and drains of a MOSFET to the gate edge.

the polysilicon lines offer too much resistance. An intermetallic silicide is grown over the top of the polysilicon. The resulting layered structure is referred to as "polycide" [5]. This chapter, however, focuses on issues associated with metal thin films employed in the "back-end" of the process line, after polysilicon definition.

The back-end of the line is composed of those processing steps required to:

- Deposit the thin metal thin films that constitute the interconnect path
- Isolate interconnect levels from each other
- Define apertures in the isolating layers to allow select regions of the interconnect to communicate with each other
- Pattern the interconnect layers
- Provide a passivating overlayer that will protect the completed circuit from scratches and environmental effects (such as corrosion)

As there are millions of devices on today's MOS or bipolar integrated circuits, connecting these spatially separate units is a major problem. The problem cannot be solved with a single layer of interconnects. "Multilayer" approaches are widely employed [1]. The first layer may occur in the "front-end" of the line, before the polysilicon layer is installed. This would be a buried layer composed of a diffused (or implanted) current path in the substrate itself. The polysilicon layer would be the second layer of interconnects and a series of isolated thin metal films complete the on-chip interconnect structure.

A modern interconnect scheme used in bipolar processing is shown in cross section in Figure 10.2 [6]. Three levels of metal (M1, M2, M3) are employed. Points of connection between metal layers are labeled "V," for "via." Points of connection between metal layers and the substrate are labeled "C," for "contact." Insulating layers used to isolate the different interconnect layers, one from another, are labeled "I." In the bipolar technology shown, n-doped polysilicon can be used to contact n-type regions on the silicon substrate. It cannot be used to form an ohmic nonblocking contact to p-type silicon. Also, since it has relatively high resistivity (20–40 Ω/cm^2), it can only be used to connect device regions that are in close proximity (i.e., it can only be used for "local" interconnects). To contact the p-type regions, or widely separate regions of the chip, metal lines are required.

Where the metal (usually an aluminum alloy) contacts the underlying silicon at the site of a shallow junction, a barrier metal is required [6].

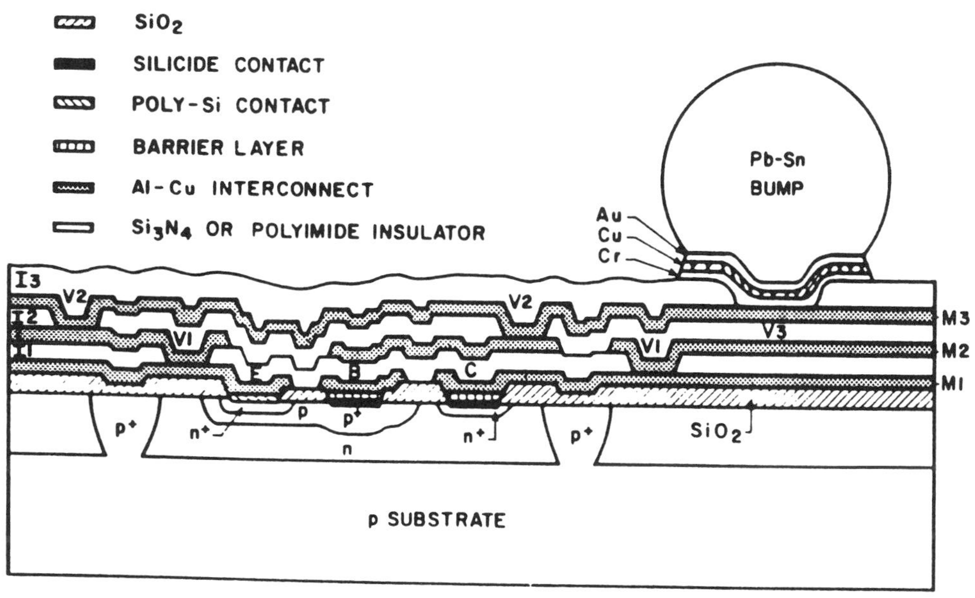

FIGURE 10.2. A multilayer metallization scheme for bipolar devices.

The barrier prevents the interdiffusion of the underlying silicon and metal line. Titanium–tungsten alloys have proved very effective in this area. Also, incorporation of 1–2% silicon into the aluminum brings the metal close to its solid-solubility limit for silicon. This impedes further interdiffusion from the underlying substrate.

Most scaling schemes used to shrink the size of integrated circuits require the formation of shallow junctions thinner than 200 nm. Any degree of alloying between the substrate and the covering metal will destroy these junctions. Most metals, though, require some high-temperature (about 400°C) "sintering" step to reduce surface oxides, elevate the local doping, and create a good electrical contact. One way around this is to form a metal silicide that provides an ohmic contact to silicon and will also contact the metal interconnect over it without adding resistance to the composite structure [5]. A metal, such as platinum or cobalt, will alloy with a fixed amount of silicon to make a stoichiometric, low-resistance contact. The depth to which the alloying proceeds depends only on the initial thickness of the silicide-forming metal. Thus we see silicides in contact areas in Figure 10.2.

Large bonding pads are used to make contact with off-chip circuitry. In the example used in Figure 10.2, a gold/copper/chrome composite is used for the bonding pad. Chrome is used to promote adhesion between isolation layer I3 and the copper. It also serves as a barrier to intermetallic formation between the copper and the aluminum contact. Gold is never placed in contact with aluminum, as these materials readily form intermetallics which are poor conductors. The intermetallics formed between aluminum and gold have a purplish hue and are known as "purple plague." Neither gold nor copper are ever placed in contact with the bare silicon surface. Both metals are rapid diffusers and they increase bulk generation in the silicon above acceptable levels. Barrier metals are frequently used to prevent such undesired alloying.

A lead solder "bump" is plated over the pad. The placement of these pads and their associated "bumps" match a pad and interconnect pattern on a circuit card. The chip is placed upside-down over the card. The pads on the chip are registered to the pads on the card. Heating the card melts the solder bumps contacting it. Many chips can be assembled on a single card in this way, creating a multichip carrier. Interconnects may be viewed as extending off-chip to the cards on which the devices are mounted.

It should also be pointed out that the processing surfaces tend to be less flat as we move away from the silicon substrate. This is a problem in a number of areas. First, steep steps are difficult to cover as a result of shadowing effects in the metal deposition process. Contact and via holes are particularly hard to cover with a continuous layer of metal. This is due to both shadowing effects and overgrowth of metal as shown in Figure 10.3. One solution to this problem comes through the use of chemical vapor deposition (CVD) technology. In CVD, reactive gases decompose and form deposited films on heated processing surfaces. In most cases, blanket deposits are accomplished, which are highly conforming to the underlying substrate. This, by itself, leads to good step

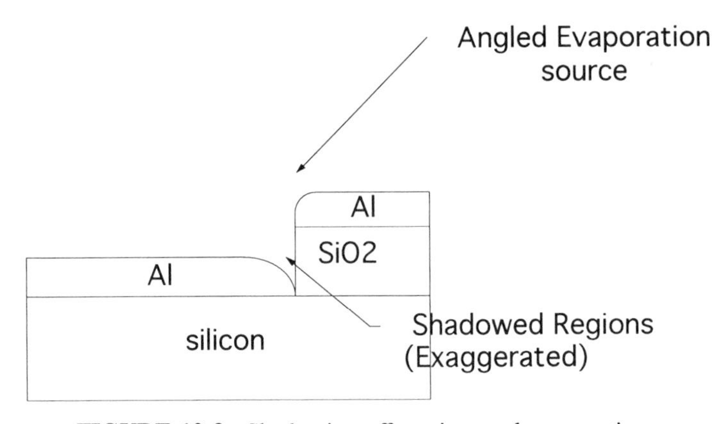

FIGURE 10.3. Shadowing effects in metal evaporation.

coverage. In some cases, it is possible for the substrate to play an active role in the deposition chemistry. In such cases, "selective" depositions are possible. For example, films may be deposited over bare silicon only, excluding deposition over adjacent regions of oxide.

Tungsten films may be deposited using CVD processes. Depending on the chemistry used, the deposition can be either of the blanket or of the selective type. If the depth of the contact window does not vary significantly across a wafer, contact windows can be selectively filled with a "plug" of tungsten. This creates a highly planar surface on which to deposit aluminum or aluminum alloys. Blanket tungsten can be grown to be thick enough to cover the whole contact in a planar fashion, as shown in Figure 10.4. The resulting layer can be patterned into an interconnect. However, for long runs, the resistivity of tungsten is too large ($>10\ \Omega/\square$) for high-speed circuitry. The tungsten outside the contact window can be "planed" down using anisotropic etches or by polishing, as shown in Figure 10.4. Lower resistivity metals can then be patterned over the filled contact or via.

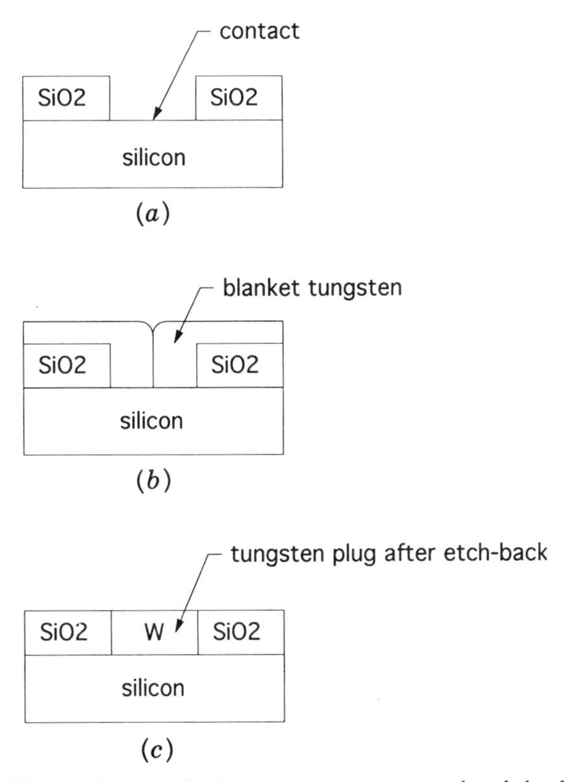

FIGURE 10.4. A blanket tungsten process and etch-back.

It should be noted that most metallization processes require some form of planarization for contacts and for interlayer isolation. High-resolution lithographic systems used to pattern metal layers have poor depth-of-focus. That is, if there is any departure from planarity, the image of the metal feature to be defined will go out of focus, destroying resolution and degrading the final interconnect quality. Submicron technology requires surface planarity of about the order of a micron.

To conclude, we have described the major features of modern-day metallization systems in this section. We have described the need for aluminum alloys, silicides, and barrier layers to impede interdiffusion and intermetallic formation, the need for advanced CVD processes to achieve step coverage, and the need for planarization to sustain high-resolution lithography. In the next section, we describe how these new systems degrade over time in use in actual circuits.

10.3 THE PHYSICS OF ELECTROMIGRATION

While a number of metallization system failure mechanisms are possible only a few of them are time dependent. It is these time-dependent mechanisms that have the greatest impact on reliability. They escape most initial screening attempts and they manifest themselves when the parts are in use. The most pronounced of these mechanisms is electromigration.

Electromigration is the formation of voids in current-carrying metallic strips. Observations of this effect date back to the 1930s [7]. It was initially thought that electric fields present in the strips forced metal ions out of their equilibrium lattice positions. This effect would be most prevalent in the vicinity of defects. Pile-up of ions would occur near defects, leading to hillock formation.

Marker studies performed in the 1950s, however, indicated that this was not the case [7]. These studies indicated that the ion motion was in the same direction as the electron current flow. That is, a small nick in a strip would appear to progress down the strip toward the biasing anode. The rate of ion migration is proportional to current density. It appears that the electron current can transfer momentum to lattice ions, dislodging them from their normally stable lattice sites, and transporting them to nucleating sites downsteam.

According to the quantum theory of electron transport in a metal, the Bloch waves representing the moving electrons can transfer energy to the lattice only by interacting with defects or through phonon–electron interactions. Thus mechanical defects in the interconnect must play a significant role in void formation. Preferential nucleation must also involve some significant defect agglomeration if hillocks are to form. Grain boundaries must also be important as defect sites and as possible transport channels.

The basis of the electron–defect interaction process is illustrated in Figure 10.5. An electron is scattered by an ion in a defect site. In the course of this scattering, the electron's momentum is reversed, causing an average change in momentum in the transport direction equal to $2m \langle \mathbf{v} \rangle$ where m is the electron mass and $\langle \mathbf{v} \rangle$ is the mean velocity of the electron in the direction of current flow. The friction force on the ion is

$$F_f = \frac{2m \langle \mathbf{v} \rangle}{\tau_{\text{col}}} \qquad (10.1)$$

where τ_{col} is a time constant representing the time between collisions. From elementary transport theory, we have

$$J_e = ne \langle \mathbf{v} \rangle \qquad (10.2)$$

where J_e is the electron current density and n is the density of electrons available for transport. Solving Equation (10.2) for $\langle \mathbf{v} \rangle$ and substituting the result in Equation (10.1) yield

$$F_f = \frac{2mJ_e}{ne\tau_{\text{col}}} \qquad (10.3)$$

We have thus derived an expression for the friction force acting on an ion in terms of the electron current transport parameters.

We now proceed to use this information to derive the ion currents. Let us assume, in a fashion analogous to collision-dominated electron

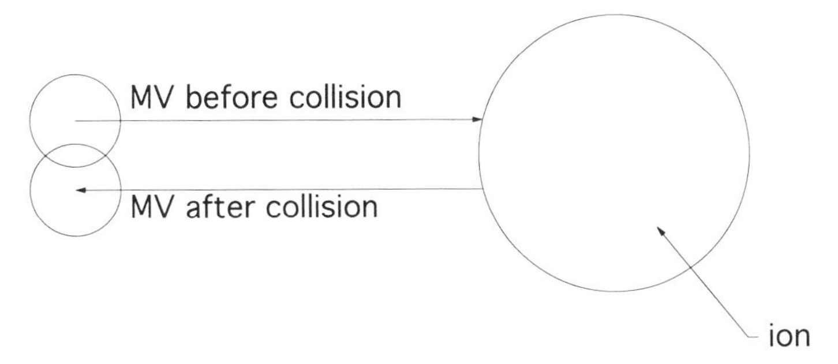

FIGURE 10.5. The electron–ion interaction in electromigration.

transport, that the speed of ion transport is simply proportional to the force applied:

$$v_i = \mu F_f \tag{10.4}$$

and

$$\mu = \frac{J_i^f}{eF_f N} \tag{10.5}$$

where v_i is the ion speed (actually, this is the mean ion velocity in the direction of transport), μ is a proportionality constant referred to as the ion mobility, J_i^f is the ion current density due to electron momentum transfer, and N is the density of ions available for transport. According to the Nernst–Einstein equation, the ion diffusion coefficient, D, and μ are related as follows:

$$\frac{D}{\mu} = \frac{kT}{e} \tag{10.6}$$

where k is the Boltzmann constant. Combining Equations (10.4)–(10.6) yields

$$J_i^f = \frac{q^2 DN}{kT} F_f \tag{10.7}$$

From Equation (10.3), we see that F_f is proportional to J_e:

$$F_f = C_1 J_e \tag{10.8}$$

Thus where C_1 is the proportionality constant derived from Equation (10.3)

$$J_i^f = \frac{q^2 DNC_1}{kT} J_e \tag{10.9}$$

From Ohm's law, this equation becomes

$$J_i^f = \frac{q^2 DNC_1}{kT} \rho_e E \tag{10.10}$$

where ρ_e is the electron resistivity and E is the electric field.

Simply accounting for the friction force driving the ion flow will not provide a full picture of mass transport. The electric field will induce an ion current, J_i^e, which is counter to the friction current. Once again, we can use the basic transport relation to describe the field-induced ion current J_i^E:

$$J_i^E = eN\mu E \tag{10.11}$$

Again, making use of the Nernst–Einstein relationship,

$$J_i^E = \frac{e^2NDE}{kT} \tag{10.12}$$

The total ion current, J_i, is given by

$$J_i = J_i^f - J_i^E \tag{10.13}$$

or

$$J_i = (eN)(C_1\rho - 1)\left(\frac{eD}{kT}\right)E \tag{10.14}$$

If we take the second term in parenthesis, $(C_1\rho - 1)$, and call it Z', Equation (10.14) takes a more easily recognizable form:

$$J_i = eZ'N\left(\frac{eD}{kT}\right)E \tag{10.15}$$

The simple interpretation of this equation is that the ion current is equal to the effective charge on the ion, multiplied by the density of ions available for transport, the ion mobility, and the electric field.

Other physical effects may give rise to net ion currents and to the ion current divergence necessary for void formation. For example, temperature gradients occurring in the conducting strip will also create the ion flux divergences responsible for open-metal device failures. This is because the ion diffusion coefficients (and hence ion mobilities) will become position dependent. Mobilities will be greater in the hotter regions. These regions will be sources, while the cooler regions will be sinks of material.

Agarwala et al. [8] have derived an expression for the diffusion coefficient gradient based on the following considerations. The diffusion coefficient is generally considered to be temperature dependent through the relationship

$$D = D_0 \exp\left(\frac{-\Delta H}{kT}\right) \tag{10.16}$$

where ΔH is the activation energy for ion transport. The temperature gradient of D is therefore

$$\frac{dD}{dT} = \frac{D_0}{kT}\left[\exp\left(\frac{-\Delta H}{kT}\right)\right]\left(\frac{-\Delta H}{kT} - 1\right). \tag{10.17}$$

Using the chain rule,

$$\frac{dD}{dx} = \frac{dT}{dx}\frac{dD}{dT} \tag{10.18}$$

we have

$$\frac{dD}{dx} = \frac{D_0}{kT}\left[\exp\left(\frac{-\Delta H}{kT}\right)\right]\left(\frac{-\Delta H}{kT} - 1\right)\frac{dT}{dx} \tag{10.19}$$

This effect is particularly important in accelerated life-testing and will be discussed in greater detail below.

Kinsbron [9] has shown that stress in the conducting strip affects electromigration in a number of ways. As material piles up at a nucleation site, a "back-pressure" is created that impedes ion migration. Deposition stresses in the film also influence the process. Compressive and tensile stresses may increase or decrease ion migration activation energies, changing D. All these factors must be accounted for in electromigration studies. In addition, as mass accumulates in a given region, there is an increased tendency toward back-diffusion [10].

A more complete equation that accounts for these effects is [11]

$$J_i = eZ'N\left(\frac{eD}{kT}\right)E - eD\,\nabla N - \frac{eDN\omega\,\nabla N}{\beta N_0 kT} \tag{10.20}$$

where ω is the atomic volume, β is the film compressibility, and N_0 is the atom density of the film. The first term accounts for friction flow, the second accounts for the concentration gradient, and the third accounts for film stress.

10.4 MATERIAL MODELS FOR TIME-DEPENDENT WEAROUT OF VLSI METALLIZATION

Microscopic examination of most deposited thin films indicates a pronounced cellular structure to the film generally referred to as the "grain structure" of the film. These "cells" arise as a result of the processes of nucleation and growth, which form the film. The presence of many nucleation sites usually leads to small grains and to rapid growth. Surface conditions and film growth parameters (e.g., substrate temperature, rate of arrival of metal atoms to the growth surface) determine grain density. Grain boundaries represent interfaces with an associated free energy of surface formation. During growth and subsequent annealing cycles, some grains may grow and others disappear in order to minimize this free energy. In any event, the grain structure boundaries play a dominant role in determining electromigration rates.

The grain boundaries represent relatively low-resistance ion conducting channels. As reported by d'Heurle and Ho [12], at standard integrated circuit operating temperatures, bulk ion migration processes are slow and the grain boundaries carry the bulk of the ion current. The bulk grain does carry the full electron current, though.

Furthermore, the properties of the grain structure enable considerable refinement of material models for electromigration. Specifically, there are three properties that have immediate impact on reliability models. They are:

- The orientation of the boundary with respect to the electric field
- The angles of the grain boundaries with respect to each other
- Changes in the number of the grains per unit area (grain density)

Each of these properties can give rise to the ion divergences necessary to create voids in metal strips. We consider each of these effects below.

Consider Figure 10.6. Here, we see a so-called grain boundary "triple point," the intersection of three grain boundaries. If the boundary to the left is parallel to the applied field, the apparent ion mobility is highest along that boundary. Migration along the two adjacent boundaries is the result of a projected field component and is lower. The ion current density divergence at the point of intersection between two grains making angles Ψ_1 and Ψ_2 with respect to the electric field is given by Berenbaum [13] as

$$\nabla \cdot J_i = \frac{eNDeZ'}{kT} \frac{\cos \Psi_1 - \cos \Psi_2 - 1}{\Delta x} \tag{10.21}$$

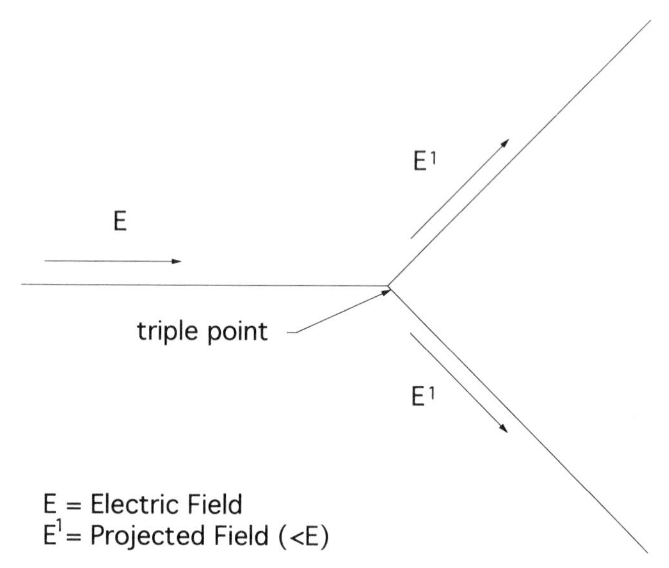

FIGURE 10.6. Ion migration at a grain-boundary "triple-point."

The grains themselves may be viewed as composed of linear arrays of dislocations characterized by some Burgers vector, **b**. The degree to which the crystallite representing the grain is misoriented with respect to the bulk crystal is given by the tilt angle, θ. If d is the separation distance of dislocations,

$$\frac{b}{d} = 2 \sin\left(\tfrac{1}{2}\theta\right) \tag{10.22}$$

The most common assumption relating grain boundary diffusion to tilt is that the ion mobility is simply proportional to the number of dislocations in the boundary [14]. Thus

$$\mu_i = A \sin\left(\tfrac{1}{2}\theta\right) \tag{10.23}$$

where A is proportionality constant.

If the grain-size changes along the strip, the density of ion conduits into and out of a region must also change. A densely grained region will channel ions out more effectively than a sparsely grained region. This creates the ion current divergence necessary to form a void. Similarly, ion pile-up can occur in regions in which the grain sizes increase in the direction of electron flow.

The degree to which a grain can conduct ions further depends on the texture of the oriented crystallite growth in the metal film. The electromigration lifetime of a strip is longer for $\langle 111 \rangle$ oriented films as contrasted with $\langle 200 \rangle$. The degree to which one or the other of these orientations is preferred is obtained from diffraction analysis through the ratio of the spot intensities, I, for each planar reflection. An emperical formula accounting for this effect was given by Vaidya et al. [15]:

$$\text{MTBF} = \frac{Rs}{\sigma^2} \log \left(\frac{I_{111}}{I_{200}} \right) \tag{10.24}$$

where s is the mean grain diameter, σ is the standard deviation of the grain diameter distribution, and R is a proportionality constant. Frequently, the right-hand side of this equation is used as a prefactor in determining ion current densities in Equation (10.15).

All these effects must be included in wearout models for metallization reliability. How these models are created is discussed in the next section.

10.5 MODELING RELIABILITY

Ideally, one would like to make a few rapid measurements on a device structure and determine, from these measurements, how long a device will last in operation. However, linear extrapolations from data taken over short time intervals rarely provided meaningful information. At the very least, a model for the wearout process is required. The early time datapoints can then be used for a more meaningful extrapolation. Also, in slow processes, such as electromigration, some form of accelerated testing is frequently required. The validity of existing models under accelerated conditions must be critically examined. Two types of reliability model are generally encountered in electromigration studies. These models are here referred to as "empirical" and "statistical." We begin by describing the empirical models. This is followed by a discussion of how these models fit into accelerated testing schemes.

The empirical approach was pioneered by Black [16,17]. He analyzed Equation (10.9) and assumed that the mean-time-before-failure (MTBF) of a conducting strip would be inversely proportional to the ion current. Ion current would, in turn, be dominated by the electron current, J_e N, the number of ions available for transport. He further assumed that the number of ions available for transport would be proportional to an

Arrhenius factor multiplied by the rate at which electrons impinged on the stationary ion in an attempt to mobilize it. The impingement rate is also proportional to the electron current. As a result of these considerations, Black presented the following equation:

$$\text{MTBF} = \frac{B}{J_e^2 \exp(-\phi/kT)} \tag{10.25}$$

where ϕ is the activation energy for mobile ion production. Subsequent researchers have found that the square exponent to which the electron current density is raised is best dealt with as an empirical parameter, n. Values of n have been reported that are as low as 1 and as high as 7. Typical activation energies are in the 0.6–0.7 eV range.

Now let us consider the statistical approaches to reliability modeling Attardo and his group [14] were early contributors to this area, and their approach is presented here. The basic idea is to consider all the ion channels into and out of a given section of strip. The ion divergence is proportional to the differences in mobilities as summed over the channels. Assuming the three grain-boundary divergence mechanisms discussed in the preceding section, we can write the following expression for current divergence along the transport direction per unit width of strip:

$$\frac{dJ_i}{dx} = \sum_{\text{paths out}} A \sin(\tfrac{1}{2}\theta_i) \cos(\Psi_i) - \sum_{\text{paths in}} A \sin(\tfrac{1}{2}\theta_i) \cos(\Psi_i) \tag{10.26}$$

where A is a proportionality constant. The MTBF for the section would be inversely proportional to the divergence times the width:

$$\text{MTBF} = \frac{GW}{\sum_{\text{paths out}} A \sin(\tfrac{1}{2}\theta_i) \cos(\Psi_i) - \sum_{\text{paths in}} A \sin(\tfrac{1}{2}\theta_i) \cos(\Psi_i)} \tag{10.27}$$

where W is the strip width and G is another proportionality constant.

Attardo et al. [14] describes in some detail a method for using Equation (10.27) to derive failure distribution statistics. Based on actual microscopic measurements, the grain boundary diameter distribution is measured. Sections of strip are constructed whose length, Δx, is the mean grain length. Using a Monte Carlo approach, grains are stacked to span the strip length. The number of channels into and out of the section are counted. Each path is assigned a Ψ and a θ, again using a Monte Carlo approach. The Ψ angle was allowed to vary randomly from 0° to 90°. Assuming a ⟨111⟩-oriented film, symmetry requires consideration of angles ranging from 0° to 60° for θ. Again, random orientation was assumed for θ. Proportionality constants were chosen to best fit experimental MTBF studies. The section of strip with the smallest MTBF, as ascertained by evaluation of Equation (10.26), could then be identified.

As a result of these simulations, it was determined that the statistical life distribution (failure rate versus time curve) best fit a log-normal distribution curve. The parameters of this curve could be fit using relatively few datapoints, greatly reducing the amount of work needed to make reasonable reliability projections. Attardo's calculations were extended in recent years by Schoen [18] to include the effects of a significant number of grain sizes exceeding the mean strip width. Scherge et al. [12] have included the back-diffusion and strain effects, which are a part of Equation (10.20).

There are many techniques for setting up experiments to parameterize the reliability models described above. Recent reviews of these techniques have been published by Ghate [19], Schreiber [20] and Vook and Vankar [21], and the reader is referred to these reviews for further details. The balance of this section deals with difficulties in extending electromigration reliability models to accelerated testing.

The most prevalent technique to accelerate electromigration failure is to increase the current density in the strip. Attempts to do this often lead to an apparent current-stress dependence of the activation energy. As Lloyd et al. [22] have pointed out, there is no physical basis for this assumption. Ion migration is responsible for electromigration failure. As discussed in Section 10.3, the activation energy relates to the ion diffusion coefficient in the grain boundary. This term is just the product of the grain boundary vacancy diffusivity and the vacancy concentration [11]. Neither of these terms is influenced by electron current density.

As Lloyd et al. [22] point out, there is a change in failure mechanism at high current densities. Incipient voids in metal lines have smaller effective areas for current transport. Local hot spots form at these sites, giving rise to thermal gradients. According to Equation (10.19), these thermal gradients will give rise to a position-dependent ion diffusivity and hence an ion current divergence. These effects become more pronounced at higher current densities.

Two approaches are possible. The easiest is to follow Lloyd's method of estimating the temperature gradient in the vicinity of an incipient failure site [22] at a given current density. The current density is kept small enough so this gradient term doesn't affect the measurement. Or one can incorporate hot spot formation in the reliability model [23].

10.6 CURRENT STATUS AND FUTURE TECHNOLOGIES

One of the earliest approaches to reducing the probability of electromigration failure was through alloying. Small amounts ($<4\%$ atomic) of copper, for example, alloyed with aluminum metal strips would dramatically increase the electromigration MTBF [24]. In the past, it was felt that copper solute would segregate at grain boundaries, blocking boundary diffusion sites. More recent work [25] indicates that the initial amount of grain boundary precipitation of copper is small. It appears that the solute is wrapped up in a supersaturate precipitate which can replenish lost material as electromigration progresses.

As interconnect linewidth shrinks down to a few micrometers, electromigration failures increase. It is easier to form a void over a smaller width of line. Local heating effects are more pronounced. However, as linewidths shrink below 2 μm there is an improvement in MTBF [26]. The reason for this is seen in Figure 10.7. As linewidth shrinks below the mean grain boundary diameter, the boundary conduction channels appear normal to the film. These channels cannot effectively funnel material, and electromigration rates reduce. These normal boundaries are referred to as "bamboo" structures.

It may be thought that metals with higher melting temperatures should exhibit higher ion-migration activation energies. This has proved to be the case for tungsten thin films. Recent studies show that tungsten exhibits a much longer electromigration MTBF. Tungsten, however, exhibits a relatively high resistivity, high stress on deposition, and poor adhesion making it less attractive for integrated circuit work. Al–W–Al sandwiches are currently under active evaluation [27]. Significant improvements in electromigration reliability have been noted using these complex metallization schemes. Furthermore, it has been reported that intermetallic compounds formed at the Al–W interface may be stable and electromigration-resistant, further improving reliability [28].

Most metal layers are coated with protective insulating layers. These layers themselves usually serve to impede electromigration failure. The mechanical stress these films generate may be sufficient to increase the "back-pressure" of ion buildup, slowing ion migration. Recent work

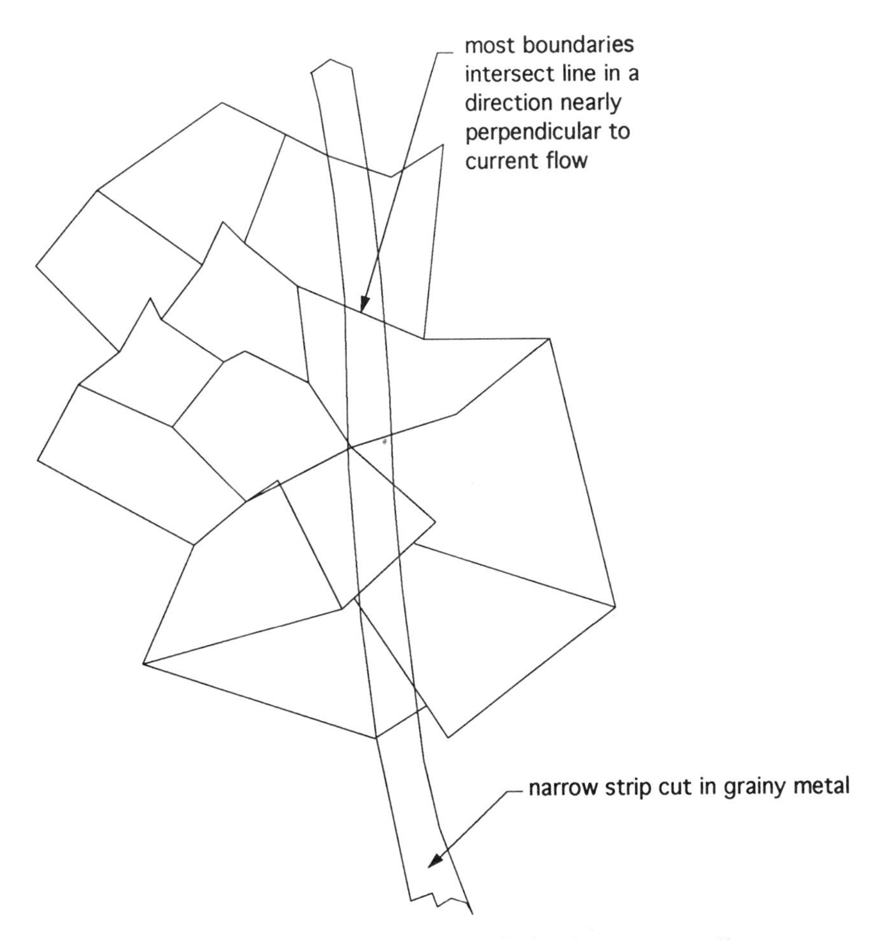

FIGURE 10.7. Development of "bamboo structure."

[22] indicates that certain passivants (polyimide, in particular) may introduce atomic species into the grain boundary that impede ion diffusion.

The introduction of dielectric protective layers may have some drawbacks under certain circumstances. These layers may be deposited under tension or under compression. The resulting stresses can be transferred to the underlying metal films, accelerating void or hillock formation. This phenomenon is known as "stress-voiding" [4]. As the difference in thermal expansion coefficients of insulators and metals may vary, the phenomenon is temperature dependent. Sullivan [29] has modeled this effect by adding a temperature-dependent strain term to the basic diffusion equation for vacancies. Of course, one solution to this problem is to minimize strain in passivating overlayers. As there is no comprehensive model relating residual strain to processing conditions, such minimization is currently done empirically.

10.7 SUMMARY AND CONCLUSIONS

In this chapter we have reviewed the basic philosophy behind the design of modern metallization systems used in large-scale integrated circuits. After reviewing the key design challenges, we focused on the issue of time-dependent wearout. The most significant of these failure mechanisms, electromigration, was discussed in detail.

Of particular interest was the way in which material properties influenced device life. The models discussed here were dominated by the grain structure of the metal. Grain size and orientation distributions had immediate impact on failure rate projections. Techniques for incorporating significant material parameters into reliability models were presented. The resulting models significantly eased the data-gathering burden for making reliability projections.

Future work will hinge on defining new material systems that are less prone to electromigration. Techniques such as the metal alloys or sandwich structures or the optimization of passivating layers discussed above all appear to significantly improve metallization lifetime.

ACKNOWLEDGMENTS

The author would like to thank Drs. O. Glembocki, C. Marrian, E. Dobisz, and S.P. Murarka for helpful discussions concerning this chapter.

REFERENCES

1. S.P. Murarka and M.C. Peckerar, *Electronic Materials: Science and Technology*, Academic Press, Cambridge, 1989.
2. J.L. Vossen and W. Kern, *Thin Film Processes*, Academic Press, Cambridge, 1978.
3. J.L. Vossen and W. Kern, *Thin Film Processes II*, Academic Press, Cambridge, 1991.
4. S.P. Murarka, *Metallization: Theory and Practice for VLSI and ULSI*, Butterworth-Heineman, Boston, 1993.

5. S.P. Murarka, *Silicides for VLSI Application*, Academic Press, Cambridge, 1986.
6. M. Wittmer, "Barrier Layers: Principles and Applications in Microelectronics," *J. Vac. Sci. Technol. A*, Vol. 2, No. 2, 273 (1984).
7. H.B. Huntington and A.R. Grone, "Current-Induced Marker Motion in Gold Wires," *J. Phys. Chem. Solids*, Vol. 20, Nos. 1–2, 76 (1961).
8. B.N. Agarwala, M.J. Attardo, and A.P. Ingraham, "Dependence of Electromigration Induced Failure Times on Length and Width of Aluminum Thin-Film Conductors," *J. Appl. Phys.*, Vol. 41, No. 10, 3954 (1970).
9. E. Kinsbron, "A Model for Width Dependence of Electromigration Lifetimes in Aluminum Thin-Film Stripes," *Appl. Phys. Lett.*, Vol. 36, No. 12, 968 (1980).
10. M. Shatzkes and J.R. Lloyd, "A Model for Conductor Failure Considering Diffusion Concurrently with Electromigration Resulting in a Current Exponent of 2," *J. Appl. Phys.*, Vol. 59, No. 11, 3890 (1986).
11. M. Scherge, V. Breternitz, and Ch. Knedlik, "Simulation of Electromigration Behavior in Aluminum Metallization of Integrated Circuits," *Microelectron. Reliab.*, Vol. 32, Nos. 1/2, 21 (1992).
12. F.M. d'Heurle and P.S. Ho, "Electromigration in Thin Films," in *Thin Films: Interdiffusion and Reactions*, J.M. Poate, K.N. Tu, and J.W. Meyers, eds., Wiley, New York, 1978, p. 243.
13. L. Berenbaum, "Electromigration Damage of Grain Boundary Triple Points in Aluminum Thin-Films," *J. Appl. Phys.*, Vol. 42, 880 (1971).
14. M.J. Attardo, R. Rutledge, and R.C. Jack, "Statistical Metallurgical Model for Electromigration Failure in Aluminum Thin-Film Conductors," *J. Appl. Phys.*, Vol. 42, No. 11, 4343 (1971).
15. S. Vaidya, D.B. Fraser, and A.K. Sinha, "Orientation Effects in Electromigration," in *Proceedings of the IEEE International Reliability Physics Symposium*, p. 165 (1980).
16. J.M. Black, "Electromigration—A Brief Survey and Some Recent Results," *IEEE Proc. Electron. Devices*, Vol. Ed-16, 338 (1969).
17. J.M. Black, "Electromigration Failure Modes in Aluminum Metallization for Semiconductor Devices," *Proc. IEEE*, Vol. 57, 1587 (1969).
18. J.M. Schoen, "Monte Carlo Calculations of Structure-Induced Electromigration Failure," *J. Appl. Phys.*, Vol. 51, No. 1, 513 (1980).
19. P.B. Ghate, "Electromigration Induced Failures in VLSI Interconnects," *Solid State Technol.*, p. 113 (1983).
20. H.U. Schreiber, "Reduced Aluminum Electromigration in Future Integrated Circuits: A Problem of Test Procedure and Threshold Mechanism," *Thin Solid Films*, Vol. 175, 29 (1989).
21. R.W. Vook and V.D. Vankar, "Microelectronic Failure Due to Electromigration Damage," *Israel J. Technol.*, Vol. 25, 23 (1989).

22. J.R. Lloyd, M. Shatzkes, and D.C. Challener, "Kinetic Study of Electromigration Failure in Cr/Al–Cu Thin-Film Conductors Covered with Polyimide and the Problem of the Stress-Dependent Activation Energy," in *Proceedings of the IEEE International Reliability Physics Symposium*, p. 216 (1988).
23. J.C. Ondrusek, A. Nishimura, H.H. Hoang, T. Suguira, R. Blumenthal, H. Kitagawa, and J.W. McPherson, "Effective Kinetic Variations with Stress Duration for Multilayered Metallizations," in *Proceedings of the IEEE International Reliability Physics Symposium*, p. 179 (1988).
24. I. Ames, F.M. D'Heurle, and R.E. Horstmann, "Reduction of Electromigration in Aluminum Thin Films by Copper Doping," *IBM J. Res. Dev.*, Vol. 14, 461 (1970).
25. M.B. Small and D.A. Smith, "Selection of Solutes for Improving Electromigration Resistance of Metals: A New Insight," *Appl. Phys. Lett.*, Vol. 60, No. 26, 3235 (1992).
26. S. Vaidya, T.T. Sheng, and A.K. Sinha, "Linewidth Dependence of Electromigration in Evaporated Al–0.5% Cu," *Appl. Phys. Lett.*, Vol. 36, No. 6, 464 (1980).
27. C.F. Dunn, F.R. Brotzen, and J.W. McPherson, "Electromigration and Structural Properties of Al–Si/Ti, Al–Si VLSI Metallization," *J. Electron. Mater.*, Vol. 15, 273 (1986).
28. H.H. Hoang, "Effects of Annealing Temperature on Electromigration Performance of Multilayer Metallization Systems," in *Proceedings of the IEEE International Reliability Physics Symposium*, p. 173 (1988).
29. T.D. Sullivan, "Thermal Dependence of Voiding in Narrow Aluminum Microelectronic Interconnects," *Appl. Phys. Lett.*, Vol. 55, No. 23, 2399 (1989).

INDEX